The River of the Mother

Aldo Leopold, 1928. (Leopold Collection, UW Archives)

The River of the Mother of God

AND OTHER ESSAYS
BY ALDO LEOPOLD

Edited by
Susan L. Flader
and J. Baird Callicott

THE UNIVERSITY OF WISCONSIN PRESS

The University of Wisconsin Press
1930 Monroe Street
Madison, Wisconsin 53711

3 Henrietta Street
London WC2E 8LU, England

Copyright © 1991
The Board of Regents of the University of Wisconsin System
All rights reserved

9   8   7   6

Printed in the United States of America

Library of Congress Cataloging-in-Publication Data
Leopold, Aldo, 1887-1948
   The river of the mother of God and other essays / by Aldo Leopold
edited by Susan L. Flader and J. Baird Callicott.
   400 pp.   cm.
   Includes index.
   1. Nature conservation. 2. Natural history—Outdoor books.
I. Flader, Susan. II. Callicott, J. Baird. III. Title.
QH81.L557   1991
333.9516—dc20   90-45491
ISBN 0-299-12760-5   ISBN 0-299-12764-8 (pbk.)

Contents

PREFACE / ix

ALDO LEOPOLD: A BRIEF CHRONOLOGY / xiii

INTRODUCTION / 3

A Tramp in November (1904) / 33
The Maintenance of Forests (1904) / 37
The Busy Season (1911) / 40
To the Forest Officers of the Carson (1913) / 41
The Varmint Question (1915) / 47
The Popular Wilderness Fallacy: An Idea That Is
   Fast Exploding (1918) / 49
Forestry and Game Conservation (1918) / 53
Notes on the Behavior of Pintail Ducks in a Hailstorm (1919) / 60
Wild Lifers vs. Game Farmers: A Plea for Democracy
   in Sport (1919) / 62
"Piute Forestry" vs. Forest Fire Prevention (1920) / 68
The Forestry of the Prophets (1920) / 71
The Wilderness and Its Place in Forest Recreational
   Policy (1921) / 78
Standards of Conservation (1922) / 82

Some Fundamentals of Conservation in the Southwest (1923) / 86
A Criticism of the Booster Spirit (1923) / 98
Pioneers and Gullies (1924) / 106
Grass, Brush, Timber, and Fire in Southern Arizona (1924) / 114
The River of the Mother of God (1924) / 123
Conserving the Covered Wagon (1925) / 128
The Pig in the Parlor (1925) / 133
Wilderness as a Form of Land Use (1925) / 134
The Home Builder Conserves (1928) / 143
Ho! Compadres Piñoneros! (1929) / 148
Report to the American Game Conference on an American
    Game Policy (1930) / 150
Game Methods: The American Way (1931) / 156
Game and Wild Life Conservation (1932) / 164
Grand-Opera Game (1932) / 169
The Virgin Southwest (1933) / 173
The Conservation Ethic (1933) / 181
Conservation Economics (1934) / 193
Helping Ourselves (1934) / 203
The Arboretum and the University (1934) / 209
Land Pathology (1935) / 212
Coon Valley: An Adventure in Cooperative
    Conservation (1935) / 218
Review of Elton, *Exploring the Animal World* (1935) / 224
Wilderness (1935) / 226
Threatened Species (1936) / 230
Means and Ends in Wild Life Management (1936) / 235
Conservationist in Mexico (1937) / 239

Chukaremia (1938)   /   245
Letter to a Wildflower Digger (1938)   /   247
Engineering and Conservation (1938)   /   249
The Farmer as a Conservationist (1939)   /   255
A Biotic View of Land (1939)   /   266
New Year's Inventory Checks Missing Game (1940)   /   274
The State of the Profession (1940)   /   276
Ecology and Politics (1941)   /   281
Wilderness as a Land Laboratory (1941)   /   287
The Last Stand (1942)   /   290
Land-Use and Democracy (1942)   /   295
The Role of Wildlife in a Liberal Education (1942)   /   301
What Is a Weed? (1943)   /   306
Conservation: In Whole or in Part? (1944)   /   310
Review of Young and Goldman, *The Wolves of North America* (1945)   /   320
The Outlook for Farm Wildlife (1945)   /   323
Review of Farrington, *The Ducks Came Back* (1946)   /   327
Adventures of a Conservation Commissioner (1946)   /   330
Wherefore Wildlife Ecology? (1947)   /   336
The Ecological Conscience (1947)   /   338

PUBLICATIONS OF ALDO LEOPOLD   /   349
INDEX   /   371

Preface

Aldo Leopold (1887-1948) is best known as the author of a slender volume of natural history vignettes and philosophical essays dealing with the relationship of people to land. Composed during the last decade of his life and published posthumously in 1949, *A Sand County Almanac and Sketches Here and There* expresses the distilled essence of Leopold's mature thought.

At first the almanac's audience was small and select, consisting primarily of professionals and dedicated amateurs in the conservation community. Then, riding the crest of the cultural self-criticism and environmental awakening of the 1960s, it grew in popularity and was reprinted in several hardbound and paper editions, eventually surpassing Rachel Carson's *Silent Spring*, Stewart Udall's *Quiet Crisis*, and Barry Commoner's *Closing Circle* as the philosophical touchstone of the modern environmental movement. Today *Sand County Almanac* is regarded as a classic of natural history literature as well as the conservationist's bible, and its author is routinely called a modern American prophet. In retrospect, one may say without exaggeration that Aldo Leopold led a generation to a new perception of nature and to a new vision of relationship with the natural environment.

However closely associated Leopold has been in the public mind with a single chef d'oeuvre, the strength of his reputation and the credibility of *Sand County Almanac* rest upon a lifetime of extraordinary professional achievement and a large body of other writing. During his tenure with the United States Forest Service from 1909 to 1928, Leopold was a leader and innovator in the young profession of forestry, especially in recreation planning, game management, and soil erosion control. He is considered the father of the national forest wilderness system and the father of the profession of wildlife management in America. From 1933 until his death in 1948, Leopold occupied the nation's first chair of game management, which was created for him at the University of Wisconsin. In the course of his life, he offered his remarkable leadership talents to more than a hundred conservation organizations, agencies, and committees. Every facet of his career left its mark in his literary estate, which consists of three books, some five hundred published articles, reports, handbooks, newsletters, and reviews, and more than five hundred unpublished essays, speeches, and policy papers.

Virtually every one of Leopold's writings is a gem of its genre, and each delighted, informed, and challenged its particular audience. But little of his work is readily available or widely known today. Perhaps best known after *Sand County Almanac* is Leopold's professional magnum opus, *Game Management*. Published in 1933, it became the standard text in a then-new field of applied science and continued as such for decades. Now available in a paperback edition, it is read and savored as a classic. (Leopold's 1931 *Report on a Game Survey of the North Central States* was influential and highly regarded in its day, but it is now long out of print and little known.) After the early success of *Sand County Almanac*, Luna B. Leopold brought out excerpts from his father's early hunting journals and ten previously unpublished essays under the title *Round River* in 1953. After *Round River* went out of print, eight of the essays were included in an enlarged edition of *Sand County Almanac* in 1966. Beyond this the public has had little access to Leopold's work.

Many of Leopold's remaining papers are mileposts on their author's intellectual journey and continue to be of general interest and lasting significance. Our purpose is to gather into a single volume the best and most representative among them. The items assembled here are drawn from two sources: Leopold's unpublished manuscripts, deposited in the archives of the University of Wisconsin-Madison; and his published work, scattered in sundry periodicals ranging from the *Carson Pine Cone* (a newsletter Leopold edited during his years as a forest officer in New Mexico) to the *Bulletin of the Garden Club of America*.

Selecting some fifty items from a universe twenty times as large involved the editors (and on occasion their collaborators) in protracted discussions of the significance of Leopold's work and the contributions individual pieces could make to understanding it. Philosopher Callicott was especially attracted to the more thought-provoking essays that illuminate the major themes for which Leopold is best known today: wilderness preservation, ecology, natural esthetics, and the land ethic. Historian Flader argued for sampling a wider range of Leopold's corpus and including selections that would represent the periods in his productive life and career, exemplify the various genres in which he wrote, and reflect his many professional activities. The result, we hope, is a volume that combines the virtues (and avoids the vices) of both approaches: a volume that illuminates Aldo Leopold—his eventful life, his multifaceted career, and his diverse writings—and one that documents the development of his most original and influential ideas.

The papers selected for publication here are presented in chronological order. Each is preceded by a brief headnote indicating something of the circumstances that inspired it, the audience to which it was addressed, and other particulars of special interest or significance. A biographical chronology and a bibliography of Leopold's publications are included in this volume as well. They too should help to place the selections in context. To elucidate the contribution of each selection to Leopold's enduring intellectual and professional legacy, the editors have also included an introduction to the book as a whole. Highlighting key essays, the introduction tracks the evolution of the principal themes in *Sand County Almanac*—conservation ecology, natural esthetics, and environmental ethics—and Leopold's thinking about wildlife management, conservation economics, agricultural land use, and wilderness preservation.

As these papers show, Leopold was keenly aware of his different audiences and varied his rhetoric accordingly. Readers of *Sand County Almanac* will recognize his inimitable style in them all. His prose is literate yet informal, reflective, and wry. It is characteristically direct, compressed, vivid, and precise. In this collection, one may also detect a subtle development in the author's style spanning all his genres and voices. There was music in Leopold's prose from the start, but in the earlier papers the higher registers and clarion tones predominate, while in the later ones the mellow baritone and *basso profundo* voices reign. Leopold's signal literary qualities are present throughout this volume; one just finds them becoming more refined as one reads from beginning to end.

These selections are presented with minimal editing. Previously published items are as they appeared in the cited source, though nonintegral illustrations have been omitted and obvious typographical errors silently corrected. Unpublished papers likewise remain as they were found, with edits in Leopold's hand retained; capitalization, spelling, and punctuation were infrequently adjusted, where it seemed that Leopold or his secretary had erred.

Preparation of this volume was greatly encouraged by Nina Leopold Bradley and Charles Bradley of Baraboo, Wisconsin. Elizabeth Steinberg and Alice Honeywell of the University of Wisconsin Press took a personal interest in the project from its inception, and Angela Ray provided skillful editing. J. Frank Cook and the staff of the University of Wisconsin Archives facilitated research in the Leopold Papers. Carolee Cote expertly typed all the material by Aldo Leopold at the University of Wisconsin-Stevens Point, while Patty Eggleston at the University of Missouri wrestled with the bibliography. Curt Meine of the University of Wisconsin-Madison and Donald Worster of the University of Kansas read the entire manuscript with care and made very helpful suggestions. To all we extend our deepest gratitude.

Aldo Leopold: A Brief Chronology

1887 Aldo Leopold born in Burlington, Iowa, on January 11, eldest of four children of Carl and Clara Leopold. Educated in public schools of Burlington until 1903.
1904 Attends Lawrenceville School in New Jersey from January 1904 to May 1905 to prepare for college.
1905 Attends Sheffield Scientific School at Yale (class of 1908).
1906 Begins coursework at Yale Forest School (Master of Forestry, 1909).
1909 Joins U.S. Forest Service (established 1905). Reports to District 3, Arizona and New Mexico territories. First field assignment as assistant on Apache National Forest in southeastern Arizona.
1911 Transferred to Carson National Forest in northern New Mexico as deputy supervisor, then supervisor. Founds and edits *Carson Pine Cone*, a forest newsletter.
1912 Marries Estella Bergere of Santa Fe on October 9. Five children: Starker, 1913; Luna, 1915; Nina, 1917; Carl, 1919; Estella, 1927.
1913 Stricken with acute nephritis in April. Recuperates until September 1914, mostly in Burlington.
1914 Assigned to district headquarters in Albuquerque in office of grazing, then in 1915 placed in charge of new work on recreation, game and fish, and publicity.
1915 Helps found game protective associations throughout the Southwest and begins editing the *Pine Cone*, bulletin of the Albuquerque (later New Mexico) Game Protective Association.
1918 After U.S. entry into World War I alters Forest Service priorities, leaves the service in January to accept a full-time position as secretary of the Albuquerque Chamber of Commerce.
1919 Rejoins Forest Service in August as assistant district forester in charge of operations, with responsibility for business organization, personnel, finance, roads and trails, and fire control for the twenty million acres of national forests in the Southwest. In December,

meets in Denver with Arthur Carhart to discuss prospects for a policy for wild and scenic areas (wilderness) in the Forest Service.

1922 Submits formal proposal for administration of Gila National Forest as a wilderness area (administratively designated by Forest Service on June 3, 1924).

1923 Completes *Watershed Handbook* (mimeographed) for district, reflecting observations on numerous inspection tours of southwestern forests.

1924 Accepts transfer to U.S. Forest Products Laboratory in Madison, Wisconsin, as assistant (later associate) director. Becomes increasingly uncomfortable with the industrial emphasis of the institution.

1927 Circulates initial chapters for book "Southwestern Game Fields," later titled "Deer Management in the Southwest" (unpublished).

1928 Leaves Forest Products Laboratory and Forest Service to conduct game surveys of midwestern states, funded by the Sporting Arms and Ammunition Manufacturers' Institute. Prepares survey reports for nine states and publishes book-length summary (1931). Delivers course of lectures on game management at the University of Wisconsin.

1930 As chairman of the Game Policy Committee of the American Game Conference, leads in formulating an American Game Policy, adopted in December.

1931 As a consulting forester, after the depression terminates his Arms Institute funding, conducts additional surveys for Iowa and Wisconsin conservation agencies and works on *Game Management* (published in May 1933).

1933 In the early months of Franklin Roosevelt's New Deal, May–June 1933, returns to the Southwest to supervise erosion control work in Civilian Conservation Corps camps for the Forest Service. In July, accepts appointment to a new chair of game management in the Department of Agricultural Economics at the University of Wisconsin.

1935 In January, assists in founding the Wilderness Society. In April, acquires the Wisconsin River farm ("the shack") that would be the setting for the almanac sketches. In autumn, studies forestry and wildlife management in Germany on a Carl Schurz fellowship. These experiences combined with insights from his 1936 trip to Mexico result in a profound reorientation of his thinking about the purposes of land management.

1936 Assists in establishing a society of wildlife specialists, by 1937 renamed the Wildlife Society. In September, makes first of two pack trips along the Rio Gavilan in Chihuahua, Mexico.

1938 Begins multi-year series of brief natural history articles for *Wisconsin Agriculturist and Farmer*.
1939 Becomes chairman of a new Department of Wildlife Management at the University of Wisconsin. Begins teaching Wildlife Ecology 118, an introductory course for liberal arts as well as wildlife students.
1941 Develops initial plans for a volume of ecological essays. U.S. entry into World War II draws off most students.
1942 Begins studies of excess deer problems in Wisconsin and nationwide.
1943 Appointed by governor to a six-year term on the Wisconsin Conservation Commission, a tenure dominated by debates over deer policy.
1944 Assembles ecological essays in book manuscript for submission to Macmillan, which rejects the project. Knopf expresses potential interest until November 1947, when it in effect rejects the book. Writing continues.
1946 After the war, students flock back to the university. Leopold develops recurring symptoms of tic douloureux, a painful condition of a facial nerve that impedes his work and finally requires surgery in September 1947. Some health problems continue.
1947 In December, submits revised book manuscript titled "Great Possessions" to Oxford University Press, which notifies him of acceptance on April 14, 1948. Published with few changes as *A Sand County Almanac* in fall 1949.
1948 Stricken by heart attack and dies on April 21 while helping to fight a grass fire on a neighbor's farm at the shack. Burial in Burlington, Iowa.

The River of the Mother of God

Introduction

Aldo Leopold's writing was remarkably wide-ranging. He addressed a broad spectrum of topics and mastered a number of literary genres. Nevertheless, almost all of Leopold's diverse interests as well as his varied professional career orbited around a stable center, land conservation, and the condensed essay was his preferred and perfected medium of expression. Throughout his life, Leopold returned periodically to topics on which he had written before, rearmed with fresh experience, new information, and critical reflection. And he was as keen a learner as he was a teacher. The papers included in this volume span a period of nearly half a century. In them one can see the author's thinking on a variety of distinct but interrelated themes evolving, becoming gradually richer, clearer, more subtle, and more profound.[1]

The chronological arrangement of this collection of papers, most of which are short essays, precludes the other natural alternative of grouping them into categories by subject. The purpose of this introduction is to identify the recurrent themes found in Leopold's hitherto less accessible literary estate and to trace their development diachronically.

The Almanac Themes and Their Evolution

Among the last things Leopold wrote before his untimely death was the foreword to *A Sand County Almanac*. In it he identified the major burdens of his book—ecology, esthetics, and ethics—and stated simply that "these essays attempt to weld these three concepts."[2]

In this collection, we can follow the evolution of Leopold's thinking on these topics, which were the concerns most important to him at the end of his life and the most familiar to his public.

Conservation Ecology

Some scientists see their work as separate from economics, politics, religion, and philosophy. Aldo Leopold could not so compartmentalize his thinking.

He consciously strove for an integrative understanding. Hence, for Leopold ecology was never just a pure science among other sciences. Nor was it for him merely an applied science, a means to an end, a tool to get more productivity from range, forest, and farm. For Leopold ecology was also a way of perceiving and comprehending that called into question the human ends as well as the means inherited from earlier epochs.

During Leopold's lifetime ecology came of age. The stages in its maturation are reflected in the evolution of Leopold's ecological worldview. To one degree or another ecology informs almost all these papers, but in some its larger implications are expressed in a particularly pure and pointed way.

The ecological science that Leopold first encountered in his forestry training at Yale was plant ecology, an essentially descriptive schema in botany developed in Henry C. Cowles's studies of plant succession on the sand dunes of Lake Michigan and Frederick E. Clements's work on "plant formations" and "climax" vegetation in Nebraska.[3] Early animal ecologists followed the lead of their botanical colleagues and sketched the fauna into the successional seres. Ecology had yet to take a functional approach to the total environment.

Thus in Leopold's early writing it is not surprising to find certain dogmas that were uncritically based upon contemporaneous ecological theory and canonized in forestry. For example, Leopold supported the tenet that forest fires should always be prevented because they set back succession in "Piute Forestry vs. Forest Fire Prevention" (1920). But the clearest indication of the limits of Leopold's early ecological thinking was his leadership in the effort to remove predators from southwestern forests and ranges in order to protect game, illustrated in "The Varmint Question." In 1915, when he began to organize game protective associations, "varmints" seemed expendable. It would be at least a decade before he started to sense that wolves and grizzlies might have a redeeming natural function and more than two decades before he would call explicitly for their preservation and restoration in "Threatened Species" (1936). Public acknowledgment of his personal intellectual transformation, which came in his review of Young and Goldman's *The Wolves of North America*, would require yet another decade.[4]

It is revealing that Leopold's most significant early advances in ecological thinking came not in game management, his chosen specialty, a field in which he was perhaps more committed to ground that he himself had previously staked out, but rather in watershed management, a field quite removed from his previous training and experience. This aspect of management dealt with what Leopold regarded as the most fundamental of all resources: the land itself.

"Grass, Brush, Timber, and Fire in Southern Arizona," published in 1924, records and summarizes the ecological discoveries that Leopold made

Leopold at Tres Piedras in Carson National Forest, New Mexico, c. 1911. Leopold's keen observations of landscape as he rode thousands of miles on horseback over the national forests of the Southwest formed the basis of his ecological understanding. (Leopold Collection, UW Archives)

during his years as chief of operations in the Forest Service's southwestern district, through which he traveled repeatedly on official inspections. The essay shows how far Leopold had advanced as a critical field ecologist since penning "Piute Forestry" and is notable for its independent thought. He asserts that "15 years of forest administration were based on an incorrect interpretation of ecological facts and were, therefore, in part misdirected." Leopold challenges the convenient assumption of both cowboys and forest rangers, which he himself had once shared, that more rather than less intense grazing by domestic stock is necessary to hold down grass that could carry fires, pointing out that southwestern watersheds had maintained their integrity despite centuries of periodic wildfire. Grass is a more effective conserver of watersheds than trees or brush, Leopold argues, challenging another dogma about the influence of forests in retarding erosion. Leopold does not reject Clements's model of succession out of hand, commenting that "the climax type is and always has been woodland." Rather, he makes a more discriminating use of it than he had before and than had most others in his profession. "The Virgin Southwest," written in 1927 and revised and presented as a lecture six years later, is less technical and formal and reaches similar conclusions based upon similar evidence.

A year before the painstaking analysis of "ecological facts" in "Grass, Brush, Timber, and Fire," Leopold had attempted a bolder stroke in "Some

Fundamentals of Conservation in the Southwest," drawing on ecology as an encompassing point of view. In "Some Fundamentals" he is concerned with the same problem, the deterioration of watersheds, but he moves quickly beyond the particulars to a discussion of basic philosophy, contrasting the organismic worldview of ecology with the prevailing "mechanistic conception of the earth." Quoting and paraphrasing the popular Russian mystic philosopher P. D. Ouspensky, Leopold suggests that we might conceive the whole earth to be a single living being. His scientific predecessors Clements, Cowles, and S. A. Forbes had pioneered organicism, regarding ecology as an extension of physiology—the "physiology" of plant and animal associations. But Leopold does not cite them in "Some Fundamentals," preferring to rely for support on Ouspensky, liberally buttressed by quotations from the Old Testament, John Muir, John Burroughs, and William Cullen Bryant's "Thanatopsis." He may not have cited any American ecologist because none had explicitly expanded the organismic model of plant and animal associations to encompass the whole earth. And certainly the supposition of "an invisible attribute—a soul or consciousness" that Leopold here speculatively attributes to the whole earth organism would transcend the limits of science altogether. Indeed, it may be significant that he never published the paper, possibly sensing that he was going too far out on a slender philosophical limb.[5]

Ecology's intensely holistic and frankly metaphysical organicism began to give way to a more pedestrian, but scientifically more palatable, social model during the 1920s.[6] The ecological "community concept" was perhaps most vividly expressed in the work of the distinguished British zoologist Charles Elton, who became a friend of Leopold in the early 1930s and whom Leopold especially admired for his ability to present the science of ecology to a lay audience, as Leopold's 1935 review of one of Elton's books testifies. According to Elton, plants and animals are links in "food chains." Each species plays a role in the "biotic community" or occupies a "profession" in the "economy of nature."[7] The ecological paradigm thus began to change from nature-as-organism to nature-as-society.

"The Arboretum and the University" (1934) directly links ecology to our most fundamental and comprehensive questions—the nature of nature and our proper human role within it. In a single sentence Leopold unites the two theoretical metaphors that were by then widely employed by ecologists: "Plants, animals, men, and soil are a community of interdependent parts, an organism." He goes on to develop the organismic ecological image and contrasts it to the Judeo-Christian "world view" (Leopold himself uses the term thus, in quotation marks) augmented by the "iron-heel mentality" associated with our age of "engines" and "machines." In "Engineering and Conservation," written four years later, Leopold expresses his belief that ecology contains the seeds of a philosophy of nature with the potential to supplant the reigning mechanical worldview epitomized by engineering: "All history shows this: that civilization is not the progressive elaboration of a single idea, but the successive dominance of a series of ideas. . . . Engineering is clearly the dominant idea of the industrial age. What I have here called ecology is perhaps one of the contenders for a new order."

The biotic community concept has become so thoroughly institutionalized in ecology that its essentially metaphoric character is often forgotten. The British plant ecologist Arthur G. Tansley, however, was acutely aware of the metaphoric (and metaphysical) character of the previous conceptual models in ecology and so proposed the term *ecosystem*, expressly inspired by physics, as a fully abstract, scientifically mature alternative in 1935.[8] Tansley called for the previously bifurcated organic and inorganic elements of the environment—plants and animals on the one hand, and soils and waters on the other—to be considered a systemic unit. Although Tansley did not emphasize energy dynamics in his seminal paper, his suggestion that ecology draw upon physics for a conceptual model invited quantification of Elton's qualitative "trophic pyramid," in which plant "producers" are eaten by animal "consumers," as solar energy, measurable at each point of capture, coursing through food chains.

Tansley's programmatic ecosystem paradigm is usually regarded as having languished until Raymond L. Lindeman fulfilled its promise in an influential paper published posthumously in 1942 (with an addendum by G.E. Hutchinson).[9] Lindeman presented lacustrine ecosystem dynamics as fundamentally a matter of energy transfer, and he is deservedly credited as the first to carry out the ecosystem model's invitation to quantification. But Aldo Leopold, three years earlier, clearly articulated its salient conceptual elements in a less technical paper, "A Biotic View of Land," which was presented to a joint meeting of the Society of American Foresters and the Ecological Society of America on June 21, 1939. In it, he begins with Elton's "biotic pyramid," as a more accurate representation of ecological relationships than the earlier image of a "balance of nature," and asserts that "land, then, is not merely soil; it is a fountain of energy flowing through a circuit of soils, plants, and animals. Food chains are the living channels which conduct energy upward; death and decay return it to the soil."

In "A Biotic View," moreover, Leopold examines ecological relationships in an evolutionary context and argues that the trend of evolution has been to diversify the biota. He posits a relationship between the complex structure of the system and the flow of energy through it—a relationship between the evolution of ecological diversity and the capacity of the land system for renewal, what he would later term "land health." The essay represents Leopold's most concerted effort to draw an ecological and evolutionary picture of the environment. Later he would extensively borrow from this essay in his effort to evoke a "mental image of land" in relation to which "we can be ethical" for "The Land Ethic."

In "Conservation: In Whole or in Part?" (1944) Leopold again forcefully sketches an ecological picture of nature, relying heavily once more on the organismic metaphor, and advocates a holistic approach to conservation, as the essay's title suggests. He compares the diversity and stability, or "health," of Wisconsin before and after European settlement, but when he turns to the implications of presently diminished diversity for future stability, he is painstakingly cautious: "The circumstantial evidence is that stability and diversity in the native community were associated for 20,000 years, and presumably depended on each other. . . . Presumably the greater the losses and alterations, the greater the risk of impairments and disorganizations."

Until recently conservationists rhetorically employed the hypothetical relationship between species diversity and biotic stability as if it were a natural law of ecology.[10] It is a tribute both to Leopold's scientific sensitivity and to his intellectual honesty that, while he believed coevolved diversity was necessary for stability in ecosystems, he recognized that the concept of stability eluded precise definition and that its dependence upon diversity was not well established empirically and therefore could not be stated categorically. In "The Land Ethic," he carefully says that "these creatures [wildflowers and songbirds] are members of the biotic community, and if (as I believe) its stability depends on its integrity, they are entitled to continuance."[11]

Natural Esthetics

Leopold's esthetic sensitivity to land was closely related to his conception of ecological systems. To be sure, his appreciation of beauty in nature antedated his ecological outlook, but as in all of his lifelong preoccupations, his esthetic tastes changed and matured as his ecological comprehension grew.

The first item in this collection, "A Tramp in November," was written in 1904 when Leopold was a seventeen-year-old schoolboy at Lawrenceville School in New Jersey. It is not *about* the esthetics of nature; that is, it does not provide an analysis of natural esthetics, as one may find here and there in parts 2 and 3 of the almanac.[12] It is, rather, an exercise in natural sensibility and expression, as pure a piece of descriptive nature writing as "Great Possessions" in the "shack sketches" of the almanac's part 1. Overwritten and self-conscious, it largely registers a painfully conventional esthetic response, featuring the picturesque view from the top of Stony Mountain, the peacock hues of the sunset which ornamented the young naturalist's homeward trek, and the cold light of the full moon on "the gray sea of a late-autumn landscape." In the last paragraph, the teenage author too clearly reveals his esthetic intentions (and his literary pretensions) when he writes, "There are some who even must go to Europe for scenery, and poets as well."

Nevertheless, read in light of the understated celebratory prose of the mature naturalist that he would become, the author's early essay reveals the seeds of the very different, ecologically informed "land esthetic" that he would later consciously articulate. He records *hearing* the "cheery voice" of a little chickadee (and *surmises* that it had eaten a "grub-worm") and the "loud rattle" of a kingfisher leaving its streamside perch. He sees the silent daylight flight of a great horned owl and *imagines* its "hunting call, a loud sonorous hooting, or on special occasions, a terrible shriek that is the most blood-curdling sound in nature." And he suggests the *function* of the owls' shriek: "frightening their prey into moving in its concealment."

Natural esthetics is a pitifully underworked topic in Western philosophical and critical studies and in Western intellectual history. This much, however, can be said with confidence: In the Church-dominated Middle Ages, sensitivity to natural beauty was regarded as vaguely sinful or at best a distraction from the soul's proper preoccupation with its spiritual pilgrimage.[13] In medieval painting the natural world, accordingly, was pictured as a symbolic backdrop for the artists' central religious motifs. After the European Renaissance, nature became a compelling subject of study for modern philosophers, as it had been for the ancients. Following suit, European painters made nature the subject of their art. Thus as the West became reattuned to nature and Western sensibilities were liberated from their medieval cloister, renascent esthetic appreciation of nature was mediated by painting. In sum, people saw landscape paintings in galleries, enjoyed an esthetic experience, and so turned to the painters' motif for similar gratification.[14] The classical Aristotelian dictum, Art imitates nature, was, in effect, inverted. Only nature that imitated art was considered worthy of esthetic contemplation. Since a painted landscape can be appreciated only by being seen, a natural "landscape," it seems generally to have been assumed, can only be appreciated in the same way. Countryside, correspondingly, was critically evaluated for its scenic and picturesque qualities.[15] Vision thus became the predominant sensory modality for experiencing natural beauty in the conventional Western natural esthetic, and natural beauty was judged by esthetic criteria originally developed for the evaluation of painting.

By contrast, in Leopold's revolutionary land esthetic all the senses, not just vision, are exercised by a refined taste in natural objects, and esthetic experience is as cerebral as it is perceptual. Most important, form follows function for Leopold as for his architectural contemporaries. For him, the esthetic appeal of country, in other words, has little to do with its adventitious colors and shapes—and nothing at all to do with its scenic and

10   Introduction

Estella and Aldo, c. 1913. Estella Bergere of Santa Fe nurtured Leopold's special esthetic sensitivity throughout their devoted marriage. (Leopold Collection, UW Archives)

picturesque qualities—but everything to do with the integrity of its evolutionary heritage and its ecological processes.

Thus, as one consequence, Leopold is able to appreciate esthetically country that he has never seen but only imagined. In "The River of the Mother of God" (1924), for example, he reflects on an unexplored river in the wilds of Amazonia. His attraction to the piñon jay in "Ho! Compadres Piñoneros!" (1929), on the other hand, derives from his sense that it represents, as he would later express this subtle esthetic idea, the *noumenon*, or imponderable essence, of the juniper foothills of the Southwest. Leopold adapted the concept of *noumena* from Ouspensky and used it to designate the vital signs, so to speak, of the health of land organisms, species without which ecosystems would lack wholeness and integrity.[16] Thus he prefers the bobwhite quail in "Grand-Opera Game" (1932) to the bigger and flashier, but artificially propagated and foreign pheasant, because the quail is a wild, native member of the American land community.

In "Land Pathology," a 1935 lecture, Leopold uses the terms *land esthetic* and *land ethic*, both apparently for the first time. Here he is concerned to reconcile and integrate "the utility and beauty of the landscape." He lampoons the conventional esthetic taste for "scenery" immured in the national parks as "an epidemic of esthetic rickets." And he stresses the fundamental importance to the conservation equation of "the revival of land esthetics in rural culture." But he does not attempt an analysis of "landscape beauty," focusing instead on economic and political policies to protect the public stake in private land use. Unfortunately for us, "land esthetics," he writes, "lies outside the scope of this paper."

Later in 1935, Leopold traveled abroad for the first time, studying forestry and wildlife management in Germany. This experience inspired the essay "Wilderness" in this volume, a handwritten draft of a speech Leopold possibly prepared for delivery while he was still in Germany. In it, he expresses an esthetic dissatisfaction not only with "what the geometrical mind has done to the German rivers" but also with the decidedly picturesque German landscape, because "to the critical eye, there is something lacking that should not be." The original complement of native species, especially predators, had been extirpated from the German *Wälder* by "cubistic forestry" and by "the misguided zeal of the game-keeper and the herdsman." Leopold laments especially the absence of "the great owl or 'Uhu,'" more often heard than seen, "without whose vocal austerity the winter night becomes a mere blackness." The Uhu, we realize, is to Leopold the *noumenon* of the German forest. Maintenance of an unnaturally high deer population by German gamekeepers, moreover, has resulted in "an illusive burglary of esthetic wealth . . . an unnatural simplicity and monotony in the vegetation of the forest floor, which is still further aggravated by the too-dense shade cast by the artificially crowded trees, and by the soil-sickness . . . arising from conifers."

In "Wilderness," as in the almanac's haunting "Marshland Elegy," Leopold fuses the evolutionary and ecological dimensions of informed perception. He expresses his displeasure with a "forest landscape . . . deprived of a certain exuberance which arises from a rich variety of plants fighting with each other for a place in the sun." But species diversity, the lack of which is discernible to the ecological eye, is an evolutionary legacy. "It is almost," he suggests, "as if the geological clock had been set back to those dim ages when there were only pines and ferns."

The following year, and again in 1937, Leopold would make some bow-and-arrow hunting trips to the Sierra Madre in Mexico. In the almanac's "Song of the Gavilan," he would distill an esthetic generalization from his esthetically positive experience in Mexico by means of an arresting musical metaphor elaborating the popular "harmony of nature" ecological conceit: "its score inscribed on a thousand hills, its notes the lives and deaths of plants and animals, its rhythms spanning the seconds and the centuries."[17] But in "Wilderness," from his esthetically negative trip to Germany, Leopold distills a complementary esthetic generalization by means of an even more powerful musical metaphor: "I never realized before that the melodies of nature are music only when played against the undertones of evolutionary history. In the German forest . . . one now hears only a dismal fugue out of the timeless reaches of the carboniferous."

Presenting his newly vested profession to a student audience in "Means and Ends in Wild Life Management" (1936), Leopold remarks that "our tools are scientific whereas our output is weighed in esthetic satisfaction rather than in economic pounds or dollars." He identifies rarity and wildness as two factors contributing to the esthetic value of wildlife species and the capacity to appreciate and foster "natural equilibria" as the mark of "good taste" in biotic communities. In "The State of the Profession," his 1940 presidential address to the Wildlife Society, he calls, in effect, for righting the relationship that Aristotle had proclaimed between art and nature and that had been turned on its head since the seventeenth century. "Is it not a little pathetic," Leopold asks rhetorically, "that poets and musicians must paw over shopworn mythologies and folklores as media for art, and ignore the dramas of ecology and evolution?" He goes on to envision a new genre of literature, the "ecological novel," and lauds scientifically informed painting and photography. Thus an autonomous natural esthetic might lead rather than follow artifactual esthetics.

Leopold's evolving natural esthetic permeates these papers, just as his mature natural esthetic subtly pervades *Sand County Almanac*. In sharp contrast with the prominent landmark papers on the way to "The Land Ethic," which stands majestically at the end of the almanac as its summit and fulfillment, those papers which have, like "Wilderness" and "Means and Ends," esthetics as their principal burden are difficult to identify. "The River of the Mother of God," "Ho! Compadres Piñoneros!" "Grand-Opera Game," "The Virgin Southwest," "Conservationist in Mexico," "The Farmer as a Conservationist," "The Last Stand," and "What is a Weed?" all document, in one way or another, Leopold's increasingly self-aware and deliberately expressed evolutionary/ecological esthetic sensibility.

Environmental Ethics

The third governing concept of the almanac, the ethical aspect of conservation, like the esthetic aspect, appears as early as Leopold's Lawrenceville years, when he wrote "The Maintenance of Forests." There is no intimation in this essay of his original and revolutionary land ethic; there is no hint, that is, of a human obligation to respect the forest itself or its trees and other constituents. Rather, perfectly reflecting its time, the Progressive era of American history, it treats forests as timber resources and condemns their destruction for private profit.[18] But the essay is certainly morally charged. The family business, the Leopold Desk Company, depended upon a sustainable supply of high-quality timber, and Leopold's father had been a pioneer in sportsmanship who impressed his outdoor ethic upon his sons. Thus, Aldo Leopold's eventual land ethic, as this early essay clearly reveals, was rooted deeply in his personality, his family heritage, and in the temper of the times of his formative years.[19]

Although "The Forestry of the Prophets" (1920) contains some hints of Leopold's search for a broader ethical consciousness, his first sustained effort to express a rationale for conservation beyond narrow utilitarian criteria came in 1923 with "Some Fundamentals of Conservation in the Southwest" (which is also liberally salted with quotations from and allusions to the Bible).[20] On the one hand, an understanding of the "delicately balanced interrelation" of water, soil, plants, animals, climate, and human technology, which Leopold expounds in "Some Fundamentals," has profound prudential implications for individual and collective human self-interest. Thus the first part of Leopold's essay summarizes his novel ecological analysis of the progressive deterioration of the organic "resources" of the Southwest and presents its upshot as an "economic issue," an issue that he further elaborates in a contemporaneous popular article, "Pioneers and Gullies." On the other hand, the same analysis brings to the fore a wholly new understanding of nature as "a coordinated whole, each part with a definite function." To Leopold's ethically oriented mind, this notion had distinctly nonanthropocentric (or, as he writes, non-"anthropomorphic") moral implications. Leopold points out that the view of nature "given us by physics, chemistry, and geology," that had ruled science since the seventeenth century, posits a "'dead' earth." But "we can not destroy the earth with moral impunity," he goes on to argue, if "the 'dead' earth is an organism possessing a certain kind and degree of life." These two aspects of "Some Fundamentals," the economic and the ethical, are united as two sides of one coin. The metal is ecology.

Leopold's next excursus into the ethical implications of ecology came exactly a decade later. The economic and ethical sides of the ecological coin are again present in "The Conservation Ethic" (1933), but they are cast in dramatically different ways. In "Some Fundamentals" and certainly in "Pioneers and Gullies," Leopold assumed that public awareness of the ruinous costs of land abuse would be sufficient to motivate reform. The responsible land use dictated by collective enlightened self-interest could be buttressed, secondarily, by a more noble motive—the ethic implicit in a fully elaborated ecological worldview. By 1933, in the midst of the worst depression in the nation's history, Leopold is not as sanguine about the effectiveness of the profit motive in promoting responsible land use. Employing examples drawn from agriculture, forestry, and migratory waterfowl, he illustrates in "The Conservation Ethic" how complex economic factors, from prices to taxes, conspire against conservation goals and expresses skepticism that any exclusively economic philosophy, however new or radical, could do the job. Hence, ethics must bear more weight.

Moreover, the ecological underpinnings of "The Conservation Ethic" are quite different from those of "Some Fundamentals." We find no mention of a whole earth organism or of a world soul. In the interval between "Some Fundamentals" and "The Conservation Ethic," ecology was shifting steadily from the Clementsian organismic model to the Eltonian community model. Leopold found in the new model, nevertheless, an equally compelling moral mandate: the necessity for *"mutual and interdependent cooperation* between human animals, other animals, plants, and soils." He also found the new model to be an equally radical departure from the prevailing mechanical paradigm, as his frequent disparaging references to "machines," "the machine age," "mechanical ingenuity," "machine-made commodities," and "better motors" amply attest.

He begins "The Conservation Ethic" with a recapitulation of the origin and development of ethics similar to that classically expressed in Charles Darwin's *The Descent of Man*. He then envisions an extension of human moral sentiments and social instincts beyond individual persons and society as a whole—stages Darwin himself had envisioned—to land, the "third element in human environment."

In "Land Pathology," written two years after "The Conservation Ethic," Leopold hints at an interesting relationship between a land esthetic and a land ethic. The former seems to be understood as a prerequisite or at least as an antecedent of the latter. From "mechanisms for protecting the public interest in private land," "remedial practices" (like reforestation and wildlife management), and "land esthetics," there "may eventually emerge a land ethic more potent than the sum of the three." And while Leopold seems to think of a land esthetic as an immediately achievable cultural accomplishment, a Darwinian evolutionary account of ethics evidently suggests to him here that the general infusion of a land ethic must await something more like a biological change: "The breeding of ethics is as yet beyond our powers. All science can do is to safeguard the environment in which ethical mutations might take place."

This statement might profitably be compared with similar remarks that Leopold makes in the almanac's closely studied "The Land Ethic." There, for example, in language appropriated without change from "The Conservation Ethic," he describes ethics as "actually a process of ecological evolution" and the land ethic as "an evolutionary possibility."[21] Thinking perhaps of human society itself as a kind of superorganism, he speculates that "ethics are possibly a kind of community instinct in the making," or, as he put it in "The Conservation Ethic," "a kind of advanced social instinct in-the-making." By 1947, in the material newly written for "The Land Ethic," Leopold

Aldo and Estella at the shack. Leopold's ethical consciousness was deepened through his personal struggles to rebuild a diverse, healthy, esthetically satisfying biota on his worn-out river-bottom farm in Wisconsin. (Bradley Study Center Files)

more incisively speaks of "the land ethic as a product of social evolution" and thus, it would seem, a cultural development, more difficult and distant of realization, perhaps, but at least a change of the same kind as required for the wide distribution of a land esthetic.[22]

Leopold worked out a major conceptual element of the land ethic in his 1939 "A Biotic View of Land," in which he posited a relationship between the evolved diversity, structural complexity, and healthy functioning of the biotic community. Ethical obligation in a system thus conceived consists in acting to preserve the health of the system by encouraging the greatest possible diversity and structural complexity and minimizing the violence of man-made changes.

The famous summary moral maxim of "The Land Ethic"—"A thing is right when it tends to preserve the integrity, stability, and beauty of the biotic community. It is wrong when it tends otherwise"—was adapted from "The Ecological Conscience," the last milepost essay on the way to "The Land Ethic."[23] The relationship between ecology and ethics is stated here perhaps more directly and succinctly than anywhere else: "Ecology is the science of communities, and the ecological conscience is therefore the ethics of community life." An ecological conscience, Leopold remarks, "is an affair of the mind as well as the heart." An ecological worldview, in other words, must inform our native social instincts for an environmental ethic to emerge as a cultural reality.

By 1947, a year before he died and the year during which "The Ecological Conscience" was written and "The Land Ethic" synthesized, Leopold had come to see economic self-interest and ecologically informed conscience as more often contradictory than complementary. Indeed, in "The Ecological Conscience" he expressly warns of "the dangers that lurk in the semihonest doctrine that conservation is only good economics." In three of the four case histories that compose the bulk of "The Ecological Conscience," economic pressures thwarted morally responsible land use. In the other, moral motives, uninformed by a biotic view of land, did more harm than good. The 1944 essay "Conservation: In Whole or in Part?" is even more emphatic about the need to decouple conservation practices from the profit motive and to link them instead with a sense of moral responsibility.[24]

Professional and Public Policy Themes

Throughout his career Aldo Leopold was engaged in conservation at all levels—from the mobilization of the local civic group or garden club to achieve some small victory for songbirds or wildflowers to the lofty overview of conservation's cognitive foundations, history, and future. Included in this collection are a number of papers that record Leopold's work with local folks for concrete and immediate conservation goals. Among the most memorable of these are "A Criticism of the Booster Spirit," "Helping Ourselves" ("coauthored" with farmer Reuben Paulson, although written in Leopold's distinctive, unmistakable style), "Coon Valley: An Adventure in Cooperative Conservation," the hard-hitting "Letter to a Wildflower Digger," and "Adventures of a Conservation Commissioner." Leopold's expansive philosophical vision grew directly out of his lifelong fieldwork and extensive hands-on experience. For all his literary productivity and wide reading, he was equally a man of action. Leopold was, in short, the consummate professional field naturalist/conservationist.

Professionalism is itself a subtle and, as it turns out, ironic thread running through these papers. It is ironic because Leopold was ultimately convinced that only ecologically literate and ethically motivated ordinary citizens could effectively conserve privately held, productive land, although he would use professional expertise to inform and assist them. The professionalization of conservation (as well as other fields from engineering and city planning to education) was a hallmark of the Progressive era.[25] The school essay "The Maintenance of Forests" foreshadows Leopold's choice of forestry as a profession. He turned his considerable powers of persuasion to upholding the ideals and expanding the province of professionalism within the Forest Service almost immediately upon joining it. His 1911 verse "The Busy Season," his 1913 open letter "To the Forest Officers of the Carson," and his 1922 "Standards of Conservation" are examples. He cared passionately about what happened on the ground, and he searched for ways to get the Forest Service, already becoming burdened with manifold administrative rules and other trappings of bureaucracy, to focus on "the Forest."

Leopold's multifaceted professional and public policy contributions developed on four major fronts, each of which he initiated during his Forest Service years and pursued with renewed vigor during his university career. Wildlife management, a new profession that he began to conceive and to shape early in Albuquerque, eventually led him out of the Forest Service so that he could pursue it full time in the Midwest. Conservation economics, a public policy interest developed in the Southwest, was honed in the University of Wisconsin's Department of Agricultural Economics. Sustainable agriculture, likewise a concern in the Southwest, also became much more focused in Wisconsin. Finally, wilderness preservation, for which Leopold campaigned vigorously in New Mexico, was a policy that he actively supported throughout his career.

Wildlife Management

The 1918 article "Forestry and Game Conservation" is a blueprint for the professionalization of game management within the administrative structure of the Forest Service. Grazing and recreation, under the Pinchot doctrine of "highest use," after all, had already expanded the mandate of the service beyond timber production and watershed protection. Game management seemed to Leopold a logical next step. The essay's governing concept is that "the history of game conservation in the United States has been exactly analogous to the history of forest conservation." Therefore, future professional game management may be modeled on past professional forestry. The essence of both is a strong scientific foundation. Point by point, Leopold explores the analogy: A game census is to game management what reconnaissance is to forestry; law enforcement against poaching is analogous to fire control, breeding stock to seed trees, license fees to stumpage rates, bag limits and closed seasons to limitation of cut, and game farm to nursery. The fundamental utilitarian/economic concepts of supply and demand, sustained yield, and market forces all apply equally to both, and so on.

Like "The Varmint Question" and "Piute Forestry," "The Popular Wilderness Fallacy" (1918) provides a starting point for measuring the extent to which Leopold's thinking was transformed by his subsequent ecological odyssey. The fallacy that Leopold explores in this essay is that we need wilderness to have game. On the contrary, Leopold argues, game penury is not an inevitable result of civilization any more than timber famine is (a proposition which in fact, appropriately qualified and in respect to particular species, he never abandoned, but which he certainly did not retain in the sanguine, laissez-faire form expressed here). The eradication of "varmints" following the conquest of wilderness will make more game available to increased numbers of hunters, fire suppression will save game and its food from destruction, flowages behind dams will replace drained and seasonally dried marshes for waterfowl, and so on. In short, everything will be coming up roses (or deer, in this case) after the complete human possession of the continent. Indeed, Leopold writes, "Nature was actually improved upon by civilization." It is as instructive to juxtapose this early piece with the rather apocalyptic "The Outlook for Farm Wildlife" of 1945 as to compare "The Varmint Question" with "Threatened Species."

While Leopold's enthusiasm in "The Popular Wilderness Fallacy" seems untempered by critical reflection, his thinking about wildlife and wildlife policy appears to be quite perspicacious in most of his other early papers. During his years in the Southwest he published numerous scientific notes and short articles detailing field observations about particular species, usually in the pages of the *Condor*. "Notes on the Behavior of Pintail Ducks in a Hailstorm" (1919) is an example. In the 1919 essay "Wild Lifers vs. Game Farmers: A Plea for Democracy in Sport," Leopold draws a sharp distinction in policy between scientific management for the perpetuation of wildlife in situ and artificial commercial propagation of game for shooting and meat. The latter was in essence the European system, notorious for its undemocratic (and thus un-American) social consequences. Here Leopold expresses preference for native over exotic species (e.g., the American heath hen over the Chinese pheasant) and he even regards the continuation of native species as an "ethical" matter transcending any "utilitarian objective." He would maintain this predilection in subsequent papers such as "Grand-Opera Game" (1932) and "Chukaremia" (1938). Although unregenerate on the question of "vermin" (to the point of openly conceding that game farmers have taken the lead in insisting upon the importance of their eradication) he recognizes "non-game" animals and "non-gunners" as appropriate objects and constituencies, respectively, of the "wild lifer's" art. Just as in forestry, in which the professional manages forest and range lands that are publicly owned, so in wildlife management, public ownership of hunting grounds seemed early on to Leopold an essential part of the conservation equation. He would later alter this emphasis significantly.

Outstanding among Leopold's many contributions to the professionalization of wildlife management was his leadership in the effort around

1930 to draft an American game policy. "Report to the American Game Conference on an American Game Policy" became Leopold's touchstone for a national campaign to implement its comprehensive vision and a positive mandate for the embryonic profession of game management. The game policy that Leopold proposed served as a blueprint for conservation measures that states would enact to maintain and enhance recreational hunting and fishing.

Its most problematic element was inducing the private landowner to perform a public service. While forest, range, and wilderness game species thrive on lands cheap enough to be publicly owned and managed on a scale large enough to be meaningful, farm game does not. The proposed policy advocated a shift in emphasis from public to private ownership and management of *land* for farm game, while retaining public ownership and access to all game *animals*. The key to the policy's feasibility was the creation of an apparatus to provide farmers, upon whose shoulders responsibility for producing farm game would thus settle, with encouragement, incentives, and assistance, short of granting them title to game that they could sell at whatever price the market would bear.

In response to curiosity and criticism, Leopold expanded the democratic philosophy animating his game policy in "Game Methods: The American Way" (1931). Exclusion of the "one-gallus" sportsman was the inevitable result of the European system, which wed hunting privileges and profits from marketing meat to management responsibilities, he explained. Such a system would be incompatible with the free-hunting American tradition. Of course free hunting was by then long past, but equal access to regulated hunting was a value of the American tradition that Leopold strongly believed could and should be preserved. Attacked from another quarter—by a preservationist who damned Leopold in the pages of the *Condor* for his apparent greater concern with producing game to shoot than with preserving wildlife for its own sake—Leopold poignantly and pointedly defended the purity of his personal ends, despite the practicality of his political and economic means, in "Game and Wild Life Conservation" (1932).

By 1940 Leopold could survey from its pinnacle the profession he had done more than anyone else to create. In "The State of the Profession," his presidential address to the Wildlife Society, he focused less on the achievements of the past (and, with characteristic modesty, not at all on his own) and more on the problems and opportunities confronting the profession as it continued to grow and mature. In addition to deeper commitment to basic ecological research, he looked forward to an "almost romantic expansion in professional responsibilities," as wildlifers would develop new extension enterprises to help landowners and sportsmen help themselves and generate new local teaching materials and university courses to reach the general student body and prospective teachers. Not surprisingly, however, Leopold's sights were set on something much more profound. The science of wildlife management was built upon a value premise, "loyalty to and affection for a thing: wildlife." Hence it might help to "rewrite the objectives of science" and "change ideas about what land is for." But to do that was "to change ideas about what anything is for."

Leopold himself had begun to reach out to landowners with a series of statewide radio talks and popular articles upon joining the faculty of the University of Wisconsin. In 1938 he began publishing how-to articles at least monthly in the *Wisconsin Agriculturist and Farmer*, some of which he eventually transformed into essays for the almanac. The 1940 "New Year's Inventory Checks Missing Game," for example, foreshadows "January Thaw." In 1939 he began offering a new wildlife course intended for liberal arts majors. "Ecology and Politics" was his riveting introductory lecture for the spring term of 1941, as the specter of total war hovered over the land. Leopold further elaborated his expansive professional vision in a paper presented at the seventh North American Wildlife Conference in 1942 entitled "The Role of Wildlife in a Liberal Education." Finally, in a midterm lecture, "Wherefore Wildlife Ecology?" probably presented in spring 1947, he mused again, for what might have been the last time, on the object of his professional quest.

Conservation Economics

Leopold's fascination with the economics of conservation paralleled both his own career changes and events in the larger world. When World War I shifted Forest Service priorities toward wholesale production of timber and livestock, he left the service for a stint as secretary of the Albuquerque Chamber of Commerce, where he hoped he could work more effectively to promote wildlife conservation. But commerce was not all conservation, and in 1919 he returned to the service, assuming responsibility for, among other things, business organization and finance. In 1924 he left the Southwest and endured four years as an administrator at the hopelessly utilitarian Forest Products Laboratory in Madison, Wisconsin, resigning in 1928 to concentrate on game management. And in 1933, in the depths of the depression, he joined the faculty of the Department of Agricultural Economics at the University of Wisconsin, where he was challenged to consider broader social and institutional issues, and began to experience firsthand the counterproductive results of conservation through government ownership of land and single-track resource management in the New Deal. Increasingly, Leopold began to emphasize the importance of personal stewardship on the part of private landowners, based ultimately upon attitudes and values that resist economic reduction. Indeed, the land ethic cannot be fully comprehended

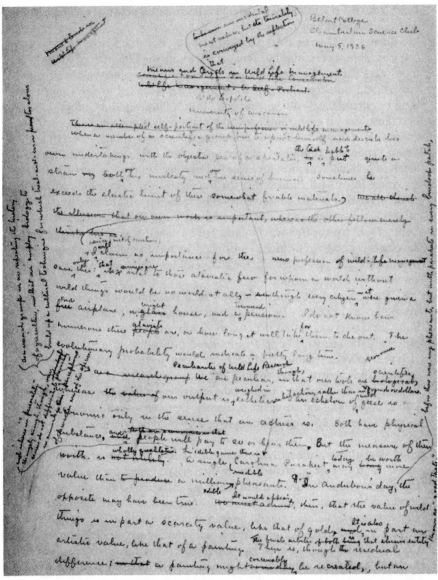

The original pencil draft of "Means and Ends in Wildlife Management," May 5, 1936. Leopold's heavy editing and search for just the right title reveal his commitment to clarity of expression and his struggle to come to terms with the deeper significance of his chosen profession. (Leopold Collection X25 2277, UW Archives)

apart from the essentially Jeffersonian reflections on political economy with which it was so closely connected.

In the 1923 article "A Criticism of the Booster Spirit," which was inspired, almost certainly, by his experience with the Chamber of Commerce four years earlier, Leopold blasted the narrow economic values of rapid growth and quick profits that substituted for sustainable development of local resources, local culture, and good taste. "Pioneers and Gullies" (1924) best illustrates his early thinking about conservation economics in regard to private lands in the Southwest. Searching for a way to make research at the Forest Products Laboratory relevant to the ordinary citizen and to his own ideals of political economy, he wrote "The Home Builder Conserves" (1928), an article filled with choices that the thoughtful and responsible consumer could make that would conserve wood by reducing unnecessary waste in its utilization. A decade and a half later, during World War II, in "Land-Use and Democracy," a piece that was far ahead of its time, Leopold would explicitly advocate boycotting products produced in an ecologically irresponsible manner and paying a premium for commodities purchased from businesses committed to conservation.

The geography of conservation is such that most of the best land will always be held privately for agricultural production. The bulk of responsibility for conservation thus necessarily devolves upon the private custodian, especially the farmer. Leopold formulated these ideas in a major address, "Conservation Economics," which he delivered to the Taylor-Hibbard Economics Club shortly after joining the university. The 1934 address is a brilliant and devastating critique of the effectiveness of conservation through public ownership and government agency. Leopold's description of conservation "experts" working at cross-purposes would be hilarious if it were not so disquieting. Here he suggests that economic incentives might reward good stewardship, primarily through differential property taxes. But good stewardship cannot be created solely by legislation and tax incentives. The private custodian's values and attitudes toward land must also change. "Conservation Economics" was undoubtedly conceived as a companion to "The Conservation Ethic" and should be read in tandem with it.

In "Land Pathology," an address delivered the day after Black Sunday, the most devastating of the Great Plains dust storms, Leopold focuses on the "vehicles for public influence on private land use" developed in "Conservation Economics" and restates his analysis "from a different angle," from an ethical and esthetic viewpoint. He diagnoses the failure of the "profit motive" to impel landowners to conserve and points out that government ownership and rehabilitation is a remedial, rather than a preventive, expedient. Insisting that finding practicable vehicles is a soluble research problem for social scientists, he does not go into detail but pleads instead "for positive

Leopold and his dog in Wisconsin farm country. Leopold's desire to retain habitat for farm wildlife broadened into concern for ecologically sensitive agricultural practices. (Bradley Study Center Files)

and substantial public encouragement, economic and moral, for the landowner who conserves the public values—economic or esthetic—of which he is the custodian."

Sustainable Agriculture

Leopold's most complete and elegant articulation and defense of the fusion of the economic and esthetic values that he only generally envisioned in "Land Pathology" is his 1939 address "The Farmer as a Conservationist." He notes here that we hold dear many things that are not profitable; as an extreme and somewhat far-fetched example he cites the unprofitable parts of our own bodies. But we are in fact amputating parts of the land organism that have no recognizable economic or utilitarian value. Then he imagines what a farmstead might be like when "self-expression" replaces "blind compliance with economic dogma."

In this beautiful essay Leopold's hope that an organic, ecological motif will replace the prevailing mechanical mentality in American agriculture resounds as an audible overtone. He notes that today "the farmer boy tending his tractor or building his own radio" is commonplace. A future farm boy "who becomes curious about why red pines need more acid than white is closer to conservation than he who writes a prize essay on the dangers of timber famine." (Was Leopold making an ironic reference to himself at age seventeen?) His description of the prototypical farmstead of a future ecologically literate America is especially poignant in view of the subsequent ever more intense mechanization of the countryside.

Leopold's interest in agriculture runs throughout these essays from beginning to end. Agriculture is a primary transformative force on the face of the land. In the early essays, which grew out of his experience as a forester in the Southwest, Leopold is acutely aware that agriculture has not only built empires but also destroyed them. In the later essays, which grew out of Leopold's experience as a professor of game management in a midwestern land-grant institution, one finds a persistent criticism of conventional agricultural practices and a clear vision of an alternative approach. In "The Outlook for Farm Wildlife," written as World War II was drawing to a close and the prospect of war-spawned technologies unleashed on the homeland loomed on the horizon, Leopold anticipates the critique of industrial agriculture by such contemporary mavericks as Wendell Berry and Wes Jackson; in this article he introduces the concept with the phrase "industrialization . . . of farm life."[26] And in "The Farmer as a Conservationist" and "A Biotic View of Land," he sketches an alternative "biotic farming," or sustainable agriculture, as presently it is called.

Wilderness

Beyond *A Sand County Almanac*, Aldo Leopold is probably best known for his advocacy of a system of wilderness preserves and his role in its eventual establishment. His wilderness papers fall into three sets, one for each of the decades—the 1920s, 1930s, and 1940s. They not only document his advocacy, but they also dramatically reveal the impact of his maturing ecological outlook on his thinking about a particular concern.

The first paper, "Wilderness and Its Place in Forest Recreational Policy," appeared in 1921 and was addressed to Leopold's Forest Service colleagues. The principal, indeed the only, reason for wilderness preservation that it advances is recreation, which should not be surprising since Leopold had earlier been in charge of recreation for the southwestern district of the national forests. Wilderness is even defined by Leopold in recreational terms, as an area "big enough to absorb a two weeks' pack trip." This definition occurs throughout the 1920s wilderness series. His rationale is explicitly,

Leopold memorial at Gila Wilderness Area, New Mexico, dedicated September 1954. A portion of the Gila was designated the Aldo Leopold Wilderness in 1980. (U.S. Forest Service Photo, Leopold Collection X25 1937, UW Archives)

and, for his audience, appropriately, utilitarian. He expressly invokes Gifford Pinchot's famous adaptation of John Stuart Mill's maxim, The greatest good to the greatest number, and Pinchot's doctrine of "highest use." He also draws a historical comparison between the present need for wilderness areas and the earlier need for the national forests themselves. Finally, Leopold appeals to the American political principle of minority rights and suggests that one distinctive area of each state, unsuited to industrial development, be set aside as wilderness. The headwaters of the Gila River was Leopold's favorite candidate in his beloved Southwest. A wilderness area plan that Leopold drafted during a 1922 inspection of the Gila Forest became the basis for administrative designation of the Gila as wilderness in 1924, setting the pattern for what would eventually become a nationwide system of wilderness areas.

In 1925 Leopold published no fewer than six pieces on wilderness (including his "skit," "The Pig in the Parlor") and would have published a seventh if the *Yale Review* had accepted "The River of the Mother of God." This publication blitz was understandably stimulated by the conspicuous failure of the National Conference on Outdoor Recreation to address the subject in 1924. Each of Leopold's articles makes essentially the same arguments, but each is cast in a distinctly different tone and vocabulary, carefully crafted to persuade its target audience.[27] "Conserving the Covered Wagon" and "The River of the Mother of God" are no less utilitarian than Leopold's earlier wilderness manifesto had been, but they take the higher ground. In the former, inspired perhaps by his new neighbor in Madison, Frederick Jackson Turner, Leopold argues that wilderness preserves for posterity the physical remnants of the pioneer episode in American history and, therefore, the distinctly American virtues forged on the frontier. He also suggests that wilderness conservation is spiritually expansive and morally ennobling. The latter is even more abstract and explores the relationship of wilderness to the human psychological need for the mystery of "Unknown Places."

In all the 1920s wilderness papers the "good roads movement" is the bête noire. In "The River of the Mother of God," roads and motors virtually become metaphysical symbols. The phrase "the Great God Motor" and other similar references in this essay reveal that the automobile represented for Leopold much more than an indiscriminate intrusion into wild country. It was emblematic of the obsolescing "mechanistic conception of the earth," to which he had contrasted an organic conception in "Some Fundamentals," written not long before.

"Wilderness as a Form of Land Use" marshals the arguments developed in the other contemporaneous wilderness papers and expresses them more formally and systematically. Leopold notes that his notion of wilderness is one of "relative condition" and "exists in all degrees." In his initial proposals he had suggested that the scattered cattle ranches in the Gila added to rather than detracted from the flavor of the country and that the necessaries of forest fire prevention—towers, telephone lines, and access trails—were acceptable improvements. These qualifications are reaffirmed here. More fully than elsewhere, Leopold here develops the national character argument, as it might be called. The distinctive attributes of American culture grew out of the interaction of Europeans and the wild North American landscape, he explains. To perpetuate these valued qualities in future Americans we must preserve the frontier environment in which they were forged. In constructing the almanac's penultimate essay, devoted to wilderness, Leopold borrowed several passages from "Wilderness as a Form of Land Use" that make this point.

Leopold's return to the wilderness theme in the mid-1930s was brief but telling. In early 1935 he helped found the Wilderness Society and wrote a frequently reprinted rationale for the organization, "Why the Wilderness Society?" The aforementioned "Wilderness," a product of Leopold's trip to Germany later that year, is about wilderness only in the sense that it comments on Germany's conspicuous lack of it. Germany lacks not only wilderness, Leopold observes, but because of a "former passion for unnecessary outdoor geometry," it also lacks a "degree of wildness" compatible with "utility." This piece suggests that for all his advocacy and defense of wilderness, Leopold was primarily interested in an optimal mix of wild and managed land, a "more important and complex task" than saving "wild remnants." He returned to the United States determined to avoid the overly artificial management of land and wildlife that so profoundly unsettled him in Germany.

Leopold looked back on his quite different Mexican adventures in the autobiographical foreword that he submitted to Alfred A. Knopf with his manuscript "Great Possessions" (which later became *Sand County Almanac*): "It was here that I first clearly realized that land is an organism and that all my life I had seen only sick land, whereas here was a biota still in perfect aboriginal health."[28] A literary record of that realization is preserved in "Conservationist in Mexico" (1937). Leopold's trips to Germany and Mexico—places at opposite ends of the spectrum of wilderness values—framed his thinking about wilderness in the 1930s.

Stimulated by these dramatically contrasting foreign excursions, he began to formulate a more subtle, essentially ecological rationale for wilderness preservation. The idea that each state should have one wilderness preserve for the affordable experience of its residents was transformed into the idea that each biome should have a wilderness preserve as a benchmark of health and stability against which ecologists might measure the disorganizing impact of land-use technologies in similar biotic communities elsewhere. Wilderness, in short, had hitherto unheralded scientific values. Complementing Leopold's deepening commitment to a nonutilitarian conservation ethic, these values provided a more profound and compelling basis for his defense of wilderness.

In the 1941 article "Wilderness as a Land Laboratory," Leopold more fully elaborated the "need of wilderness as a base-datum for problems of land-health." In addition to several examples of "land-sickness" that it reviews, this paper is significant for the persistent presence of the unmixed organismic ecological metaphor that governs it. Leopold refers not only to the familiar health and sickness tropes of the organismic metaphor but also to the "physiology of Montana." One can only wonder if his use of "physiology" is a conscious allusion to Clements and Cowles. The essay "Wilderness" in the almanac was, like "The Land Ethic," partly constructed from earlier essays. "Wilderness as a Land Laboratory" was prominent among them.

Conclusion

Recent years have witnessed a renewed appreciation of Aldo Leopold's contribution to American letters and science, to the cause of wilderness preservation and wildlife management, and to philosophy (at least as philosophy once was understood—as disciplined reflection on how the world works, how we fit into it, and what is ultimately worthwhile). Aldo Leopold's popular reputation has rested until now upon a slender literary footing, almost exclusively upon *A Sand County Almanac*. This volume of Leopold's forgotten essays, we hope, will deepen public understanding of his multifaceted contribution.

These essays do not show us a new Aldo Leopold. They do, however, provide a different mix and a stronger dose of the Leopold with whom we are fondly familiar. Here Leopold the activist and advocate is more prominent along with Leopold the innovative and ever-growing conservation professional. Here too is a clearer view of Leopold's practical work with both soil and game on the ranges and forests of the Southwest and the farms and marshes of Wisconsin. We also see his struggle with the bureaucratic, political, and economic dimensions of conservation. Perhaps the greatest value of these essays is the insight they afford into the development of Leopold's thought on a variety of topics, including natural esthetics, environmental ethics, wildlife management, conservation economics, agriculture, and wilderness, all driven by his evolving systemic worldview which matured apace with the science of ecology.

Notes

1. For further discussion of Leopold's life and thought, and the biographical context of most of the papers included here, see Susan L. Flader, *Thinking Like a Mountain: Aldo Leopold and the Evolution of an Ecological Attitude toward Deer, Wolves, and Forests* (Columbia: University of Missouri Press, 1974); and Curt Meine, *Aldo Leopold: His Life and Work* (Madison: University of Wisconsin Press, 1988).

2. Aldo Leopold, *A Sand County Almanac and Sketches Here and There* (New York: Oxford University Press, 1949), ix.

3. See Frederick E. Clements, *Research Methods in Ecology* (Lincoln: University Publishing Co., 1905); and Robert P. McIntosh, *The Background of Ecology: Concept and Theory* (Cambridge: Cambridge University Press, 1985).

4. See Flader, *Thinking Like a Mountain*, for a study of the evolution of Aldo Leopold's attitude toward predators. Donald Worster, *Nature's Economy: The Roots of Ecology* (San Francisco: Sierra Club Books, 1977), suggests that attitude toward predators is a litmus test for an ecologically informed outlook: "The story of the varmint's changing reputation is thus the story of the movement in American conservation toward an ecological point of view" (261). In Worster's opinion, "Leopold was rather slow to switch to this new attitude; but when he did, he came over with an elegance and credibility that quickly made him one of the leaders of the new ecological element" (274).

5. For further discussion of Leopold's papers on southwestern watersheds see Susan L. Flader, "Leopold's 'Some Fundamentals of Conservation': A Commentary,"

Environmental Ethics 1 (Summer 1979), 143-148; and Flader, *Thinking Like a Mountain*.

6. See Worster, *Nature's Economy*; and McIntosh, *Background of Ecology*.

7. Charles Elton, *Animal Ecology* (London: Sidgwick and Jackson, 1927).

8. Arthur G. Tansley, "The Use and Abuse of Vegetational Concepts and Terms," *Ecology* 16 (1935), 284-307.

9. Raymond L. Lindeman, "The Trophic Dynamic Aspect of Ecology," *Ecology* 23 (1942), 399-418, with an addendum by G. E. Hutchinson. McIntosh, *Background of Ecology*, writes, "The term *ecosystem* lay fallow for several years until it was integrated with the trophic-dynamic concept stimulated by progress in the study of food cycles in lakes by Raymond Lindeman (1942)" (98).

10. The coup de grace to the diversity-stability hypothesis appears to have been delivered by Daniel Goodman, "The Theory of Diversity-Stability Relationships," *Quarterly Review of Biology* 50 (1975), 237-266. A more recent and temperate review is provided by Thomas Zaret, "Ecology and Epistemology," *Bulletin of the Ecological Society of America* 65 (1984), 4-7.

11. Leopold, *Sand County Almanac*, 210.

12. For discussions see J. Baird Callicott, "Leopold's Land Aesthetic," *Journal of Soil and Water Conservation* 38 (1983), 329-332, and "The Land Aesthetic," in *Companion to "A Sand County Almanac": Interpretive and Critical Essays*, ed. J. Baird Callicott (Madison: University of Wisconsin Press, 1987), 157-171.

13. See Alfred Biese, *The Development of the Feeling for Nature in the Middle Ages and in Modern Times* (London: G. Routledge and Sons, 1905).

14. See Eugene C. Hargrove, *Foundations of Environmental Ethics* (Englewood Cliffs, N. J.: Prentice Hall, 1989), for a discussion of the link between landscape painting and conventional Western natural taste.

15. A device of the period, the Claude-glass, named for the seventeenth-century French landscape artist, Claude Lorrain, suggests how thoroughly derivative the newly emerged European natural esthetic was from its artifactual antecedent. Natural esthetes carried the rectangular, slightly concave, tinted mirror with them into the countryside. Upon finding a suitably picturesque prospect, they turned their backs to it and looked at its image in the mirror. Thus framed in the Claude-glass, the natural landscape appeared to be almost (but of course not quite) as pretty as a picture. For a discussion, see Christopher Hussey, *The Picturesque: Studies in a Point of View* (London: G. P. Putnam's Sons, 1927); and J. Baird Callicott, "The Land Aesthetic," *Orion Nature Quarterly* 3 (Summer 1984), 16-23.

16. See "Guacamaja" in Leopold, *Sand County Almanac*, 138. See Callicott, "Land Aesthetic," in *Companion to "A Sand County Almanac*," for a discussion of Leopold's esthetic concept of *noumena* and its relation to Kant's concept of *noumena*.

17. Leopold, *Sand County Almanac*, 149.

18. See Samuel P. Hays, *Conservation and the Gospel of Efficiency: The Progressive Conservation Movement* (Cambridge: Harvard University Press, 1959), and Stephen Fox, *John Muir and His Legacy: The American Conservation Movement* (Boston: Little, Brown and Company, 1981). Worster, *Nature's Economy*, provides a discussion of Leopold's historical relationship to progressivism.

30   Introduction

19. For a discussion of all three, see Flader, *Thinking Like A Mountain*; Meine, *Aldo Leopold*; and Boyd Gibbons, "A Durable Scale of Values," *National Geographic* 160 (1981), 681-708. See also Frederic Leopold, "Aldo's School Years: Summer Vacation"; Sharon Kaufman, "Built on Honor to Endure: Evolution of a Leopold Family Philosophy"; and Carl Leopold, Estella Leopold, Luna Leopold, Nina Leopold Bradley, and Frederic Leopold, "Reflections and Recollections"; all in *Aldo Leopold: The Man and His Legacy*, ed. Thomas Tanner (Ankeny, Iowa: Soil Conservation Society of America, 1987), 145-151, 153-160, 165-175.

20. The oft-repeated characterization of Leopold himself as a "prophet" appears traceable to Roberts Mann, "Aldo Leopold: Priest and Prophet," *American Forests* 60 (Aug 1954), 23, 42-43. The epithet was picked up, apparently, by Ernest Swift, "Aldo Leopold: Wisconsin's Conservation Prophet," *Wisconsin Tales and Trails* 2 (Sept 1961), 2-5. It was given academic imprimatur by Roderick Nash in his chapter "Aldo Leopold: Prophet," in *Wilderness and the American Mind* (New Haven: Yale University Press, 1967), 182-199. Ron Myers's *A Prophet for All Seasons* has transferred the prophet cachet to the medium of film. The biblical and prophetic associations evoked in readers by the almanac and its author may be grounded in Leopold's style, which is often biblical in its allusions and phrasing; for a discussion see John Tallmadge, "Anatomy of a Classic," in *Companion to "A Sand County Almanac,"* ed. Callicott, 110-127. See Donald Fleming, "Roots of the New Conservation Movement," *Perspectives in American History* 6 (1972), 7-91; Clay Schoenfeld, "Aldo Leopold Remembered," *Audubon* 80 (May 1978), 28-37; and Wallace Stegner, "Living on Our Principal," *Wilderness* 48 (Spring 1985), 5-21, for the claim that *A Sand County Almanac* is the "bible" of the modern conservation/environmental movement.

21. Leopold, *Sand County Almanac*, 202, 203.

22. Ibid., 225.

23. Ibid., 224-225.

24. For more detailed discussion of "The Land Ethic" and its antecedents, see Curt Meine, "Building 'The Land Ethic,'" and J. Baird Callicott, "The Conceptual Foundations of the Land Ethic," both in *Companion to "A Sand County Almanac,"* ed. Callicott, 172-217; Susan Flader, "Aldo Leopold and the Evolution of a Land Ethic," in *Aldo Leopold*, ed. Tanner, 3-24; and J. Baird Callicott, *In Defense of the Land Ethic: Essays in Environmental Philosophy* (Albany: State University of New York Press, 1989).

25. See Hays, *Conservation and the Gospel of Efficiency*.

26. Wes Jackson, Wendell Berry, and Bruce Coleman, eds., *Meeting the Expectations of the Land: Essays in Sustainable Agriculture and Stewardship* (San Francisco: North Point Press, 1984), is dedicated in part to Aldo Leopold. See Wendell Berry, *The Unsettling of America: Culture and Agriculture* (San Francisco: Sierra Club Books, 1977), and *The Gift of Good Land: Further Essays Cultural and Agricultural* (San Francisco: North Point Press, 1981); and Wes Jackson, *New Roots for Agriculture* (San Francisco: Sierra Club Books, 1980), and *Altars of Unhewn Stone: Science and the Earth* (San Francisco: North Point Press, 1987). For a discussion, see

Curt Meine, "The Farmer as Conservationist: Leopold on Agriculture," in *Aldo Leopold*, ed. Tanner.

27. For a discussion, see Craig W. Allin, "The Leopold Legacy and American Wilderness," in *Aldo Leopold*, ed. Tanner, 25-38; Susan L. Flader, "'Let the Fire Devil Have His Due': Aldo Leopold and the Conundrum of Wilderness Management," in *Managing America's Enduring Wilderness Resource*, ed. David W. Lime (St. Paul: Minnesota Extension Service, 1990), 88-95; and Curt Meine, "The Utility of Preservation and the Preservation of Utility," in *The Wilderness Condition: Essays in Environment and Civilization*, ed. Max Oelschlaeger (San Francisco: Sierra Club Books, 1991). For the history and current status of the Rio Madre de Dios (the River of the Mother of God), see Susan L. Flader, "The River of the Mother of God: Introduction," *Wilderness* 54:192 (Spring 1991), 18-22.

28. Aldo Leopold, "Foreword," in *Companion to "A Sand County Almanac,"* ed. Callicott, 285-286.

A Tramp in November [1904]

During his school years in Iowa and later in New Jersey and Connecticut, Leopold took hikes into the countryside several times a week, a practice he continued throughout his life. He described many of his tramps in letters to his mother, but one he used as the topic for a classroom essay at Lawrenceville School in New Jersey, in which he enrolled in January 1904. The essay reveals his keen observation of landscape and wildlife and his youthful command of a fine, if somewhat overblown, narrative style. Found in an undated brown notebook, it is handwritten and contains several minor edits and comments by a teacher (omitted here). The essay was probably written in November 1904 when Leopold was seventeen.

On a certain morning of that good old month, November, the sun was rising with even more than his usual alacrity. For old Sol had been accused of oversleeping of late, and wanted to justify himself; moreover he likes work like any good man, and Jack Frost had been giving him a plenty of it. Nor was it missing on this particular morning, for lo! no sooner had his rays overtopped the southeastern hills, but the whole land sparkled with the frost crystals it had gathered during the night. But old Sol only went to work the more merrily, for the clear, cold air was in his favor, and the task a pleasant one.

Now it is a hard thing, even for an honest man, to see another begin a pleasant task, and then to go about his own daily round without envy in his heart. So on this particular morning, when the writer arose (late, as is the manner of men) to resume his studies, and looking out upon the diamond-bedecked Campus, saw the sun setting about his work so cheerfully, he vowed a great vow, and it was this: "I will tramp to Stony Mountain and back today, if it take till midnight to do it." Then he thought that this was surely the fairest county in all Jersey, and that after all it is a great thing to be alive. For a breath of the glorious morning strayed through the open window and blew all envy away.

How those morning hours of work did fly! For at noon the sun seemed

A Tramp in November.

In a certain morning of that good old month, November, the sun was rising with even more than his usual alacrity. For old Sol had been accused of oversleeping of late, and wanted to justify himself; moreover he likes work like any good man, and Jack Frost had been giving him a plenty of it. Nor was it missing on this particular morning, for lo! no sooner had his rays overtopped the southeastern hills, but the whole land sparkled with the frost crystals it had gathered during the night. But old Sol only went to work the more merrily, for the clear, cold air was in his favor, and the task a pleasant one.

Now it is a hard thing, even for an honest man,

First page of "A Tramp in November," c. November 1904. (Leopold Collection X25 2223, UW Archives)

hardly to have completed his frost-melting, when clothed and shod for travel I started out. To be sure it had become warmer, for a little Chicadee back in the orchard was piping at the top of his cheery voice, before I even emerged from the village, that a grub-worm had ventured forth and fallen his victim. I left him behind still piping, and crossing the Trolley, set off due southward over the hills. It was a sore temptation to break into a run wherever a particularly long hillside or a flat stubble-field stretched away before, but it was without that indulgence that I reached the Stony Brook Divide, whence the first sight of Stony Mountain is to be had. Even in the clear air, for the sky was of almost an April-blue, the mountain appeared to be a great dark wave far away over the sunny ripples of the hills between. So hastening on through the fields and woodlots of the valley, I soon reached Stony Brook itself, just below the quaint old "Upper Mill" with its tree-grown dam. Wading across the cold water, where a patient Kingfisher leaves his perch with a loud rattle and is away down the stream, I climbed the opposite bank with the thought that one third of the distance was left behind. Thence it was off over the fields again, on and on, so that at half past three I was at the border of the wooded mountain-side, with the long climb ahead.

What great satisfaction there is in plowing through the rich brown autumn leaves of the woods on a fine sunny day! But add to that the pleasure of setting one's toes through them into the mountain-side, and with mighty pushes such as only the bracing air enables one to give, to scale it to the rocky heights above—and you have what it was like to reach the summit that November afternoon. Then it was grateful to sit on a mighty boulder of granite that the glaciers may have brought all the way from Greenland just for the occasion, and breathing the rare ozone of the heights, enjoy the view.

And such a view! For twenty miles in every direction were wooded hills and peaceful valleys. In a great semicircle the Princeton Hills swept off to the northeastward till they were blue in the distance. Ahead were merely hill and dale, to the far east shading off into the sand hills of the coast. And to the southwest, with a great majestic dip, lay the valley of the Delaware, with the opposite bluffs lying like low clouds far away. Everything lay peaceful in the evening hour (for the sun was low, and the crows beginning to fly in silent bands to the westward, e're I awoke from my reflections), and it was with regret that I took a last look and began the descent.

The way down the mountain lay through a little copse of Red Cedars, and I was passing through the midst of them, when not twenty paces away a Great Horned Owl flew from their protecting needles and noiselessly flapped away through the woods. It is marvellous how so great a bird (for this one must have spread at least five and a half feet) can pass in absolute silence through the densest tangle of trees and branches. And equally mysterious are their ways of frightening their prey into moving in its concealment by sounding the hunting call, a loud sonorous hooting, or on special occasions, a terrible shriek that is the most blood-curdling sound in nature. But suffice to say that I was much pleased at seeing the Owl, it being a hitherto unobserved species for me in this state.

On the way back I soon fell into that pace which only a cold evening can inspire. Yet there was time to look back from time to time at the magnificent and ever changing sunset. At first only great lights of amber, yellow, and liquid green, it gradually changed to masses of purple and rose against a background of the most delicate lilac. Then, as if inspired with new vigor, the whole heavens became flushed with bars and undulations of fiery red, shading overhead into blurred crimson and then maroon and dusky purple. And when the first stars began to twinkle, it had again receded to a warm glow of embers between black clouds, gradually fading and sinking away.

Hardly had the glowing orb of the sun disappeared when the full moon, at first a mere white area against the dark sky began to shed its light over the fields. How much more wan and cold the moonlight is on the gray sea of a late-autumn landscape, than on the dark green grass of only a few weeks before! Nevertheless there is a certain added cheer, and many new beauties brought to view. For instance, what is a simpler yet more beautiful picture than the graceful branches of a beech tree seen against the moonlight? How the gray, smooth trunk, the dismantled twigs, with here and there a few leaves, fluttering sere and dry, and the safely covered yet bare buds, seem to express the season of the year! There are those who say that only Spring is beautiful, and hie themselves to a warmer climate for the winter months. There are others, who, without the means of fleeing from the beauties about them at all seasons, waste their happiness with complaining. And there are some who even must go to Europe for scenery, and poets as well. But give me my native land at all hours of the day, all seasons of the year, and for all the years of my life; because its beauties, its interests, and its ennobling influences are intended for its sons above all others. And have no fear, all who may doubt, that anyone by trying may get far more of use from a stroll over its hills, than even I did on this momentous 'Tramp in November.'

The Maintenance of Forests [1904]

This essay concerning the mindless destruction of forests and the need for rational policies reflects Leopold's outlook on his chosen profession before he began to study forestry at Yale and even before the U.S. Forest Service was established in 1905. The version here is from a handwritten revision, dated November 1904, of a paper written the previous June. Leopold presented it at Lawrenceville School's annual contest in original speaking in December.

It has become a generally recognized fact that wood is, and will continue to be, one of the necessaries of life. In spite of the progress and the possibilities of science, all who have given the subject consideration, say that wood is indispensible to our welfare, and that no substitute is likely to be found for it. Furthermore we know that the lumber supply of our country, once believed to be inexhaustible, is now almost used up; two decades, it is estimated, will see its end. This present supply is confined mainly to the northern and western states, and even there an average of only about ten percent remains of the original stand of valuable lumber.

Two agents have brought about this condition. They are, first, the lumberman, second, the forest fire. To see how these act upon a lumber region, let us follow up the history of a given tract of timber which has been exposed to their ravages. In a primeval state its condition is largely stationary; trees grow up, die of old age, and are followed by another generation. But when settlers come in, they make permanent clearings. If the land is arable, well and good, but if the soil is not fit for farming, as is generally the case in places where those trees, valuable for lumber, grow, the land is forced through a process which we are pleased to call "lumbering." Lumbering a region ought to mean "gathering its forest crop," but only too truly it generally does mean "destruction of the forest crop." This is the modern method which has caused the squandering of our timber resources, and which is everywhere employed.

Cutters go over a region felling every tree large enough to make a sawlog, and then every remaining tree large enough to help build the "skid" over which the saw-logs are rolled to some road or stream, whence to be taken to the saw-mill. The "skid" is made, saw-logs rolled out and taken to the mill, and the region abandoned for new fields of destruction. Abandoned to what? Need one ask? Look at the ground, a continuous mass of dry tree-tops and skidding logs, and the answer is simple. Abandoned to the tender mercies of the forest-fire. It needs only a spark from an engine, or the like, to kindle a great tide of destruction which may sweep for miles and miles over the land and leave behind it only smouldering remains. The very soil is often consumed down to bare and sterile rocks. The slopes, deprived of the binding roots, soon become eroded and cut into deep gullies, and no vegetation save the hardiest weeds can exist. Where was yesterday a bountiful land, is today a barren, lifeless, waste, destined so to remain for years to come or perhaps forever.

Fires such as this often range over hundreds of square miles, and their deplorable results are constantly in evidence in every timbered region. As has been shown, they destroy not only the standing timber but also all opportunity of rendering the land useful in the future. And it is seldom that land, lumbered by present methods, escapes them.

It is now reasonable to ask, "How can this useless destruction be prevented?" The problem has been solved by some of the European nations, and can be solved by the United States at will.

The whole fault lies in careless and unnecessary methods in handling forest lands. The lumberman, in the mad race for his dollar, is blind to the future; too blind to see that a rational policy will not only bring him as much present profit as his destructive policy, but more than this, that that same profit will be continued indefinitely by it. "What is a Rational Forest Policy?" It is to cut no tree that is not to be used, and to so heap up the branches and tops of the trees that are used that fire cannot spread from one to another. Even from a local standpoint what can the lumberman lose by such a policy? Obviously nothing. And what does he gain? A forest crop for all years, instead of the same crop for one, and only one year.

If these statements are true, it seems incredible that in this age of enlightenment such methods should continue. But they do continue, and in common with many other evils of today, have one underlying cause, namely: the passion for "Money, and money quick!" And the lumbermen are not the only ones who bring this cause to bear. It is the people, speaking indirectly through the government, who are to blame and to whom one must look for a remedy. What can we expect, if our government collects the same taxes from forest land as it does from farms in the same region! The owners of forests quickly get what they can by indiscriminate cutting and then cede the land back to the government to avoid the taxes. The government holds today hundreds of thousands of square miles of such land, rendered a desert for years to come. And all for a few dollars taxes for a single year, or perhaps two! This is the "Prosperity" of the lumber business, the modern method of increasing the yield of the land.

Nor is it for a supply of wood alone that the forests of our country must be maintained. Their destruction involves the overturning of Nature's balance, and gradually but inevitably changes in climate will follow. Look to fallen Spain for an example. She was once the most powerful nation on the face of the earth. The fertility of her fields was second to none; her people were prosperous. But in her greed for gain the mountain forests were destroyed, just as ours are being destroyed today. And now this same Spain lies blistering under the heat of the tropical sun, a rainless desert. Nature's balance is a thing not to be meddled with. It has taken from the Creation to establish it, yet man can undo it only too quickly and too easily. It is for us, then; and for this our nation, to guard and maintain a condition so indispensible to our future welfare.

The Busy Season [1911]

From time to time Leopold tried his hand at verse, especially early in his career when he was stimulated by fellow rhymesters in the fledgling Forest Service. The *esprit* and camaraderie of the service as well as its utilitarian emphasis are illustrated by this bit of doggerel that Leopold inserted unsigned into a mimeographed monthly staff bulletin that he initiated in 1911, when he transferred from the Apache to the Carson National Forest. It is the first of four poems from the *Carson Pine Cone*, 1911–1914, that John D. Guthrie, Leopold's first supervisor on the Apache—where this verse may have been written—attributed to Leopold when he reprinted them in his book *The Forest Ranger and Other Verse*.

There's many a crooked, rocky trail,
    That we'd like all straight and free,
There's many a mile of forest aisle,
    Where a fire sign ought to be.

There's many a pine tree on the hills,
    In sooth, they are tall and straight,
But what we want to know is this,—
    What will they estimate?

There's many a cow-brute on the range,
    And her life is wild and free,
But can she look at you and say,
    She's paid the grazing fee?

All this and more,—it's up to us—
    And say, boys, Can we do it?
I have but just three words to say,
    And they are these: "TAKE TO IT."

To the Forest Officers of the Carson [1913]

Leopold was supervisor of the Carson National Forest in New Mexico when he was stricken in April 1913 with acute nephritis, the result of exposure during severe weather on a field trip in a remote district. He returned to his parents' home in Burlington, Iowa, for what would ultimately extend to eighteen months of recuperation. From there he penned the first of four letters to his staff that were printed between July 1913 and February 1914 in the *Carson Pine Cone*. Although overwritten, like a number of his writings at the time, the letter provides an early statement of a key element of Leopold's philosophy of management, a determination to judge people's decisions and actions by their effects on the land rather than by their adherence to policies, procedures, or abstract principles.

Burlington, Iowa
July 15, 1913

To the Forest Officers of the Carson—
Greetings:

We take it that the well known proverb, "Troubles never come singly," is indeed but a dilution of that modern, but more heartfelt saying, "Everything comes in bunches!" Albeit in this case, not *quite* everything *came*. There was one exception—the Supervisor. He went!

We make bold to assume that the above at least roughly approximates the feelings of your esteemed Deputy Supervisor, who on April 26 last, with the *Pine Cone* ten days overdue, and your humble servant saying "goodbye everybody" from the observation platform of the Chile Flyer, he calmly gazed at the galaxy of "things due," and said things, gently but softly, into the April twilight.

A great executive once said, "Do not make excuses, nor take them." In my opinion, your esteemed deputy is prone to follow in his footsteps. I therefore take it upon myself (not without a guilty smile) to remind you that among the things that "came in bunches" were working plans many and diverse, annual plans manifold, statistical reports and financial statements,

Blueprint cover of *Carson Pine Cone* with illustration by Aldo Leopold. (Leopold Collection, UW Archives)

and circulars—yea, even unto seven generations! And flocks and herds had their allotments in a thirsty land, and there were fires on the face of the deep. All this and more "came in bunches." By the same token, the *Pine Cone* came not at all.

Obviously, the writer might long ago have been expected to write a *Pine Cone* himself—to help out. But the doctor, at least by implication, had respect unto the proverb:

> Even a fool, when he holdeth his peace, is counted wise:
> When he shutteth his lips, he is esteemed as prudent.

Now, however, I am at least partly back on my feet, and it gives me pleasure to contribute a few lines to the July number.

After many days of much riding down among thickets of detail and box canyons of routine, it sometimes profits a man to top out on the high ridge leave without pay, and to take a look around. Most of us always *have* envied the Lookouts, anyhow. When your "topping out" is metaphorical and prescribed by the doctor—that is a circumstance which merely augments your envy, without decreasing your profit. But be that as it may, I will crave your indulgence while I attempt to describe what I now see from my point of vantage. Peradventure I may generalize a little, but only with the object of arriving at, and pinning down, what seems to me a final specific truth.

National Forest Administration has for its object the actual, concrete, specific application of the well known principles of conservation, to the resources within the National Forests. We are entrusted with the protection and development, through wise use and constructive study, of the Timber, water, forage, farm, recreative, game, fish, and esthetic resources of the areas under our jurisdiction. I will call these resources, for short, "The Forest." Our agencies for this development are: first, the Forest Users; second, our own energies, labor, and example; and third, the funds placed at our disposal.

It follows quite simply, that our sole task is to increase the efficiency of these three agencies. And it also follows that the sole measure of our success is the *effect* which they have on the Forest.

In plainer English, our job is to sharpen our tools, and make them cut the right way.

Now in actual practice, we are confronted, surrounded, and perhaps sometimes swamped, with problems, policies, ideas, decisions, precedents, and details. We ride in a thicket. We grapple with difficulties; we are in a maze of routine. Letters, circulars, reports, and special cases beset our path as the logs, gullies, rocks, and bog-holes and mosquitoes beset us in the hills. We ride—but are we getting anywhere? To that question I here propose an answer. I here offer a 66 foot chain wherewith to measure our progress. My measure is THE EFFECT ON THE FOREST.

This sounds simple, of course, but after all, day by day, how often do we apply this acid test? Suppose we lay out a few sample plots, and see. I select these as they come to my mind, without effort or guidance, in the hope they may thus attain some degree of fairness.

Plot # 1. A question arises as to whether the contract in a two million foot sale shall compel the utilization of White Fir. The Ranger, Supervisor, and District office exchange letters and memoranda, and two men make a field inspection.

Without doubt the resulting decision has a direct, tangible effect on the Forest.

Plot # 2. The District Office issues a circular call for estimates of type acreages. The Rangers report, the Supervisor corrects and tabulates, and submits as called for. What is the effect on the Forest?

The data is embodied in the Forester's annual report, which enlightens a limited portion of the public. The final effect on the Forest is very indirect and hard to measure.

Plot # 3. A trespass report on ten head of cattle involves heavy field and office work by four men, and a trip to court. The forage consumed and judgement obtained are in themselves negligible. What other effect on the Forest? An example to Users—an indirect effect—doubtless good; but no one knows whether, if balanced against the outlay, the final net effect is a profit or a loss.

Plot # 4. A ranger reports on a special use case. He submits only one copy of Form 964. The report is returned and a letter is written requesting two, and the case delayed.

There is a loss of time, fees, forage, and increment on stock; a direct effect. This is balanced against an indirect effect on other future cases, for which the Ranger submits two copies as required.

Plot # 5. Hundreds of letters, conferences, and discussions are exchanged concerning the relative merits of #9 or #12 telephone wire.

A wise decision directly affects the Forest in the resulting efficiency of our administrative agency, the telephone.

Plot # 6. A mining company is dumping poisonous tailings into a stream. The ranger inspects and reports to the office, and preventive action results.

The effect is direct and tangible.

Plot # 7. The experiment station through study and observations evolves a system of silvicultural treatment for a certain type.

Without doubt the effect is tangibly impressed on subsequent sale acres.

Plot #8. A fire burns over two acres and is extinguished by the ranger. The effect is direct, tangible, and good.

What do these random plots indicate? Briefly as follows:

(a) Effects are direct and indirect.

(b) With the single exception of fire, all effects are applied to our agencies, rather than to the Forest itself.

(c) Effects are good and doubtful, the "doubtful" ones being cases where we can not easily tell that the *net* result is beneficial.

Let us attempt a tabular classification of our plots:

Plot No.	Good Effect		Doubtful Effect	
	Direct	Indirect	Direct	Indirect
1	X			
2				X
3				X
4				X
5	X			
6	X			
7	X			
8	X			
Totals	5	0	0	3

From the above, is it possible to assert that good effects are always *direct* and that doubtful effects are always *indirect*? Perhaps that is on shaky ground, but we can at least assert that fire prevention is the most direct of all our activities, and hence also susceptible of developing the greatest relative efficiency.

A further analysis of our plots will show that indirect effects are mostly under the head of what we call "routine," and further, that most routine results from a striving after *uniformity*. If a ranger in Idaho did not have to handle a grazing case according to the same procedure as his New Mexican fellow-ranger, much routine would of itself be eliminated. No man doubts the wisdom of a policy of uniformity, but do we always remember that uniformity is simply a *policy of Operation* and not a *Conservation Principle*, and as such is not an *end* but a *means* only?

To come back to our original question, how many of us, when we write a letter, talk to a permittee, call for or make a report, recommend an improvement, or decide on our day's work—how many of us stop to try to figure out what will be the *effect on the Forest* of each separate action we take? Are we not more or less in a rut? If so, what puts us there? Principally, the *Operative Policy of Uniformity*, which I have just mentioned. But here is the point: the Policy of Uniformity is meant simply to guide our daily task, and it is *not* meant to confine our minds. And the Forest Officer who lets it do so is burying his talents.

I now come to the point I have been trying to make in the foregoing discussion. I have tried to point out the necessity for clear, untrammelled, and independent thinking on the part of Forest Officers. On this point I wish to submit two propositions:

First: The continued progress of the Service lies in the hands of the men who will thus think.

Second: The men in the best position to observe the faults and merits of our present work, to discern most clearly its effect on the Forest, and to study out the best means for improving that effect, are the *Forest Rangers*.

The Ranger is the man on the ground. He lives there. He is in the position to see the effects of our work at all seasons and under all circumstances, and it is those effects and nothing else, that count. His is the task of applying our principles in detail, and it is not until they are applied in detail that they have any effects. His is the opportunity to apply and measure and hence to study and improve. The Supervisor sees a larger area, but less closely. He may have a better chance at correlating data, and putting together disjointed observations, but most of the *first hand* and hence most vital progress must originate on the Ranger District.

In the foregoing pages I have tried to present a kind of analysis which points out the objects of our work, our agencies for performing it, the circumstances which tend to obscure our vision in trying to improve its effect, and the logically most important source of that improvement. Doubtless my propositions have some weak joints, and the location of my sample plots may be questioned—I hope that they will be. In fact, why not let the *Pine Cone* and our Ranger Meetings be the occasion for constructive discussions by all who "get an idea"?

In the next number—weather permitting—I hope to offer a few thoughts on the evolution of the Service since the "old days"; the resulting good changes to be fostered, and less good changes to be offset; and some outlines of a few specific current Service problems. A man feels better to "get shet" of these things once in a while.

The "Rest Cure," like greatness, is desired by some, while others have it thrust upon them. I wish I may soon be excused from the latter class.

My best wishes are extended to every man on the Carson.

Very sincerely yours,
(Signed) Aldo Leopold
Forest Supervisor

The Varmint Question [1915]

After returning to active duty in district headquarters in Albuquerque following his long recuperation, Leopold began to develop a new line of Forest Service work—promoting management of game for recreational hunting and organizing local game protective associations throughout the Southwest. This piece on "varmints" clearly reveals his initial hostility to wolves, lions, and other predatory animals and his strategy of winning the support of ranchers for game protection through a vigorous campaign against predators. Though unsigned, the article is consistent with Leopold's position at the time and refers specifically to cooperation with sheepman E. M. Otero, his brother-in-law. It appeared in the first issue of the *Pine Cone*, published during December 1915 by the Albuquerque Game Protective Association (thereafter the bulletin was published by the New Mexico Game Protective Association), of which Leopold was secretary. Leopold edited the *Pine Cone* and apparently wrote most of the articles that appeared in the bulletin until 1924, when he left the Southwest. Few articles, however, carry bylines.

For some unfathomable reason, there appears to have been a kind of feeling of antagonism between men interested in game protection and between some individuals connected with the stock growing industry. There have been some very notable exceptions to this rule, particularly among the stockmen themselves. It would, for instance, be a fair statement to say that certain individual stockmen have saved the antelope for New Mexico. But speaking generally the statement is true.

It seems never to have occurred to anybody that the very opposite should be the case, and that the stockmen and the game protectionists are mutually and vitally interested in a common problem. This problem is the reduction of predatory animals.

It is well known that predatory animals are continuing to eat the cream off the stock grower's profits, and it hardly needs to be argued that, with our game supply as low as it is, a reduction in the predatory animal population is bound to help the situation. If the wolves, lions, coyotes, bob-cats, foxes, skunks, and other varmints were only decreasing at the same rate as our game is decreasing, it might at least be said that there was no serious occasion for worry, but that they are not so decreasing is an established fact in the mind of every man familiar with conditions. Whatever may have been the value of the work accomplished by bounty systems, poisoning, and trapping, individual or governmental, the fact remains that varmints continue to thrive and their reduction can be accomplished only by means of a practical, vigorous, and comprehensive plan of action.

How, how is this action to be obtained? How, for instance, is the Biological Survey to receive a larger appropriation for the excellent work they have begun? How, for instance, is a more satisfactory bounty law to be enacted? How, for instance, is trapping to be made attractive to real trappers? Obviously by a united and concerted demand for these things. The stockmen alone have been demanding these things for years, and while they have accomplished a great deal, they have not accomplished enough. Why should the organized game protectionists not join with the stockmen in making these demands, and would not their added weight possibly give the necessary added effectiveness? Would not the manifestation of a sincere desire on our part to co-operate to the limit of our ability also remove the last vestige of feeling between us and the stock associations? Would not everybody, except the varmints, be benefited by such a move?

There is nothing connected with a properly conducted stock-growing operation which is going to operate against our game program. Conversely, there is nothing in our game program which is going to hurt the stock industry, or deprive any stock of their established range. Why, then, should we not get together?

Plans to have our Association actively develop this idea are already well under way. We have had an informal conference with Mr. Ligon, Predatory Animal Inspector of the Biological Survey, and obtained his personal ideas as to what ought to be done. We have had an informal conference with Mr. E. M. Otero, President of the New Mexico Wool Growers, and he was much pleased with the idea of co-operation. Shortly we shall formally tender our co-operation to the Executive Committee of the Wool Growers and, as soon as suitable occasions arise, to every other stock growing body in the Albuquerque region. If our offers are accepted, we will confer with these bodies on the question of ways and means, and there is every reason to hope that we can arrive at a mutual agreement which will bring the desired results.

It can hardly be gainsaid that we need the aid and co-operation of the powerful stock growing associations. It seems equally obvious that they could make good use of such help as we are able to give them. It is therefore to be hoped that we can get together.

The Popular Wilderness Fallacy: An Idea That Is Fast Exploding [1918]

Leopold wrote his first essay about wilderness a year before he began to think about wilderness as itself a resource to be preserved. In "The Popular Wilderness Fallacy," he argues that the demise of wilderness need not spell the doom of wildlife and makes a case for the obligation of human beings in a civilized society to practice conservation. In January 1918, the same month that this article appeared in *Outer's Book—Recreation*, Leopold left the Forest Service, which was preparing for increased wartime production of timber and livestock, to accept a post as secretary of the Albuquerque Chamber of Commerce, where he hoped he would be able to forward the cause of wildlife conservation.

When the pioneer hewed a path for progress through the American wilderness, there was bred into the American people the idea that civilization and forests were two mutually exclusive propositions. Development and forest destruction went hand in hand; we therefore adopted the fallacy that they were synonymous. A stump was our symbol of progress.

We have since learned, with some pains, that extensive forests are not only compatible with civilization, but absolutely essential to its highest development.

The same fallacy that characterized our idea of forests was bred into our attitude toward game and wild life, and unfortunately it has not yet disappeared. There are still millions of people whose opinions on wild life conservation, if they have any, are based in some degree on the assumption that the abundance of game must bear an inverse ratio to degree of settlement, and that the question of how long our game will hold out must be measured by the time it will take for man to completely occupy the land.

It is the writer's belief that this assumption is not only incorrect, but that it is exerting an incalculably mischievous influence against the progress of the movement for wild life conservation. To let the public think that economic progress spells the disappearance of wild life, is to let them believe that wild life conservation is ultimately hopeless.

It is true that the settlement and economic development of the United States has inevitably brought into operation many factors inimical to wild life, but a careful analysis of each will almost invariably reveal an accompanying counter-influence decidedly beneficial to wild life perpetuation.

Hunting, for instance, destroys millions of game animals and birds yearly. But at the same time hunting has destroyed millions of natural enemies of game. The destruction of predatory animals should enable man to take yearly, for his pleasure and food, at least part of the enormous amount of game which, in a state of nature, were the food of "varmints."

Agriculture, for instance, has usurped a large part of the former haunts of game. But to some extent at least it has replaced the natural coverts with artificial ones. At least in the case of small game, a square mile of farming country may have as great a capacity for raising wild life as a square mile of wilderness. In some cases, agricultural development has done more than this. It has literally created a game supply. In eastern New Mexico, for example, there is a large area which has recently been homesteaded by dry farmers. It formerly had little or no small game. Today it is abundantly stocked with prairie chickens. They came by natural migration, attracted by the stubblefields, and are rapidly spreading westward with agriculture. Nature was actually improved upon by civilization.

Artificial drainage has destroyed many marshes and lakes which were formerly the feeding and breeding grounds of myriads of wildfowl. But at the same time man is building yearly hundreds of artificial lakes. A good example is the huge swamp created by the Keokuk dam across the Mississippi. This is now one of the best shooting grounds in the Central West and by way of becoming an important breeding place for wood duck. Moreover, it is permanent, whereas the natural swamps along the Mississippi often dried up. In the end, we will probably give the wildfowl as many waters as we are taking from them.

Overgrazing of the public domain of the West no doubt was a powerful factor in the destruction of the antelope, mountain sheep, and other range game. But grazing ranges are coming to be dotted with thousands of artificial reservoirs for watering stock. Millions of acres of "dry range"—waterless deserts almost devoid of life—are being made usable for stock, and incidentally, for game. And at least on the National Forests and on the holdings of progressive stockmen, overgrazing has ceased and the ranges are recovering. Without any doubt, intelligent use of the western ranges for grazing purposes can be made an improvement on nature with respect to all our western big game except the buffalo. While there was no excuse for the wholesale extermination of this splendid animal, the ultimate disappearance of the huge migratory herds was inevitable.

Forest fires and prairie fires incident to the early stages of settlement burned up or starved out a great deal of game. But today these destroyers are on the wane. Under the influence of man they are approaching zero, whereas many fires ravaged the aboriginal wilderness, set by lightning and by the Indians. The time will come when there will be much less destruction by fires than took place before the coming of the white man. A sign of the times is found in the National Forests, where except in extremely dry years fires are being held down to the almost negligible average of a few acres each.

A few diseases of game are supposed to have been induced by settlement—though probably to a lesser extent than is generally believed. Scabies is alleged to have spread from domestic to mountain sheep, but the latest researches indicate that this is not the case. It is also doubtful whether the actinosis (lumpy jaw) of antelope was derived from domestic cattle. The enemies of the game have probably suffered from imported disease just as badly as the game itself. Western coyotes have died by thousands of rabies. Moreover man is to an increasing degree able to check the ravages of natural epidemic among game. He is studying wild duck disease. He has prohibited the importation of diseased quail from Mexico. Some day civilization will prevent more diseases of wild life than it introduced.

Finally, a great and ever-increasing area of game range has been actually occupied by railroads, dwellings, and cities. In part, such areas are lost to wild life. But our native birds, at least, have readily adapted themselves to the new conditions. Dozens of species like the chimney swift, the house finch, and the eastern robin now actually prefer the haunts of man. It has become a common achievement to have 20 pairs of birds make their home on a half-acre city lot. Who shall say that a square mile of suburban dwellings with modern methods of bird feeding and housing, cannot raise more wild life, though of a partly different kind, than the original wilderness?

Notable instances are not lacking in which the abundance of game seems to bear a direct rather than an inverse ratio to the degree of settlement. Compare, for example, the National Forests of New Mexico—comparatively speaking a wilderness—with the thickly settled New England States. In 1915, according to estimates prepared by the New Mexico Forest rangers, 656 deer were killed on 13,000 square miles of mountain forests; roughly, one per twenty square miles. Maine, Vermont, Michigan and New York, in the same or immediately preceding years, averaged roughly one deer killed per five square miles—a preponderance in productiveness of four to one. Who shall say that only a wilderness can raise deer?

It has often been contended that the fencing of the western ranges has destroyed the antelope. As far as concerns the occupation of the range by solid blocks of homesteads, this is probably true. But this is not the whole story. A recent analysis of figures collected by the Forest Service on the remnants of antelope in New Mexico show that of the thirty-eight herds now existing, thirty-two are found on the open range and six in large fenced pastures. The open-range herds average 30 head each, but the pasture herds 127 head each. Three-fourths of the herds reported as increasing are in fenced pastures, and all but one of the herds reported as decreasing are on the open range. The pasture herds average four times as large and several times as thrifty as the open range herds. Who shall say that only an unfenced wilderness can raise antelope?

Even those who do not believe that a wilderness is essential to a supply of game are likely to assume that where there is a real wilderness left, the game supply is comparatively safe. Alaska, the greatest remaining wilderness in North America, is by way of contradicting even this assumption. It appears to be a fact that even in the remotest region of Alaska indiscriminate slaughter is spelling the doom of the game supply. No wilderness seems vast enough to protect wild life, no countryside thickly populated enough to exclude it.

It seems safe to call a fallacy the idea that civilization excludes wild life. It is time for the American public to realize this. Progress is no longer an excuse for the destruction of our native animals and birds, but on the contrary implies not only an obligation, but an opportunity for their perpetuation. American wild life is confronted by only one unmitigated menace—indiscriminate slaughter. Its future as a part of our permanent national environment is in our hands—or its blood will be on our hands—as the case may be.

Forestry and Game Conservation [1918]

This is Leopold's classic article making the case for game conservation, following the paradigm of forest conservation. Published in the *Journal of Forestry*, the article addressed professional foresters, whom Leopold hoped would take the lead in developing a science of game management modeled on forest management. It may be viewed as a companion piece to "The Popular Wilderness Fallacy," which addressed a more general audience of sportsmen and recreationists.

The technical education of the American forester aims principally to teach him how to raise and use timber. This is obviously proper. Handling timber lands is his major function.

But when the forester begins actual work on a forest he is called upon to solve a much broader problem. He is charged with the duty of putting land to its highest use.

When foresters took charge of the National Forests in 1908 they were not slow to see that they were responsible for the regeneration and development of the Forest ranges. The fact that large areas were overgrazed was considered no reason for letting them remain so. The fact that selfish interests stood in the way of reorganization and progress was considered no obstacle against going ahead. The fact that nobody had ever heard of scientific range management was considered no reason for the continuance of an obsolete system. Foresters undertook to regenerate the ranges on a scientific basis, and succeeded. Today National Forest range management has the opportunity of becoming the most efficient in the world.

When foresters took charge of the National Forests they found their game resources, like their grazing resources, to be in a depleted and nonproductive condition. But instead of pushing the work of regenerating the game, they have, by and large, met the situation only as it has pushed them. Today the game resources of the National Forests are, in general, in just as depleted a condition as they were ten years ago, when the Forests were established. Such real work as has been done has been based almost solely on

53

the hand-to-mouth policy of preserving a little sport for the immediate morrow. There has been little vision—little effort to lay broad foundations for an aggressive game policy dovetailing into the policy for the development of timber and range and the recreational needs of the public. In short, the job bears all the earmarks of defensive instead of offensive tactics.

There are, of course, reasons for this anamolous bit of history—reasons no doubt already on the tongues of those who see no cause for dissatisfaction. But it is beginning to be suspected that these reasons will no longer hold water.

First, there is the real and mutual handicap of dual authority over National Forest game. The Federal Government owns the land and has the men on the ground, but the State owns the game and makes the laws. This has resulted in the plan of dual administration under co-operative agreements. It has grown to be almost a habit to consider these agreements fundamentally defective, and in the same breath to assume the absence of any remedy except the complete recession of the one party or the other, in either event raising the dread specter of "States rights." This assumption of three ultimate alternatives, all hopeless, has naturally blasted initiative. But the writer believes the assumption to be incorrect. In the course of a series of future articles, it is hoped to present a new method of co-operative administration, supplemented by some very simple Federal legislation, which will afford a permanently workable plan.

Second, foresters have lacked the stimulus of a strong local demand for better game administration, or, stated negatively, they have to some extent encountered local opposition to any administration at all. They have waited ten years for this demand to grow of its own accord before realizing that it is quite possible to deliberately go out and create it. In New Mexico this has been done, as indeed it was done long ago in connection with grazing administration, where the "will to do" was not lacking.

Third, foresters have labored under the vague fear that a real crop of game might interfere with both grazing and silviculture, as if grazing and silviculture might not also interfere with each other! The principle of "highest use" has evidently been more talked about than understood.

These, then, are the three reasons for the lack of an aggressive game policy on the National Forests. Assuming that they have so far made effective action impossible, they have certainly not made constructive thought impossible. Yet the lack of constructive thought seems, to the writer, to have been the greatest single obstacle to progress in this field. It is significant to recall that such really constructive ideas as that of a correlated system of National Forest Game Refuges were conceived by outside parties. It is the purpose of this paper to urge on foresters their special responsibility and special fitness for supplying constructive thought, and later an actual solution of the game problem on the Forests.

Foresters must consider themselves especially responsible for handling the game problem because: (1) Their work gives them the opportunity to be better acquainted with game conditions than any other class of men. (2) Their training in forestry especially fits them for the work. (3) They are the only large body of scientifically trained men on the ground. (4) If they do not devise means of saving the game, the recreational value of the Forests will be permanently and seriously reduced.

That foresters ought to know game conditions is not open to argument—it is obvious. That a forester's training especially fits him to supply the greatest present needs of the game problem, the writer hopes to establish by showing what those needs are.

In the first place, it is strikingly true that the history of game conservation in the United States has been exactly analogous to the history of forest conservation. Any new movement starts out as a "cause," and the first few years consist mostly of propaganda in furtherance thereof. Forestry started out as a cause, and the first ten years of its history is a story of forestry propaganda. Game conservation has started out as a cause, and as such has about run its allotted course. What, judging from the history of forestry, is the next move?

In answer, it may be well to state that the propaganda stage of forestry was concerned with the question of *whether* our forests should be conserved. The people having answered that question in the affirmative, forestry immediately entered its second stage. In this stage it was concerned with the question of *how* our forests should be conserved. Here the *science* of forestry took the floor, *prepared to cope with the situation*. Foresters had anticipated the need and had developed at least the rudiments of American forest management.

Reverting again to the game question, we may venture the statement that the American people have already answered, in a vigorous affirmative, the question of *whether* our game shall be conserved. Game conservation is ready to enter its second stage, and even the layman is beginning to ask *how* it shall be accomplished. Witness game refuges, game farming, and countless new departures in game laws. The time has come for *science* to take the floor, *prepared to cope with the situation*. But has the need been anticipated, and at least the rudiments of American game management developed? The writer believes it has not.

If it is true that the country is confronted with the eleventh-hour necessity of developing the science of game management, what can the new science borrow from the science of forestry? In the opinion of the writer, a great deal. The following brief analogy, which for the sake of simplicity deals primarily with big game, is self-explanatory:

The first step in undertaking the administration of a tract of game range is to make a game census. This corresponds to timber estimates or reconnaissance. A proper game census should give us the number of head by species (stand estimates), a game distribution map (type map), data by unit areas on predatory animal damage (fire and insect damage), data on water, cover, and foods (soils and site qualities), and figures by unit areas on past annual kill (old cuttings).

The next step is, install a system of patrol against illegal killing and predatory animal damage (timber-trespass and fire damage). This is for the purpose of safeguarding needed breeding stock (growing stock), the loss of which would seriously impair the productiveness of our forest.

We may assume in this analogy that we have an unlimited permanent demand for all the hunting we can furnish (timber market). This being the case, good management *demands* the immediate adoption of a system of regulation of annual kill (annual cut), with the aim of sustained annual kill (sustained annual yield). To determine the amount of breeding stock (growing stock) which is to serve as a basis of sustained yield, we first segregate as game refuges the ranges of rare or threatened species (protection forests). We also segregate as game refuges areas chiefly valuable for recreation and scenery (recreation forests). Next we take the breeding stock on the remaining hunting grounds and determine our annual limitation of kill (limitation of cut).

From here on the analogy is suggestive rather than absolute. For present purposes it will suffice to point out that in the determination of annual kill we use kill factors, which will be explained in future articles. They are calculated empirically and are analogous to yield tables. We may also adopt a killable age of game roughly analogous to a rotation. We also make use of game refuges, bag limits, and a limitation in the number of hunting permits, the combined effect of which is analogous to a combination of area and volume regulation of cut in forestry.

Given a market, we make game laws (sale contracts) specifying certain license fees (stumpage rates). We may adopt limitations on age and sex of animals to be taken, which are analogous to marking rules.

Before cutting begins we decide on the system of regeneration. In most instances this will be natural increase (natural reproduction), but this may be supplemented by artificial restocking (planting). If so, we may establish a game farm (nursery). It is again important to point out that in no case where we cannot restock should we allow even the local removal of more than a fixed minimum of breeding stock (seed trees).

It is the purpose of the foregoing paragraphs to suggest, not to explain, the analogy between game management and forestry. A full explanation must necessarily transcend the scope of this paper. It must be apparent to the reader, however, that the prime necessity for stock-taking, protection

Forestry and Game Conservation   57

against damage, and management for sustained production under a fixed system of regeneration is common to both. Especially so is the principle of guarding at all costs against the depletion of the normal breeding stock (normal stand).

The skeptic may promptly rebut the foregoing analogy. American foresters, he will say, have preached the principles of silviculture, notably sustained yield, but have as yet been unable to practice them. How, then, could they have practiced them with game? True; but why? *Because of lack of a demand* for inferior grades and remote stumpage. Because of our old bugbear—inaccessibility. Does game management labor under the same handicap? Emphatically it does not. *There is a demand for every head of killable big game in the United States*, wherever it may be. Five million sportsmen are looking for hunting grounds, and many in vain. Indeed, it may be said that, as far as a market is concerned, we are more ready to practice game management than to practice forest management.

The next question is: To what extent have the principles of forestry been applied to game?

The big outstanding fact which confronts us here is that absolutely no volume limitation has been applied to the annual kill except bag limits. Having failed to regulate the number of bags, we have really applied no volume limitation at all. There has been applied (but often not enforced) a *time* limitation (hunting seasons), and we have begun to discuss an area limitation (game refuges), but that is all. Hunting seasons and bag limits are essential, but they do not go far enough. They have necessarily failed to prevent depletion of the breeding stock; consequently we are raising little game. What would we expect of a Forest wherein every millman who could pay a license fee were turned loose to cut *ad libitum* from September 15 to November 1, provided he did not haul to market more than 50,000 feet in any one day? Obviously mighty little. Such practice would end by stripping the accessible parts, culling the most desirable species, no matter how badly needed for seed; leaving other areas untouched, and in general creating the antithesis of a productive forest. Yet just so have we created the antithesis of a productive game supply.

Granting that the present system applies no adequate limitation of annual kill, how are we to make good the deficiency? The following is a brief outline of a system of Federal hunting permits which would supply the necessary limitation of kill on the National Forests.

First, take our unit area (in the National Forests this will be a ranger district) and find out what is left in the way of stock. On the basis of game killed (figures, heretofore little used, happily on file for years past), and with due allowance for gradually bringing the stock back to normal, figure out for each species how many animals can be safely removed for next year. Multiply by two for unsuccessful hunters, and advertise the result as the maximum number Federal hunting permits which will be sold for that district for that year, no permit to be sold except on presentation of a proper State license. Sale will take place at a specified time and place, first come first served. Each permit will bear tags, which must be attached to carcasses during possession and later mailed to the Forest officer for cancellation. Possession of carcass without tag, or failure to turn in tags, will be grounds for refusal of permit during ensuing years. The canceled tags and a recensus will form the basis for next year's limitation. The local game protective association (an adjunct to our administration just as necessary as the stockmen's advisory board) should help determine the number and allotment of permits.

The foregoing paragraph is only a suggestive sketch of a system which will be developed in detail in a separate article. The legal basis for putting it into effect would consist of a simple Federal law authorizing the issuance of hunting permits. (Authority to issue such permits, by the way, is already vested in the United States, but it would require an act of Congress instructing the Secretary of Agriculture to exercise it.) It will be seen that no right, title, or interest of the State is in any way interfered with. The whole process consists in "raising the price" on big game, and thus creating a Federal fund to assist the State in its protection. A revision of the co-operative agreement with the State game department would complete the necessary machinery.

It should be noted that the proposed system of Federal hunting permits leaves the fixing of open and closed seasons to the State, as heretofore. The lack of Federal authority in the fixing of seasons has been one of the stock arguments against the present co-operative plan of administration. But with Federal volumetric control of kill, who cares what the open season is, as long as it be within reason? With the annual kill under regulation, open seasons are no longer a vital factor. The necessity for long closed seasons is done away with. The whole question of seasons becomes a mere matter of expediency and convenience.

We started out to prove that the undeveloped science of game management can borrow its framework from the developed science of forestry. It is hoped that the analogy between the two sciences and the sample system in which the analogy is concretely applied will throw some light on the writer's contention that foresters can meet the big needs of the game problem by simply applying the principles with which they, and they only, are familiar.

In view of the impending extermination of certain valuable game species, it seems advisable, before concluding this paper, to make a further comparison between forestry and game management in the matter of selection of species. In the writer's opinion, this is a point which cannot receive too much emphasis.

Forestry may prescribe for a certain area either a mixed stand or a pure one. But game management should always prescribe a mixed stand—that is, the perpetuation of every indigenous species. Variety in game is quite as valuable as quantity. In the Southwest, for instance, we want not only to raise a maximum number of mule deer and turkey, but we must also at least perpetuate the Mexican mountain-sheep, big-horn, antelope, white-tail deer, Sonora deer, elk, and javelina. The attractiveness, and hence the value of our Forests as hunting grounds, is easily doubled by retaining our extraordinary variety of native big game. This variety also adds enormously to their attractiveness for the summer camper, the cottager, and the fisherman. The perpetuation of interesting species is good business, and their extermination, in the mind of the conservationists, would be a sin against future generations.

Forestry does not face so acute a problem. Black walnut or yellow poplar may have become commercially defunct in our hardwood forests, but they are not extinct and never will be. We may destroy them with fire and axe, and burn off the soil of their native habitat to the uttermost extremity of abuse, but some day, somehow, we can always have a walnut or a poplar forest if the demand is sufficiently urgent. But not so with most game. White-tail deer and rabbits seem to have an immunity to extinction, but the great majority of big-game species may quite conceivably become extinct. One species of big game is extinct, two species have already been exterminated from the Southwest, and five more are even now threatened with extermination.

Foresters are quite properly concerned over the threatened commercial extermination of chestnut by blight and white pine by the blister rust. But how much concern is felt over the impending extermination of mountain-sheep and antelope on the National Forests? I am afraid, very little. Yet a good stock of mountain-sheep alone would add millions of dollars to the capital value of National Forest resources. Men go to Tibet to hunt the argali. Surely they would come to New Mexico to hunt *Ovis mexicanus*—if we had any left to hunt.

In conclusion, the writer is sensible of the fact that in arraigning the profession of forestry for a passive attitude toward the game problem, he speaks from the standpoint of a game conservation enthusiast. But why, indeed, should not more foresters likewise be enthusiasts on this question? They should—in fact, they must be, if they are to act as leaders in launching the new science of game management. Enthusiasm for forest conservation was a conspicuous attribute of foresters until long after the propaganda stage of forestry—and a very necessary one. Without it the tremendous first obstacles to launching the new science of forestry would not have been overcome. Without it we shall not overcome the first obstacles to making American game a major forest product.

Notes on the Behavior of Pintail Ducks in a Hailstorm [1919]

Leopold's lifelong bent toward natural history and his acute powers of observation in the field, so evident in *Sand County Almanac*, were illustrated early in his career in a series of short notes and articles he published in the *Condor*, the journal of the Cooper Ornithological Society. Leopold recast many of these items, such as this note on pintail ducks, from his field journals, where he carefully recorded each of his hunting expeditions and other outings.

On October 20, 1918, I was hunting ducks on the Rio Grande south of Los Lunas, New Mexico. I was sitting in my blind on a sandbar, with some dead ducks set out as decoys, when a very severe hailstorm set in. During the thick of the storm I discovered that a flock of about forty Pintail Ducks (*Dafila acuta*) had settled among my decoys not twenty yards distant. Each bird was facing toward the storm, and each had his head and bill pointed almost vertically into the air. The flock presented a very strange appearance, and I was puzzled for a moment as to the meaning of the unusual posture. Then it dawned on me what they were doing. In a normal position the hailstones would have hurt their sensitive bills, but pointed up vertically the bill presented a negligible surface from which hailstones would naturally be deflected. The correctness of this explanation was later proven by the fact that a normal position was resumed as soon as the hail changed into a slow rain.

Has any other observer noted a similar performance in this or other species of ducks, or in any other birds?

AL, H.T.J, E.E.Bliss   Oct 20, 1918
Rio Grande 1 mi. S. Los Lunas   Cloudy, warm, rainy, followed
                                              by very heavy hailstorms.
                                                    7 - 3 PM

2 mallard ♂ ------ 2 3/4# ----- These had speckled heads & dark breast-penciling
2 mallard ♀ ------ 2 7/16# ----- Very few drakes.
1 Black mallard -- 2 13/16# ---- Heavier than greenheads
   Sprig ♀ -------- 1 1/2 - 1 3/4# - These small plain colored sprigs predominated
3 Sprig ♂ -------- 1 3/4 - 2# --- Specimen No. 1 Saw a few getting white breasts
5 Greenwing ------ 11/16 - 3/4# - All still very plain plumage
1 Spoonbill ♂ ---- 1 9/16# ----- Very dark breast for young bird.
1 Gadwall -------- 1 3/4# ------ Orange & black bill; under parts all speckled
──   Speculum small & dim. No brown on shoulder
20 ---- 32 shells. 4 shells on cripples.

   Shooting light in morning on pass — I had 2, Johnson 4, Bliss 3 by noon. They then went home while I went up river and found gathering of several hundred on sandbar runs. Built small blind and set out dead ducks. All species decoyed & called readily, especially sprigs which predominated. By 2 PM had 15 and stopped shooting anything but big ducks. By 3 PM had hunt. Two very heavy hailstorms and one shower - hail 2" deep,- cut gullies in bars by runoff.

   Another man below had poor luck with decoys. They seem to work on gathering places but not on flyways. Not much movement from sloughs.

   Used open barrels and could notice no difference. Missed only one shot all day except a few second barrels.

   Saw blue herons, raven, flocks of killdeer, and some very small sandpipers.

   In early morning quite a few high flocks going south.

   Ducks were full of fleas - a large light-colored flea with faint dark markings. Did not leave ducks when dead and cold.

   Observation on Ducks in Hailstorm Flock 40 sprigs lit in decoys during heavy hailstorm and while storm was on faced wind with bills pointed almost straight up into hail — evidently to avoid pounding of hail on sensitive bill. After hail ceased normal position resumed.

Entry from Leopold's "New Mexico Journal" on Oct. 20, 1918, showing the field observations that Leopold reported in "Notes on the Behavior of Pintail Ducks in a Hailstorm." (Leopold Collection X25 2276, UW Archives)

Wild Lifers vs. Game Farmers: A Plea for Democracy in Sport [1919]

As one who believed in the possibility of managing game in the wild in order to provide free, "democratic" recreation for the general public, Leopold was dismayed by the move toward artificial management and commercialization of sport promoted by some segments of the conservation community. He made his case for a democratic American rather than an "unsocial" European system of game management in a paper prepared for a national convention of the American Game Protective Association and published in its *Bulletin*.

In the general field of American sportsmanship three things are certain:
1. There has been a general and growing scarcity of game all over the United States.
2. This decrease has not so far been checked on any considerable scale, except in the single case of waterfowl. Waterfowl have shown a perceptible response to the Federal Migratory Bird Law.
3. The annual drain on the game supply will greatly increase after the war. The return of the soldiers, the resumption of a normal amount of recreation by the whole population, and the increase in the number of motor vehicles, good roads, and modern guns, will all make for a greater annual kill.

The three foregoing facts can lead to but one conclusion: we face a dwindling supply and a growing demand for game. What are we going to do about it?

"There is nothing to do about it. The country is settling up and I guess the game must go." This could have been the answer of the average citizen twenty years ago.

Must Have the Game

"I don't know just how we will go about it, but a fair supply of game must be maintained." This is the answer of the average citizen today.

Do we fully appreciate the difference between these two answers? We have not solved the problem, but we have resolved to tackle it. This nationwide determination that "something must be done" is the greatest single achievement of the past generation of sportsmen. It is a big achievement. Extreme pessimists will do well to ponder it carefully.

It remains for this generation to evolve a general plan of action, and then execute it. We are now groping for such a general plan. Countless new departures in game laws, thousands of voluntary organizations devoted to game protection, hundreds of publications theorizing as to ways and means—all these are evidence of the fact that the great American public is seeking an answer to the question, "How shall we perpetuate the game?"

What next? This is easy to predict. Out of this maze of gropings will emerge two or more factions, or schools of thought. To these will flock the radicals, the extremists. They will begin forthwith to attack one another's opinions. This fight will be a test of strength, of which the great mass of moderate-minded men will be interested spectators. When one side or the other has developed a preponderance of plausible arguments, of actual demonstrations, and of influence, the moderates will join that side. Its plan will be put into effect—with a toning down of its extreme features, and the incorporation of a few strong points developed by the losing side.

If we may thus predict these probable steps in the evolution of a plan for managing our game, it becomes both interesting and profitable to figure out just what part of the process we are going through at the present time. It is the purpose of this paper (a) to point out that two extreme factions are now emerging; (b) to define their respective claims; and (c) to appraise those claims from a more or less new and, I believe, an important viewpoint. The two factions I will call the "Game Farmers" and the "Wild Lifers." The new viewpoint is that which regards the game not as meat, nor as sport, nor yet as a set of zoological specimens, but rather as a source of democratic recreation, a human source, a social asset.

What are the Game Farmers? Since the Herdes Powder Company started to advertise them two years ago, the country has had little opportunity to forget them. In general, the Game Farmers propose to supplement wild game with, or substitute for it, a supply produced under artificially regulated conditions. Radical Game Farmers tend to regard restrictive game laws as eventually hopeless and ineffective.

What are the Wild Lifers? They are the advocates of restrictive game laws; the scarcer the game the more restrictions. Long or even permanent closed seasons on threatened species are a logical corollary of their doctrine. The name, "Wild Lifers," is one recently and sarcastically donated by certain radical Game Farmers.

Both Partially Right

The easy-going (and sometimes shallow-thinking) moderate is apt to see no real conflict between these two factions. "They are both cranks," he says, "but both partially right. Why not conserve the game as well as we can with laws, and then breed it, too! Let game farming be a supplement to instead of a substitute for game laws, and, presto! everything will be lovely."

This is coming to be the view of many moderate-minded men, and the policy of many official state game departments. Will the compromise hold water? This is an important question, to which there is, I believe, no sweeping answer. A comparative analysis and appraisement of the two factions, however, will throw some light on the question.

A first and fundamental distinction between the two is that the game farmer seeks to produce merely something to shoot, while the Wild Lifer seeks to perpetuate, at least, a sample of all wild life, game and non-game. The one caters to the gunner, the other to the whole outdoors-loving public. Inasmuch as the camera man, the sporting naturalist, and other non-gunners are coming to comprise a considerable percentage of the national fraternity of sportsmen, it must be admitted at the start that the Wild Lifer represents, or attempts to represent, the interests of a larger group of citizens than his opponent.

Secondly, the Game Farmer, so far, at least, is purely materialistic as to what his "something to shoot" consists of. If Chinese pheasant is cheaper and easier to raise than the American heath hen, and is equally good game, then, he says, let the heath hen go hang! This may sound like a prejudiced statement, but certainly there is little in his published propaganda that shows any concern about the perpetuation of native species as such. On the other hand, the Wild Lifer regards the perpetuation of native species as an end in itself, equal if not greater in importance than the perpetuation of "something to shoot." It may be safely concluded that as to this point the Wild Lifer enjoys the advantage of an ethical as well as of an utilitarian objective.

Natural Enemies of Game

Thirdly, the Game Farmer makes a great point of "vermin." Natural enemies, he says, destroy more game than guns. On the other hand, many Wild Lifers have been practically silent on this point. This particular point is not one of theory, but of fact, and every keen observer must admit that the Game Farmer is right. Of course, when this point is carried to extremes—as when some Game Farmers argue that as long as vermin remain, restrictive game laws are a useless sacrifice, it becomes an absurdity that will deceive no thinking sportsman.

It may here and now be said that the two extreme points of view on the three foregoing points are by no means beyond reconciliation. The proper care of game does not necessarily imply neglect of non-game. The breeding of "efficient" species does not necessarily imply the neglect of weak native species. The advisability of controlling vermin is plain common sense, which nobody will seriously question. Any reasonable man can and will select and advocate the good points on both sides of the argument. But now we come to the points wherein the "dynamite" lies concealed. To wit:

Markets Must Not Be Opened

Game Farming is a business. It must return a profit. Part of the profit comes from marketing all or a part of the game as meat. Game markets, after a twenty-year fight, are closed. The Game Farmer wants them opened—to his products. Some of his publications even hint broadly that they must eventually be opened wide to all game, in order to do away with the restrictions, necessarily cumbersome, incident to making a distinction between the sale of wild game and game artificially raised. The Wild Lifer, on the other hand, regards with apprehension the opening of markets, even to artificially raised game, while the idea of a wide-open market fills him with horror.

Is it possible to reconcile these two points of view? Yes and no. It is conceivable that limited markets—that is, markets for tagged meat raised on licensed game farms—might be indefinitely maintained and even largely expanded, without resulting in serious violation of the laws prohibiting the sale of wild or public game. But the Game Farmer, under such a system (which is in effect the one now being adopted in most states) must forever bear the cost incident to restrictions, such as closed seasons, tagging, licensing, etc. It seems almost axiomatic, however, that a return to wide-open markets would spell the certain doom of wild game, and even edible non-game. The radical game breeder's hints for a wide-open market, therefore, are a wide-open challenge to choose between Game Farming and public game. With a wide-open market, public game could not exist, and game laws would become useless and unnecessary.

There is a second point wherein the doctrine of Game Farming cannot be reconciled with the doctrine of the Wild Lifer. Game Farming, whether conducted by public agencies or private persons, costs money. It cannot succeed unless the hunting is sold for money. It, therefore, inevitably implies the sale or lease of shooting privileges, and in so far as it supplants the present system, it means the end of free hunting. It means that the well-to-do will raise their own game, while the farmers will breed game and sell their shooting to the highest bidder. Even should the farmer not actually produce his game under hens, and even in the case of migratory game, which in the nature of things must remain public, the general spread of Game Farming would soon result in the general spread of commercialized shooting privileges, and the poor man would be left with a few navigable rivers, and the freedom of the seas for his hunting. We may even develop riparian hunting rights!

The writer is aware that even the most extreme proponents of Game Farming have provided, on paper, for the interests of the impecunious sportsman. They tell him of the fine overflow hunting to be picked up outside the poacher-proof fences of game preserves. This is well and good, for those who enjoy the idea of feasting on crumbs, and until the last "outside" coverts have ceased to exist. A wholesale commercialization of shooting privileges, however, would soon leave nothing open but public lands.

Conclusions

We can immediately draw one conclusion from the foregoing discussion of the proposed commercialization of game-meat and hunting privileges, and that is that to grant the wishes of the radical Game Farmers would be tantamount to adopting the European system of game management. A wide-open market, almost universal game farming, commercialized shooting privileges, and some incidental overflow shooting for the poor man—is this not the sum and substance of the European system? It is. And the European system of game management is undemocratic, unsocial, and therefore dangerous. I assume that it is not necessary to argue that the development of any undemocratic system in this country is to be avoided at all costs.

I am well aware that in the foregoing discussion I have avoided two points which add weight to the Game Farmer's argument. Let us consider them separately.

The first is the fact that posting of farm lands, theories of democracy to the contrary notwithstanding, is in some places fast rendering free hunting a thing of the past. This is a fact, not a theory, and we must face it as such. Moreover, with increasing values of lands and livestock, and increasing numbers of careless hunters, the proportion of posted lands is bound to rapidly increase and will eventually include all intensively developed private holdings.

The second is the fact that even without strictly artificial Game Farming, an abundant supply of wild public game cannot be restored without more or less winter feeding, patrol, vermin-control, maintenance of coverts and food-plants, and other special measures, all of which cost something in money, labor, land or attention. These measures, whether you call them

Game Farming or not, cost something and will have the same effect as Game Farming in commercializing hunting privileges.

From the foregoing discussion it seems logical to draw the following conclusions:

1. Free hunting of upland game in intensively developed farming regions will gradually disappear. Farmers will breed and encourage game and lease their hunting privileges to clubs or individuals. (Hunting privileges donated as personal favors are for all intents and purposes commercialized.)
2. Free hunting of migratory waterfowl in intensively developed farming regions will follow the same course, but more slowly. Ducking grounds on private lands will be artificially improved and leased. Ducking grounds on public waters will remain open.
3. Free hunting in poor or forested private lands will diminish, but will probably survive indefinitely. Control of vermin, winter feeding and maintenance of food plants, and other necessary semi-artificial measures will remain the duty of public agencies and will become of increasing importance.
4. Free hunting on public lands or waters must and will be preserved and developed at public expense. State forests and parks, national forests and parks, navigable rivers and lakes, and public reservations of all kinds, must, with the exception of necessary game refuges, be left open as the one sure resource of the man who cannot afford to buy his shooting. It is imperative that exclusive privileges be debarred from such lands absolutely, as is now done on the National Forests.
5. Restrictive game laws must continue to be developed and enforced on all lands and all game.
6. Markets must be kept closed to all except tagged and licensed game artificially produced, and this principle must be guarded as the very keystone of the whole structure of game management.
7. Closed seasons and special public game refuges must be maintained for native species of weak recuperative capacity, and such species must be preserved at public expense.
8. Where conditions are such that free hunting will probably disappear over large areas and there are no public lands or waters left for those who cannot buy their hunting, states and municipalities could start to acquire special public hunting grounds, in which a game supply may be developed at public expense. This suggestion is no more radical than the municipal golf courses and tennis courts proposed ten years ago and now actually available to the public near many cities.

"Piute Forestry" vs. Forest Fire Prevention [1920]

When Leopold rejoined the Forest Service in 1919 as assistant district forester, one of his responsibilities was fire control of twenty million acres of forests in the Southwest. This article, published in *Southwestern Magazine*, parallels similar articles by Henry Graves, chief forester of the United States, and his successor, William Greeley, which condemned the practice of "light-burning," or "Piute Forestry," as they derisively termed it, advocated by the Southern Pacific Railroad and other timber interests to reduce flammable brush in stands of mature timber. Within two years, as a result of careful observations he made during official inspections of the forests, Leopold would begin to rethink many of the arguments advanced here concerning the evils of fire. For an indication of how far he had moved from this position by 1924, see "Grass, Brush, Timber, and Fire in Southern Arizona."

Southwestern mining men and southwestern foresters have many common interests. The National Forests supply timber and water used by miners, while the mines are one of the important sources of prosperity in the National Forest communities. For this reason southwestern mining men may be interested in a new problem which now confronts the administrators of the forests, namely, the propaganda recently started by the Southern Pacific Railroad and certain timber interests in favor of "light-burning," or the so called "Piute Forestry."

"Light-burning" means the deliberate firing of forests at frequent intervals in order to burn up and prevent the accumulation of litter and thus prevent the occurrence of serious conflagrations. This theory is called "Piute Forestry" for the alleged reason that the California Indians, in former days, deliberately "light-burned" the forests in order to protect them against serious fires.

Foresters generally are strenuously opposing the light-burning propaganda because they believe that the practice of this theory would not only fail to prevent serious fires but would ultimately destroy the productiveness of the forests on which western industries depend for their supply of timber.

The whole principle involved may be put in a nutshell by stating that under certain conditions mature forests are not killed outright by light-burning, and can be kept alive in spite of surface fires, provided, however, that no other actual drain is made upon their productive capacity. In other words, certain forests will withstand light-burning provided they do not have to withstand anything else. But this proviso is the very negation of the fundamental principle of forestry, namely, to make forests productive not only of a vegetive cover to clothe and protect our mountains, but also of the greatest possible amount of lumber, forage, and other forest products.

Forests can not be productive under light-burning because:
1. Light-burning destroys most of the seedling trees necessary to replace the old stand as it is removed for human use.
2. Light-burning gradually reduces the vitality and productiveness of the forage.
3. Light-burning destroys the humus in the soil necessary for rapid tree growth, (that is, rapid lumber production, and germination of the seed.)
4. Light-burning, by inflicting scars, abnormally increases the rots which destroy the lumber, and increases the resin which depreciates lumber grades and intensifies subsequent fires.
5. Light-burning, in most cases at least, increases the destructive effects of wood-boring insects.

In other words, you can maintain some forests under light-burning, but you can not maintain them efficiently.

The light-burning fallacy ought to be of particular interest to the mining industry for the reason that it would destroy future timber supplies in an attempt to preserve present standing timber. The mines of the Southwest are not going to be exhausted during the present generation, and they are going to need timber in the future as well as at present. The heavy cost of importing this timber from distant sources is already known to most mining men. The mining industry would therefore suffer an economic handicap of no mean proportions were the future productiveness of our mountain timberlands seriously jeopardized.

The Forest Service policy of absolutely preventing forest fires insofar as humanly possible is directly threatened by the light-burning propaganda. It is up to the public and especially the users of the forests to decide whether they wish that policy continued or whether they wish to try "Piute Forestry." If the public desires the continuance of forest fire prevention, now is the time to put the quietus on the agitation for light-burning.

It is, of course, absurd to assume that the Indians fired the forests with any idea of forest conservation in mind. As is well known to all old-timers, the Indian fired the forests with the deliberate intent of confusing and concentrating the game so as to make hunting easier. It appears to be a fact that when deer or other game animals smell a smoke they stand stupefied with their heads in the air until actually singed by the heat or flames. They then make a wild break in any direction and largely lose their usual caution and ability to escape their human enemies. A bunch of deer with their heads in the air waiting for a fire presented an easy mark, even to the Indian's bow and arrow, and it was this fact, and not any desire for fancied forest conservation, which caused the Indians to burn the forests.

The destructive effects of Piute Forestry can readily be seen in California and in many areas of the Southwest. It can be stated without hesitation that a large percentage of the chaparral or brush areas found in the Southwestern states were originally covered with valuable forests, but gradually reverted to brush after repeated light-burning had destroyed the reproduction. In fact, the remains of large old stumps and pitchy roots can be found in these brush areas in many places. It is probably a safe prediction to state that should light-burning continue for another fifty years, our existing forest areas would be further curtailed to a very considerable extent. It is also a known fact that the prevention of light-burning during the past ten years by National Forest administration has brought in growth on large areas where reproduction was hitherto largely lacking. Actual counts show that the 1919 seedling crop runs as high as 100,000 per acre, and in some cases 400,000 per acre. It does not require any very elaborate argument to show that these tiny trees, averaging only two inches high, would be completely destroyed by even a light ground fire.

It is an old saying that fire is a good servant but a poor master. It is the opinion of foresters generally that the light-burners who propose to make fire their servant will find that forest fires are not so easily subjugated. In fact, it would be in practice absolutely impossible to fire the Forests without destroying the young growth, not to mention the constant risk of the fire breaking out of bounds and destroying buildings, fences, and mature timber. Most light-burning propagandists undoubtedly know this already, but do not care whether the young growth is destroyed or not. Their investments are in mature timber, and the less young growth there is left in the country the greater their chance for speculative profits. But the public does care, because wholesale destruction of young growth means timber famine at some time in the future. To actually control light-burning would cost more than the entire present system of Forest administration (about three cents per acre per year, gross) of which fire prevention is but a fraction. Why pay such a price for the privilege of burning up our future timber supplies?

The Forestry of the Prophets [1920]

Though Leopold did not profess adherence to any organized religion, he had participated in a Bible study group at Yale, and early in his career he occasionally quoted from the prophets. Here he studies the books of the Old Testament prophets for evidences of their understanding of forests, forest fires, wood utilization, and tree growth. The article was published in the Journal of Forestry.

Who discovered forestry? The heretofore accepted claims of the European nations have of late been hotly disputed by the Piutes. I now beg leave to present a prior claim for the children of Israel. I can hardly state that they practiced forestry, but I believe it can be shown that they knew a lot about forests. (Also, if any of them set fires, they knew better than to admit it.) The following notes, gleaned from a purely amateur study of the Books of the Prophets of the Old Testament,[1] may be of interest to other foresters, and may possibly suggest profitable fields of research for competent Hebraists and physiographers.

The most interesting side of forestry was then, as it is now, the human side. There is wide difference in the woodcraft of the individual prophets—the familiarity with which they speak of forests, and especially the frequency with which they use similes based on forest phenomena. It appears that in Judaea, as in Montana, there were woodsmen and dudes.

Isaiah was the Roosevelt of the Holy Land. He knew a whole lot about everything, including forests, and told what he knew in no uncertain terms. He constantly uses the forest to illustrate his teachings, and in doing so calls the trees by their first names. Contrast with him the sophisticated Solomon, who spoke much wisdom, but whose lore was city lore—the nearest he comes to the forest is the fig tree and the cedar of Lebanon, and I think he saw more of the cedars in the ceiling of his palace than he did in the hills. Joel

---

1. Quotations are from Moulton's Reader's Bible, which is based on the Revised English Version.

knew more about forests than even Isaiah—he is the preacher of conservation of watersheds, and in a sense the real inventor of "prevent forest fires." David speaks constantly and familiarly about forests and his forest similes are especially accurate and beautiful. Ezekiel was not only a woodsman and an artist, but he knew a good deal about the lumber business, domestic and foreign. Jeremiah had a smattering of woods lore, and so did Hosea, but neither shows much leaning toward the subject. Daniel shows no interest in forests. Neither does Jesus the son of Sirach, who was a keen business man, a philosopher, and a master of epigram, but his tastes did not run to the hills. Strange to say the writer of the Book of Job, the John Muir of Judah, author of the immortal eulogy of the horse and one of the most magnificent essays on the wonders of nature so far produced by the human race, is strangely silent on forests. Probably forests were his background, not his picture, and he took for granted that his audience had a knowledge of them.

Forest Fires in the Holy Land

Every forester who reads the Prophets carefully will, I think, be surprised to see how much they knew about fires. The forest fire appealed strongly to their imagination and is used as the basis for many a simile of striking literary beauty. They understood not only the immediate destructive effects of fires, but possibly also the more far reaching effects on watersheds. Strangely enough, nothing is said about causes of fires or whether any efforts were ever made toward fire suppression.

The book of Joel opens with an allegory in which the judgment of God takes the form of a fire.[2] This is perhaps the most convincing description of fire in the whole Bible. "Alas for the day!" says Joel. "The herds of cattle are perplexed, because they have no pasture; Yea, the flocks of sheep are made desolate. O Lord, to thee do I cry, for a fire hath devoured the pastures of the wilderness, and a flame hath burned all the trees of the field. Yea, the beasts of the field pant unto thee, for the water brooks are dried up. Blow ye the trumpet in Zion, and sound an alarm in my holy mountain; let all the inhabitants in the land tremble! For . . . a fire devoureth before them; and behind them a flame burneth: the land is as a garden of Eden before them, and behind them a desolate wilderness!"

Joel's story of the flames is to my mind one of the most graphic descriptions of fire ever written. It is "a day of clouds and thick darkness," and the fire is "like the dawn spread upon the mountains." The flames are "as a great people, set in battle array," and "the appearance of them is as horses, and as horsemen, so do they run. Like the noise of chariots on the tops of the

---

[2]. Parts of Joel 1 and 2 have been used in printed matter issued by the Southwestern District as fire prevention propaganda.

mountains do they leap, . . . they run like mighty men; they climb the wall like men of war; and they march every one on his way. They break not their ranks: neither doth one thrust another; they march every one in his path. . . . They leap upon the city; they run upon the wall; they climb up into the houses; they enter in at the windows like a thief. The earth quaketh before them: the heavens tremble: the sun and the moon are darkened, and the stars withdraw their shining."

Joel is evidently describing a top fire or brush fire of considerable intensity. Is there at the present time any forest cover in Palestine of sufficient density to support such a fire? I do not know, but I doubt it. If not, it is interesting to speculate whether the reduced forest cover is a cause or an effect of the apparent change in climate.[3] Isaiah (64-1) adds some intensely interesting evidence as to the density of forest cover in Biblical times when he says: "when fire kindleth the brushwood, . . . the fire causeth the waters to boil." Have there been any fires in this country, even in the Northwest or the Lake States which caused the waters to boil? One writer, who had to take refuge in a creek during one of the big fires in the Northwest in 1918, states that falling brands caused the temperature of the creek to rise "several degrees," which sounds very tame in comparison with Isaiah's statement. In fact, Isaiah's statement seems almost incredible. Was he telling fish stories? Or is there some special explanation, such as a resinous brushwood producing great heat, or drainage from a sudden rain on a hot fire, or a water hole containing bitumen or oil from a mineral seep? I will leave this question for some one personally familiar with the country.

That top fires actually occurred in the Holy Land is abundantly proven by many writers in addition to Joel. Isaiah says (10-19) that a fire "shall consume the glory of his forest, and of his fruitful field . . . and the remnant of the trees of his forest shall be few, that a child may write them." "It kindleth in the thickets of the forest, and they roll upward in thick clouds of smoke." The individual tree at the moment of combustion he likens most effectively to a "standard-bearer that fainteth." Those who have actually seen the "puff" of the dying tree, as the fire rushes up through the foliage, will not miss the force of this simile. Ezekiel says (20-46): "A fire . . . shall devour every green tree . . . and every dry tree: the flaming flame shall not be quenched."

Surprisingly little is said about how fires started. Man-caused fires were no doubt frequent, as were to be expected in a pastoral community. Tobacco fires were of course still unknown. (Samuel Butler says the Lord postponed the discovery of tobacco, being afraid that St. Paul would forbid smoking.

---

3. Prof. Ellsworth Huntington's book, *The Pulse of Asia*, contains some very readable and convincing material on climatic cycles in Asia Minor.

This, says Butler, was a little hard on Paul.) Lightning was no doubt the principal natural cause of fire. Very heavy lightning seems to have occurred in the mountains. David, in the "Song of the Thunderstorm" (Psalm, 29), says: "The God of glory thundereth, . . . the voice of the Lord breaketh the cedars; Yea, the Lord breaketh in pieces the cedars of Lebanon." His voice "cleaveth the flames of fire . . . and strippeth the forest bare." It is not entirely clear whether this refers to lightning only, or possibly also to subsequent fire.

How much did the prophets really know about the effects of fires? Joel has already been quoted as to the effects on streamflow, but there is a possibility that he meant that his "water-brooks" dried up, not as the ultimate effect of fires, but as the immediate effect of a drouth prevailing at the time of the particular fire which he describes. David (Psalm, 107) plainly states that changes in climate occur, but no forest influences or other causes are mentioned. I think it is quite possible that the effect of forests on streamflow was known empirically to a few advanced thinkers like Joel, but it is quite certain that their knowledge went no further or deeper. The habit of thinking of natural phenomena as acts of God instead of as cause and effect prevails to this day with a majority of people, and no doubt prevailed at that time in the minds of all. But even if the prophets were ignorant of science, they were wise in the ways of men. "Seemeth it a small thing unto you to have fed upon the good pasture, but ye must tread down with your feet the residue of your pasture? And to have drunk of the clear waters, but ye must foul the residue with your feet?" (Ezekiel 24-18.) Here is the doctrine of conservation, from its subjective side, as aptly put as by any forester of this generation.

Forest Utilization in the Holy Land

The old Hebrew used both saws and axes in cutting timber. Isaiah (10-15) says: "Shall the axe boast itself against him that heweth therewith? Shall the saw magnify itself against him that shaketh it?" "Shaking" the saw is a new bit of woods vernacular that leads one to wonder what the instrument looked like. Here is more woods vernacular: " . . . he shall cut down the thickets of the forest with iron, and Lebanon shall fall by a mighty one." While I am not competent to go behind the translation, the word "iron" seems to be used here in much the same way as our modern engineers used the word "steel," that is, to indicate certain manufactured tools or articles made of steel.

Very close utilization of felled timber seems to have been practiced. Solomon (Wisdom, 13-11) tells how a woodcutter sawed down a tree, stripped off the bark, carved the good wood into useful vessels, cooked his dinner with the chips, and used the crooked and knotty remainder to fashion a graven image. Expertness in whittling then, as now, seems to have been a trait of the idle, for Solomon says the wood-cutter shaped the image "by the diligence of his idleness, and . . . by the skill of his indolence." Isaiah (44-14) also tells how a man plants a fir tree, and after the rain has nourished it, he cuts it down and uses a part to warm himself, a part to bake bread, a part to make utensils, and a part to fashion a graven image. Graven images, if one is to believe the prophets, must have been an important product of the wood using industries of that day.

Here is an unsolved mystery in woods practice: "The carpenter . . . heweth him down cedars, and taketh the holm tree and the oak, and *strengtheneth for himself* one among the trees of the forest" (Isaiah, 44-14). What is meant by "strengtheneth for himself?" Some process of seasoning? Some custom of individual branding such as is practiced on bee trees? Some process of lamination in wood-working to give strength and lightness?

Ezekiel (27-4) records some interesting data on the sources and uses of timber in his satire on the glories of Tyre. "They have made all thy planks of fir trees from Senir: they have taken cedars from Lebanon to make a mast for thee. Of the oaks of Bashan have they made thine oars; they have made thy benches of ivory inlaid in boxwood, from the isles of Kittim." Isaiah (2-18) also mentions "the oaks of Bashan." Oak would seem to be a bit heavy for the long oars used in those days.

Who made the first cedar chest? Ezekiel (27-24) says that "chests of rich apparel, bound with cords and made of cedar" were an article of commerce in the maritime trade of Tyre. The use of cedar chests for fine clothing seems to be nearly as old as the hills. Solomon's palanquin was also made of cedar. Here is his own description of it, as taken from the Song of Songs (3-9): "King Solomon made himself a palanquin of the wood of Lebanon. He made the pillars thereof of silver, the bottom thereof of gold, the seat of it of purple, the midst thereof being inlaid with love from the daughters of Jerusalem." (I doubt whether Solomon "made himself" this palanquin. He does not give the impression of a man handy with tools. No doubt he had it made by the most cunning artificers of his kingdom.)

Cedar construction in Biblical days seems to have been a kind of mark of social distinction, as mahogany is today. (Witness also the marble-topped walnut of our Victorian forbears!) Solomon's bride boasts (Song of Songs, 1-16): "Our couch is green. The beams of our house are cedars, and our rafters are firs." Jeremiah (22-14) accuses Jehoiakim of building with ill-gotten gains "a wide house . . . with windows . . . ceiled with cedar, and painted with vermillion." "Shalt thou reign," exclaims Jeremiah, "because thou strivest to *excel in cedar?*"

The cedar seems to have grown to large size. Ezekiel, in a parable (31), says of one tree: "The cedars in the garden of God could not hide him; the fir trees were not like his boughs, and the plane trees were not as his branches." This cedar was Pharaoh, and the Lord "made the nations to shake at the sound of his fall."

The close utilization which seems to have been practiced at least in some localities, the apparently well developed timber trade of the coast cities, and the great number of references to the use and commerce in cedar, would lead to the surmise that the pinch of local timber famine might have been felt in the cedar woods. That this was actually the case is indicated by Isaiah (14-7). After prophesying the fall of Babylon, he tells how all things will rejoice over her demise. "Yea, the fir trees rejoice at thee, and the cedars of Lebanon: 'Since thou art laid down, no feller is come up against us.'" This impersonization of trees is characteristic of the Biblical writers; David (Psalms, 96) says, "Then shall all the trees of the wood sing for joy."

The relative durability of woods was of course fairly well known. Isaiah (9-10) says: "The bricks are fallen, but we will build with hewn stone; the sycamores are cut down, but we will change them into cedars." Ecclesiasticus (12-13) likens the permanency and strength of wisdom to "a cedar in Libanus, and . . . a cypress tree on the mountains of Hermon."

Fuel wood was evidently obtained not only from cull material, as already indicated, but by cutting green timber. Ezekiel (39-9) predicts that after the rout of the invading army of Gog, "they that dwell in the cities of Israel shall go forth, and make fires of the weapons and burn them, . . . and they shall make fires of them seven years; so that they shall take no wood out of the field, neither cut down any out of the forests." It would seem that Biblical fuel bills were either pretty light, or else Gog left behind an extraordinary number of weapons.

Hebrew Silviculture

There are many passages in the books of the Prophets showing that some of the rudimentary principles of silviculture were understood, and that artificial planting was practiced to some extent. Solomon (in Ecclesiastes 2-4) says that he planted great vineyards, orchards, gardens, and parks, and also "made me pools of water, to water therefrom the forest where trees were reared." Isaiah (44-14) speaks of a carpenter who planted a fir tree, and later used it for fuel and lumber. The context gives the impression that such instances of planting for wood production were common, but probably on a very small scale. Isaiah (41-9) seems to have had some knowledge of forest types and the ecological relations of species. He quotes Jehovah in this manner: "I will plant in the wilderness the cedar, the acacia tree, and the myrtle, and the oil tree; I will set in the desert the fir tree, the pine, and the box tree together." He also makes the following interesting statement (55-13) which possibly refers to the succession of forest types: "Instead of the thorn shall come up the fir tree, and instead of the brier shall come up the myrtle tree."

Some of the peculiarities of various species in their manner of reproduction are mentioned. Isaiah (44-4) says: "They shall spring up among the grass as willows by the watercourses." He also speaks of the oak and the terebinth reproducing by coppice (6-12). Job (14-7) also mentions coppice, but does not give the species. Ezekiel (17) in his parable of the Eagles and the Cedar, tells about an eagle that cropped off the leader of a big cedar and planted it high on another mountain, and it brought forth boughs, and bore fruit, and was a goodly tree. I do not know the cedar of Lebanon but it sounds highly improbable that any conifer should grow from cuttings. I think this is a case of "poetic license."

Isaiah (65-22) realized the longevity of some species in the following simile: "They shall not build, and another inhabit; they shall not plant, and another eat; for as the day of a tree shall be the day of my people, and my chosen shall long enjoy the work of their hands." Isaiah disappoints us here in not telling the species. Unlike Solomon and Daniel and Ecclesiasticus, he is not given to calling a tree just "a tree."

Miscellaneous

Barnes has written a very interesting article on grazing in the Holy Land, and there is much additional material on this subject which would be of interest to foresters.[4] One matter which some entomologist should look up occurs in Isaiah (7-18). Isaiah says: "And it shall come to pass in that day, that the Lord shall hiss for the fly that is in the uttermost part of the rivers of Egypt, and for the bee that is in the land of Assyria. And they shall come, and shall rest all of them in the desolate valleys, and in the holes of the rocks, and upon all thorns, and upon all pastures." What fly is referred to? The Tetse fly, or the Rinderpest?

There is also considerable material on game and fish in the Old Testament, and additional material on forests in the historical books, both of which I hope to cover in future articles.

In closing, it may not be improper to add a word on the intensely interesting reading on a multitude of subjects to be found in the Old Testament. As Stevenson said about one of Hazlitt's essays, "It is so good that there should be a tax levied on all who have not read it."

4. *National Wool Grower*, February, 1915.

The Wilderness and Its Place in Forest Recreational Policy [1921]

Not long after he returned to the Forest Service in 1919, Leopold had occasion to discuss a new concept of land management with Arthur Carhart, a colleague involved in recreational planning in the Rocky Mountain district, and from this collaboration grew the notion that the Forest Service might be induced to preserve representative portions of some forests as wilderness. This 1921 essay is Leopold's first published articulation of a proposed wilderness area policy, addressed to foresters through their professional periodical, the *Journal of Forestry*, and rationalized within the context of the Forest Service's principle of "highest use."

When the National Forests were created the first argument of those opposing a national forest policy was that the forests would remain a wilderness. Gifford Pinchot replied that on the contrary they would be opened up and developed as producing forests, and that such development would, in the long run, itself constitute the best assurance that they would neither remain a wilderness by "bottling up" their resources nor become one through devastation. At this time Pinchot enunciated the doctrine of "highest use," and its criterion, "the greatest good to the greatest number," which is and must remain the guiding principle by which democracies handle their natural resources.

Pinchot's promise of development has been made good. The process must, of course, continue indefinitely. But it has already gone far enough to raise the question of whether the policy of development (construed in the narrower sense of industrial development) should continue to govern in absolutely every instance, or whether the principle of highest use does not itself demand that representative portions of some forests be preserved as wilderness.

That some such question actually exists, both in the minds of some foresters and of part of the public, seems to me to be plainly implied in the recent trend of recreational use policies and in the tone of sporting and outdoor magazines. Recreational plans are leaning toward the segregation of certain areas from certain developments, so that having been led into the wilderness, the people may have some wilderness left to enjoy. Sporting magazines are groping toward some logical reconciliation between getting back to nature and preserving a little nature to get back to. Lamentations over this or that favorite vacation ground being "spoiled by tourists" are becoming more and more frequent. Very evidently we have here the old conflict between preservation and use, long since an issue with respect to timber, water power, and other purely economic resources, but just now coming to be an issue with respect to recreation. It is the fundamental function of foresters to reconcile these conflicts, and to give constructive direction to these issues as they arise. The purpose of this paper is to give definite form to the issue of wilderness conservation, and to suggest certain policies for meeting it, especially as applied to the Southwest.

It is quite possible that the serious discussion of this question will seem a far cry in some unsettled regions, and rank heresy to some minds. Likewise did timber conservation seem a far cry in some regions, and rank heresy to some minds of a generation ago. "The truth is that which prevails in the long run."

Some definitions are probably necessary at the outset. By "wilderness" I mean a continuous stretch of country preserved in its natural state, open to lawful hunting and fishing, big enough to absorb a two weeks' pack trip, and kept devoid of roads, artificial trails, cottages, or other works of man. Several assumptions can be made at once without argument. First, such wilderness areas should occupy only a small fraction of the total National Forest area—probably not to exceed one in each State. Second, only areas naturally difficult of ordinary industrial development should be chosen. Third, each area should be representative of some type of country of distinctive recreational value, or afford some distinctive type of outdoor life, opportunity for which might disappear on other forest lands open to industrial development.

The argument for such wilderness areas is premised wholly on highest recreational use. The recreational desires and needs of the public, whom the forests must serve, vary greatly with the individual. Heretofore we have been inclined to assume that our recreational development policy must be based on the desires and needs of the majority only. The only new thing about the premise in this case is the proposition that inasmuch as we have plenty of room and plenty of time, it is our duty to vary our recreational development policy, in some places, to meet the needs and desires of the minority also. The majority undoubtedly want all the automobile roads, summer hotels, graded trails, and other modern conveniences that we can give them. It is already decided, and wisely, that they shall have these things as rapidly as brains and money can provide them. But a very substantial minority, I think, want just the opposite. It should be decided, as soon as the existence of the demand can be definitely determined, to provide what this minority wants. In fact, if we can foresee the demand, and make provision for it in advance, it will save much cash and hard feelings. It will be much easier to keep wilderness areas than to create them. In fact, the latter alternative may be dismissed as impossible. Right here is the whole reason for forehandedness in the proposed wilderness area policy.

It is obvious to everyone who knows the National Forests that even with intensive future development, there will be a decreasing but inexhaustible number of small patches of rough country which will remain practically in wilderness condition. It is also generally recognized that these small patches have a high and increasing recreational value. But will they obviate the need for a policy such as here proposed? I think not. These patches are too small, and must grow smaller. They will always be big enough for camping, but they will tend to grow too small for a real wilderness trip. The public demands for camp sites and wilderness trips, respectively, are both legitimate and both strong, but nevertheless distinct. The man who wants a wilderness trip wants not only scenery, hunting, fishing, isolation, etc.—all of which can often be found within a mile of a paved auto highway—but also the horses, packing, riding, daily movement and variety found only in a trip through a big stretch of wild country. It would be pretty lame to forcibly import these features into a country from which the real need for them had disappeared.

It may also be asked whether the National Parks from which, let us hope, industrial development will continue to be excluded, do not fill the public demand here discussed. They do, in part. But hunting is not and should not be allowed within the Parks. Moreover, the Parks are being networked with roads and trails as rapidly as possible. This is right and proper. The Parks merely prove again that the recreational needs and desires of the public vary through a wide range of individual tastes, all of which should be met in due proportion to the number of individuals in each class. There is only one question involved—highest use. And we are beginning to see that highest use is a very varied use, requiring a very varied administration, in the recreational as well as in the industrial field.

An actual example is probably the best way to describe the workings of the proposed wilderness area policy.

The Southwest (meaning New Mexico and Arizona) is a distinct region. The original southwestern wilderness was the scene of several important chapters in our national history. The remainder of it is about as interesting, from about as large a number of angles, as any place on the continent. It has a high and varied recreational value. Under the policy advocated in this paper, a good big sample of it should be preserved. This could easily be done by selecting such an area as the headwaters of the Gila River on the Gila National Forest. This is an area of nearly half a million acres, topographically isolated by mountain ranges and box canyons. It has not yet been penetrated by railroads and to only a very limited extent by roads. On account of the natural obstacles to transportation and the absence of any considerable areas of agricultural land, no net economic loss would result from the policy of withholding further industrial development, except that the timber would remain inaccessible and available only for limited local consumption. The entire area is grazed by cattle, but the cattle ranches would be an asset from the recreational standpoint because of the interest which attaches to cattle grazing operations under frontier conditions. The apparent disadvantage thus imposed on the cattlemen might be nearly offset by the obvious advantage of freedom from new settlers, and from the hordes of motorists who will invade this region the minute it is opened up. The entire region is the natural habitat of deer, elk, turkey, grouse, and trout. If preserved in its semi-virgin state, it could absorb a hundred pack trains each year without overcrowding. It is the last typical wilderness in the southwestern mountains. Highest use demands its preservation.

The conservation of recreational resources here advocated has its historic counterpart in the conservation of timber resources lately become a national issue and expressed in the forestry program. Timber conservation began fifteen years ago with the same vague premonitions of impending shortage now discernible in the recreational press. Timber conservation encountered the same general rebuttal of "inexhaustible supplies" which recreational conservation will shortly encounter. After a period of milling and mulling, timber conservation established the principle that timber supplies are capable of qualitative as well as quantitative exhaustion, and that the existence of "inexhaustible" areas of trees did not necessarily insure the supply of bridge timber, naval stores, or pulp. So also will recreational resources be found in more danger of qualitative than quantitative exhaustion. We now recognize that the sprout forests of New England are no answer to the farmer's need for structural lumber, and we admit that the farmer's special needs must be taken care of in proportion to his numbers and importance. So also must we recognize that any number of small patches of uninhabited wood or mountains are no answer to the real sportsman's need for wilderness, and the day will come when we must admit that his special needs likewise must be taken care of in proportion to his numbers and importance. And as in forestry, it will be much easier and cheaper to preserve, by forethought, what he needs, than to create it after it is gone.

Standards of Conservation [1922]

Leopold's frequent inspections of forest resources and the effectiveness of field personnel performed in his capacity as assistant district forester led him to puzzle over the problem of setting meaningful objectives for managing the forests. This unfinished penciled essay contains a blank space for a number to be added later and ends in midsentence. It was probably drafted while Leopold was on an inspection of the Prescott National Forest in Arizona in 1922. Here Leopold gropes toward measures of what should be accomplished on the ground, in contrast to the usual "machinery" standards of the service. This concern had troubled him nearly a decade earlier in his letter to the forest officers of the Carson and is still being debated by land managers today.

The setting of standards to correlate methods and practices has now become a familiar and successful feature of administration on the National Forests. Such standards have proven a simple and effective means of detecting and ironing out the discrepancies in the intensiveness with which similar work is done in separate places, and the relative emphasis given various lines of work.

So far, however, standards have been applied principally to the machinery of administration. What would be the probable result if they were applied to the objectives of administration, as distinguished from the machinery with which those objectives are to be attained?

It is believed such an application of standards would result in certain fundamental and beneficial changes, the nature of which it is the purpose of this paper to discuss.

At the outset, it may be well to give examples of the two classes of standards. When an administrative officer is directed to spend at least 40 days a year on grazing work or to make at least two general inspections per year of each unit of range, there is set up a *machinery standard* (heretofore vaguely called administrative standard, or standard of performance). On the other hand when there is set up as an objective of administration that a

Standards of Conservation 83

certain unit of range shall be brought to an .8 density of grama grass capable of carrying 1 head per 20 acres, there is established a *standard of conservation* for that unit.

Before discussing the possible effects of standards of conservation, it may be well to answer the question of why they need be set. Is it not axiomatic that every resource should be conserved as far as possible? To be sure, but natural resources are a complex affair, and few men agree on what is possible. For example: three administrators were examining a piece of range, having about .5 oak brush and .1 grama grass, with a very few old fire killed Junipers. It developed that one was looking forward to a .5 oak brush and .3 grass objective, another to a .5 oak and .9 grass objective, and the third to a stand of Juniper and Piñon woodland with a little brush and grass mixed in. Each objective was probably obtainable, but the method of setting about it radically different in each case. How could any man administer this area intelligently without knowing which of the three he was to work toward?

Another example: A certain area was withdrawn for protection of a reclamation project watershed. Previous overgrazing had thinned the grass and begun to let in a little Juniper reproduction, whereas previous to the grazing Juniper had been kept out by grass fires, as evidenced by charred stumps. One man examining the area wanted to reduce the grazing and restore the grass as watershed cover. Another wanted to increase the grazing to fill out the catch of Juniper reproduction as watershed cover. How could either administer the area intelligently without knowing which kind of cover he was to work toward?

Another example: A certain area on a "watershed" Forest was covered with vigorous even aged pine saplings, with a scattering ground cover of non-palatable Ceanothus brush and a few weeds. In the course of an inspection it developed that the ranger had striven for years to stock the area as heavily as possible with cattle, with a view to forcing them to browse the Ceanothus and thus reduce the fire hazard. The inspector noticed that this heavy grazing was destroying all the willows in the water-courses, causing heavy silting by tearing out of banks, in spite of the excellent protection of the watershed by the young pines. He wanted to risk the fire hazard and prevent the silting. Here were two men, both anxious to conserve, but with opposite ideas as to what most needed conserving, and hence with opposite plans of administration.

It can be safely said that when it comes to actual work on the ground, the objects of conservation are never axiomatic or obvious, but always complex and usually conflicting. The adjustment of these conflicts not only calls for the highest order of skill, but involves decisions so weighty in their consequence, and so needful of permanence and correlation, that only the highest authority should make them. And until they are made by competent authority, the local administrators cannot possibly apply their efforts effectively. On the other hand such adjustments when once decided on, constitute standards of conservation, toward which generations of local administrators may work with their eyes open, often devising their own methods or machinery and unifying their efforts toward the attainment of a clearly defined ultimate goal.

To return now to the possible effects of a change of emphasis in standards. It is first necessary to ask: why do we standardize machinery at all?

Some standardization of methods, practices, procedure, customs, forms, and other machinery is so obviously necessary, especially in any big organization, that the point need not be discussed. But is there not a lot of it done for the purpose of regulating, correlating, and stimulating effort toward objectives, when a precise statement of the objective itself would be a more effective means of accomplishing the same purpose? And is not such standardization of machinery apt in time to become merely a subterfuge to cover up the absence of scientific thinking in analyzing the objectives of conservation, willingness to take the responsibility of deciding them, and skill in concisely defining the decisions?

To illustrate: The Forest Service has _____ pages of grazing manual, a large part of which is devoted to setting up a standardized machinery for range conservation. Many thousands of dollars and the efforts of many men are invested each year in the job of keeping this machinery up to date. At this time also the Forest Service, pursuant to the wishes of the Congressional Committee, is conducting an appraisal of the National Forest ranges, which, to speak broadly, will result in the setting up of figures expressing the kind, quality, and quantity of forage now growing on each small unit of range of each Forest. To get these figures requires an examination and study of each unit by skilled men. During such examination, why not also set up figures expressing the kind, quality, and quantity of forage which *should* grow on each unit? If this were done, we should have a standard of conservation toward which local administrative effort could be intelligently directed for years to come. And what is more, such effort, having a definite goal, would not need nearly so much prodding in the form of machinery standards. And the time that now goes into establishing and maintaining machinery standards, could be diverted into the technical education of field men to make their efforts constantly more intelligent. "We don't know where we're going, but we're on our way" is a laudable sentiment only up to the point where it becomes scientifically possible to state where we ought to go.

When does it become scientifically possible to state where we ought to go? The big development of recent years in range management knowledge, coupled with the timely opportunity of range appraisal, has already been cited as one example. In forest management, the management plan offers a complete opportunity of expressing standards of conservation; in fact it largely constitutes a standard. Coupled with it would be standards of loss by fire (already broached at Mather Field). In lands management, the standard could be expressed by

Some Fundamentals of Conservation in the Southwest [1923]

This important essay was the first in which Leopold attempted an integrated statement of his conservation philosophy, grounded in an analysis of the deterioration of organic resources in the Southwest—especially the problem of soil erosion—and concluding with a discussion of "conservation as a moral issue." It was found in his papers as a typescript with comments from several colleagues attached, but it is not clear for what purpose he intended the essay or why he chose not to publish it. It contains several blank spaces for numbers to be added. More than three decades after Leopold's death, it was finally published in the first volume of a new journal, *Environmental Ethics*.

Introduction

The future development of the Southwest must depend largely on the following resources and advantages:
    Minerals: chiefly copper and coal.
    Organic: farms, ranges, forests, waters and water powers.
    Climatic: chiefly health and winter resort possibilities.
    Historic: archaeological and historical interest.
    Geographic: on route to California and Mexico.
    This discussion is confined to the two first named. While the last three are of great value, the Southwest should hardly be satisfied to build its future upon them. They are what might be termed *unearned advantages*.
    Excluding these, it is apparent that all of the remaining economic resources are of such a nature that their permanent usefulness is affected more or less by that idea or method of development broadly called *conservation*. It is the purpose of this paper to discuss the extent to which this is true, and the extent to which unskillful or nonconservative methods of exploitation threaten to limit or destroy their permanent usefulness.

A brief statement of some of the salient facts about each of these resources is first necessary as a background:

Minerals. Of six of the leading Arizona copper mines, the average life in sight is twenty-two years. This is a short life. Undoubtedly, our mineral wealth will be expanded from time to time by new processes, better transportation, discovery of additional ore bodies, demand for rare minerals, exploitation of gross minerals such as sulphur and salt, and possible discovery of oil. But the fact remains that, with the exception of coal, the mineral wealth of the Southwest, from the standpoint of an economic foundation for society, is exhaustible. Our coal will probably always be handicapped by long hauls and absence of water transport.

Farms & waters must be considered together. The late drouth ought to have sufficiently redemonstrated that generally speaking, dry farming, as a sole dependence for a livelihood, is a broken reed. The outstanding fact that we can never change is that we have roughly twenty million acres of water-producing or mountain area and fifty million acres of area waiting for water. Most of the latter is tillable. But, it takes say four feet of water per year to till it, whereas less than two feet falls on the mountains, of which only a very small percent runs off in streams in usable form. Therefore, if we impounded all the non-flood runoff and had no evaporation (both impossibilities), we should still have scores of times more land than water to till it. This is partially offset by underground storage of part of the water which does not run off, but, nevertheless, we still have an overwhelming shortage of water as compared with land. Therefore, the term *irrigable land* actually represents a combination of natural resources which is really very rare and accordingly vital to our future. By artificially impounding water we are steadily adding to our irrigated area, but these gains are being offset by erosion losses in the smaller valleys, where water was easily available simply through diversion. Broadly speaking, no net gain is resulting. We are losing the easily irrigable land and "replacing" it by land reclaimed at great expense. The significant fact that is not understood is that this "replacement" is no replacement at all, but rather slicing at one end of our loaf while the other end sloughs away in waste. Some day the slicing and sloughing will meet. Then we shall realize that we needed the whole loaf.

Water powers. Erosion and silting are likewise deteriorating our water powers, though the silting of a reservoir is not so destructive to its power possibilities as to its use for irrigation. Also, the water powers not dependent upon storage are not yet badly damaged. It is obvious, however, that anything which damages the regularity of stream flow and interferes with storage of waters is depreciating the value of our power resources.

Forests. While ____% of our area bears trees, only ____ of this bears saw timber, of which the present stand is thirty-five billion feet. Most of this sawtimber land is in the national forests. The management plans of the Forest Service indicate that if handled under proper methods these sawtimber lands will sustain indefinitely a cut of 300 million feet per year. A larger cut will be possible temporarily because of the excess proportion of mature stands.

New Mexico and Arizona now consume about 450 million feet per year. The salient fact about our forests, therefore, is this: that in the long run the timber yield will only partly suffice to sustain our own agriculture, cities, and mines. Its conservation for these purposes is, of course, absolutely essential in order that we may not have to depend on expensive importations. But, even with good forestry, the Southwest cannot figure on timber export as a future source of wealth.

Ranges. Arizona and New Mexico are carrying about three million sheep and two million cattle. About one-fourth of these are on the national forests. It would not mean anything to try to state in figures the original carrying capacity of the two states because much of the virgin range was without water. Great progress has been made in developing water, but a wholesale deterioration in both the quality and quantity of forage has also taken place. On certain areas of national forests and privately owned range this deterioration has been checked and the productiveness of the forage partially restored through range improvements and conservative methods of handling stock. The remainder continues to deteriorate under the system of competitive destruction inherited from frontier days but now perpetuated by the archaic land policy of the government and some of the several states. It is safe to state that the condition of our range forage has depreciated 50% and is still going down-hill.

This overgrazing of our ranges is chiefly responsible for the erosion which is tearing out our smaller valleys and dumping them into the reservoirs on which our larger valleys are dependent. The significant element in this situation is that cessation of overgrazing will usually not check this erosion.

Summary. All of our organic resources are in a rundown condition. Under existing methods of management our forests may be expected to improve, but our total possible farm areas are dwindling and our waters and ranges are still deteriorating. In the case of our ranges, deterioration could be easily checked by conservative handling, and the original productiveness regained and restored. But the deterioration of our fundamental resources—land and water—is in the nature of permanent destruction, and the process is cumulative and gaining momentum every year.

Erosion and Aridity

The task of checking the ravages of erosion and restoring our organic resources to a productive condition is so intricate and difficult a problem that we must know something about causes before we can well consider remedies.

Is our climate changing? The very first thing to know about causes is whether we are dealing with an "act of God," or merely with the consequences of unwise use by man. If this collapse of stable equilibrium in our soil and its cover is being caused or aggravated by a change in climate, the possible beneficial results of conservation might be very limited. On the other hand, if there is no change of climate going on, the possible results of conservation are limited only by the technical skill which we can train upon the problem and the public backing available to get it applied.

In discussing climatic changes, a clear differentiation between the geological and historical viewpoints is essential. The status of our climate from the geological viewpoint has nothing to do with the question in hand. Any such changes that may be taking place would be too slow to have any bearing on human problems.

Historically speaking, our climate has recently been checked back with considerable accuracy to 1390 A.D. through study of the growth rings of yellow pine in Arizona, and to 1220 B.C. through the growth rings of sequoia in California. These studies, conducted by Douglas and Huntington, demonstrate convincingly that there has been no great increase or decrease in aridity of the Southwest during the last 3,000 years.

Yellow pines recently excavated at Flagstaff show very large growth rings, indicating a wetter climate during some recent geological epoch, but, as previously stated, that has a merely academic bearing on our problem.

Ancient Indian ditches and ruins in localities now apparently too dry for either irrigation or dry farming would seem to contradict the conclusion derived from tree rings, but little is known of the age of these relics or the habits of the people who left them. Archaeologists predict that we may soon know more about the age of the cliff culture through possible discoveries connecting it with the now accurately determined chronology of the Maya culture in Central America.

Long straight cedar timbers found in some Southwestern ruins likewise might be taken to indicate a process of desiccation, but other and unknown factors are involved, such as the effect of fire on our forests and the distances from which the timbers were transported.

Changes in the distribution of forest types likewise might be interpreted to throw a little light on the recent tendency of our climate. In the brush forests of southern Arizona there is strong evidence of an uphill recession, such as would accompany desiccation, former woodland now being occupied by brush species and former yellow pine by woodland and brush. At the same time, in the region of the Prescott and Tusayan Forests, there is an indisputable encroachment of juniper downward into former open parks. A similar encroachment of yellow pine is taking place in the Sitgreaves and Apache Forests. Most of these changes, however, can be accounted for through purely local causes such as fire and grazing. This fact, and the fact that any attempt at a climatic theory of causation would result in contradictory conclusions, makes it seem logical to regard these phenomena either as shedding no light on the question of climate, or as possibly somewhat substantiating the conclusion derived from tree rings.

In general, there thus far appears to be no clear evidence of desiccation during any recent unit of time small enough to be considered from an economic standpoint, but at least one line of pretty clear evidence as to the general stability of our climate during the last 3,000 years.

Drouth cycles and their effect. While science has shown that there is no general trend in our climate either for better or for worse, it has shown most conclusively that there are periodic fluctuations which vitally affect our prosperity and the methods of handling our resources. The same tree rings which assure us that Southwestern climate has been stable for 3,000 years warn us plainly that it has been decidedly unstable from year to year, and that the drouth now so strongly impressed on every mind and pocketbook is not an isolated or an accidental bit of hard luck, but a periodic phenomenon the occurrence of which may be anticipated with almost the same certainty as we anticipate the days and the seasons. It is cause for astonishment that our attitude toward these drouths which wreck whole industries, cause huge wastes of wealth and resources, and even empty the treasures of commonwealths should still be that of the Arkansan toward his roof—in fair weather no need to worry, and in foul weather too wet to work.

The tree rings show, in short, that about every eleven years we have a drouth. Every couple of centuries this eleven-year interval lengthens or shortens rather abruptly, running as low as nine and as high as fourteen. The drouths vary a little in length and intensity, and usually there is a "double crest" to both the high and low points, i.e., a better or worse year interlarded between the bad or good extremes. But always, and as sure as sunrise, "dust and a bitter wind shall come."

In addition to the eleven-year cycle, there is a curious chop or "zigzag" (two-year cycle), and probably a long or low groundswell measured in centuries, but the amplitude of all these is too low to have any great practical present economic significance. The eleven-year wave is the one that swamps the boats.

If there be those who doubt whether the tree rings tell a true story, let them be reminded that history supports their testimony. The great flood of the Rio Grande in 1680 is recorded in the trees. The famines of 1680-90 are there—possibly the drouth that produced them had something to do with the Pueblo Rebellion that sent De Vargas to Santa Fe. The great drouths of 1748, 1780, and 1820-23 are all concurrently reported by trees and historians. And in the last century came the weather records, which likewise concur. Douglas has tied in the eleven-year cycle of tree rings with sunspots. Munns has tied in the sunspots with lightning, and with forest fires as recorded in old scars. The chain of evidence as to the existence of the drouth cycle is very complete, and its bearing on economic and conservation problems is obvious. But like much other scientific truth, it may remain "embalmed in books, which are interred in university libraries, and then, long after, worked out by rule of thumb, by practical politicians and businessmen."

The point is that, if every eleven years we may expect a drouth, why not manage our ranges accordingly? This means either stocking them to only their drouth capacity, or arranging to move the stock or feed it when the drouth appears. But instead, we stock them to their normal capacity, and, when drouth comes, the stock eat up the range, ruin the watershed, ruin the stockman, wreck the banks, get credits from the treasury of the United States, and then die. And the silt of their dying moves on down into our reservoirs to someday dry up the irrigated valleys—the only live thing left!

Equilibrium of arid countries. To complete a background for the understanding of natural laws and their operation on our resources it is necessary to consider briefly the so-called "balance of nature."

There appears to be a natural law which governs the resistance of nature to human abuse. Broadly speaking, the law is this: the degree of stability varies inversely to the aridity.

Of course, this concept of a "balance of nature" compresses into three words an enormously complex chain of phenomena. But history bears out the law as given. Woolsey says that decadence has followed deforestation in Palestine, Assyria, Arabia, Greece, Tunisia, Algeria, Italy, Spain, Persia, Sardinia, and Dalmatia. Note that these are all arid or semiarid.[1] What well watered country has ever suffered serious permanent damage to all its organic resources from human abuse? None that I know of except China. It might be reasonable to ascribe this one exception to the degree of abuse received. Sheer pressure of millions exerted through uncounted centuries was simply too much.

A definite causal relation has long been believed to exist between deforestation and decline in productiveness of nations and their lands. But it

1. T. S. Woolsey, *Studies in French Forestry* (New York: John Wiley & Sons, n.d.).

strikes me as very curious that a similar causal relation between overgrazing and decadence has never to my knowledge been positively asserted. All our existing knowledge in forestry indicates very strongly that overgrazing has done far more damage to the Southwest than fires or cuttings, serious as the latter have been. Even the reproduction of forests has now been found to be impossible under some conditions without the careful regulation of grazing, whereas fire was formerly considered the only enemy.

The relative seriousness of destructive agencies may be illustrated by an example. Take the Sapello watershed, which forms a major part of the GOS range in the Gila National Forest. Old settlers state that when they came to the country the Sapello was a beautiful trout stream lined with willows. Yet old burned stumps show beyond a doubt that great fires burned in the watershed of the Sapello for at least a century previous to settlement. These fires spoiled the timber, but they did not spoil the land. Since then we have kept the fires out, but livestock has come in. And now the watercourse of the Sapello is a pile of boulders. In short, a century of fires without grazing did not spoil the Sapello, but a decade of grazing without fires ruined it, as far as the water courses are concerned.

Now the remarkable thing about the Sapello is that it has not been overgrazed. The GOS range is pointed to with pride as a shining example of range conservation. The lesson is that under our peculiar Southwestern conditions, any grazing at all, no matter how moderate, is liable to overgraze and ruin the watercourses. And the wholesale tearing out of watercourses is sufficient to silt our irrigation reservoirs, whether or not it is followed by wholesale erosion of the range itself.

Of course, this one example does not prove that grazing is the outstanding factor in upsetting the equilibrium of the Southwest. It is rapidly becoming the opinion of conservationists, however, that such is the case, and that erosion-control works of some kind are the price we will have to pay if we wish to utilize our ranges without ruining our agriculture.

Examples of destruction. The effect of unwise range use on the range industry, or of unwise farming on the land, are all too obvious to require illustration. What we need to appreciate is how abuses in one of these industries in one place may unwittingly injure another industry in another place.

A census of thirty typical agricultural mountain valleys in the national forests of the Southwest shows four ruined, eight partly ruined, fifteen starting to erode, and only three undamaged. Of the twenty-seven valleys damaged, every case may be ascribed to grazing or overgrazing, supplemented more or less by the clearing of cover from stream banks, fire, and the starting of washes along roads or trails.

A detailed survey of one mountain valley (Blue River, Arizona) shows 3,500 acres of farm land washed out, population reduced two-thirds, and half a million paid for a road over the hills because there was no longer any place to put a road in the valley. This entire loss may be ascribed to overgrazing of creek bottoms and unnecessary clearing of banks.

A special study of one cattle ranch showed that the loss of sixty acres of farm land through erosion imposed a permanent tax of $6 per head on the cost of production. This was a herd of 860 head.

Data on reservoirs shows that Elephant Butte and Roosevelt Lake must probably be raised prematurely because of silting. One of the big Pecos dams (Lake MacMillan) is said to have silted up 60% in fifteen years. A detailed report on the Zuni Reservoir, which may be considered typical of a smaller class, shows that its life will be twenty-one years, twelve of which have passed.[2] Silting is forcing the amortization of this half-million dollar investment at the rate of $7 per irrigated acre per year. Raising the dam is necessary to extend its life. In all these cases the silting seems to have been faster than was calculated, and is tending to force the amortization of the investments during alarmingly short periods. Inexpensive desilting methods have not yet been devised. What will be left of the Southwest if silting cuts down our already meager facilities for storage of an already meager water flow?

Here are some typical flood figures: Cave Creek, which flooded Phoenix in 1921, destroyed $150,000 in property and forced construction of a dam costing $500,000. The Pueblo flood of 1921 cost $17,000,000. A little flood in Taos Canyon in 19___ destroyed a new road costing $15,000. Every year it costs $40,000 to clear the diversion plants below Elephant Butte of silt from side washes. These are merely random examples. Undoubtedly we always had floods, but all the evidence indicates that they usually spent themselves without damage while our watercourses were protected by plenty of vegetation, and such damage as occurred was quickly healed up by the roots remaining in the ground.

Summary. Our organic resources are not only in a rundown condition, but in our climate bear a delicately balanced interrelation to each other. Any upsetting of this balance causes a progressive deterioration that may not only be felt hundreds of miles away, but may continue after the original disturbance is removed and affect populations and resources wholly unconnected with the original cause. Erosion eats into our hills like a contagion, and floods bring down the loosened soil upon our valleys like a scourge. Water, soil, animals, and plants—the very fabric of prosperity—react to destroy each other and us. Science can and must unravel those reactions, and government must enforce the findings of science. This is the economic bearing of conservation on the future of the Southwest.

---

2. H. F. Robinson, "Silt Problem of the Zuni Reservoir," *Transactions of the American Society of Civil Engineers* 83 (1919-20), 868-93.

Conservation as a Moral Issue

Thus far we have considered the problem of conservation of land purely as an economic issue. A false front of exclusively economic determinism is so habitual to Americans in discussing public questions that one must speak in the language of compound interest to get a hearing. In my opinion, however, one can not round out a real understanding of the situation in the Southwest without likewise considering its moral aspects.

In past and more outspoken days conservation was put in terms of decency rather than dollars. Who can not feel the moral scorn and contempt for poor craftsmanship in the voice of Ezekiel when he asks: *Seemeth it a small thing unto you to have fed upon good pasture, but ye must tread down with your feet the residue of your pasture? And to have drunk of the clear waters, but ye must foul the residue with your feet?*

In these two sentences may be found an epitome of the moral question involved. Ezekiel seems to scorn waste, pollution, and unnecessary damage as something unworthy—as something damaging not only to the reputation of the waster, but to the self-respect of the craft and the society of which he is a member. We might even draw from his words a broader concept—that the privilege of possessing the earth entails the responsibility of passing it on, the better for our use, not only to immediate posterity, but to the Unknown Future, the nature of which is not given us to know. It is possible that Ezekiel respected the soil, not only as a craftsman respects his material, but as a moral being respects a living thing.

Many of the world's most penetrating minds have regarded our so-called "inanimate nature" as a living thing, and probably many of us who have neither the time nor the ability to reason out conclusions on such matters by logical processes have felt intuitively that there existed between man and the earth a closer and deeper relation than would necessarily follow the mechanistic conception of the earth as our physical provider and abiding place.

Of course, in discussing such matters we are beset on all sides with the pitfalls of language. The very words *living thing* have an inherited and arbitrary meaning derived not from reality, but from human perceptions of human affairs. But we must use them for better or for worse.

A good expression of this conception of an organized animate nature is given by the Russian philosopher Ouspensky, who presents the following analogy:

> Were we to observe, from the inside, one cubic centimetre of the human body, knowing nothing of the existence of the entire body and of man himself, then the phenomena going on in this little cube of flesh would seem like elemental phenomena in inanimate nature.

He then states that it is at least not impossible to regard the earth's parts—soil, mountains, rivers, atmosphere, etc.—as organs, or parts of organs, of a coordinated whole, each part with a definite function. And, if we could see this whole, as a whole, through a great period of time, we might perceive not only organs with coordinated functions, but possibly also that process of consumption and replacement which in biology we call the metabolism, or growth. In such a case we would have all the visible attributes of a living thing, which we do not now realize to be such because it is too big, and its life processes too slow. And there would also follow that invisible attribute—a soul, or consciousness—which not only Ouspensky, but many philosophers of all ages, ascribe to all living things and aggregations thereof, including the "dead" earth.

There is not much discrepancy, except in language, between this conception of a living earth, and the conception of a dead earth, with enormously slow, intricate, and interrelated functions among its parts, as given us by physics, chemistry, and geology. The essential thing for present purposes is that both admit the interdependent functions of the elements. But "anything indivisible is a living being," says Ouspensky. Possibly, in our intuitive perceptions, which may be truer than our science and less impeded by words than our philosophies, we realize the indivisibility of the earth—its soil, mountains, rivers, forests, climate, plants, and animals, and respect it collectively not only as a useful servant but as a living being, vastly less alive than ourselves in degree, but vastly greater than ourselves in time and space—a being that was old when the morning stars sang together, and, when the last of us has been gathered unto his fathers, will still be young.

Philosophy, then, suggests one reason why we can not destroy the earth with moral impunity; namely, that the "dead" earth is an organism possessing a certain kind and degree of life, which we intuitively respect as such. Possibly, to most men of affairs, this reason is too intangible to either accept or reject as a guide to human conduct. But philosophy also offers another and more easily debatable question: was the earth made for man's use, or has man merely the privilege of temporarily possessing an earth made for other and inscrutable purposes? The question of what he can properly do with it must necessarily be affected by this question.

Most religions, insofar as I know, are premised squarely on the assumption that man is the end and purpose of creation, and that not only the dead earth, but all creatures thereon, exist solely for his use. The mechanistic or scientific philosophy does not start with this as a premise, but ends with it as a conclusion, and hence may be placed in the same category for the purpose in hand. This high opinion of his own importance in the universe Jeanette Marks stigmatizes as "the great human impertinence." John Muir, in defense of rattlesnakes, protests: ". . . as if nothing that does not obviously make for the benefit of man had any right to exist; as if our ways were God's ways." But the noblest expression of this anthropomorphism is Bryant's "Thanatopsis":

> . . . The hills
> Rock-ribbed and ancient as the sun—the vales
> Stretching in pensive quietness between;
> The venerable woods—rivers that move
> In majesty, and the complaining brooks
> That make the meadows green, and, poured round all
> Old oceans gray and melancholy waste—
> *Are but the solemn decorations all*
> *Of the great tomb of man.*

Since most of mankind today profess either one of the anthropomorphic religions or the scientific school of thought which is likewise anthropomorphic, I will not dispute the point. It just occurs to me, however, in answer to the scientists, that God started his show a good many million years before he had any men for audience—a sad waste of both actors and music—and in answer to both, that it is just barely possible that God himself likes to hear birds sing and see flowers grow. But here again we encounter the insufficiency of words as symbols for realities.

Granting that the earth is for man—there is still a question: what man? Did not the cliff dwellers who tilled and irrigated these our valleys think that they were the pinnacle of creation—that these valleys were made for them? Undoubtedly. And then the Pueblos? Yes. And then the Spaniards? Not only thought so, but said so. And now we Americans? Ours beyond a doubt! (How happy a definition is that one of Hadley's which states, "Truth is that which prevails in the long run"!).

Five races—five cultures—have flourished here. We may truthfully say of our four predecessors that they left the earth alive, undamaged. Is it possibly a proper question for us to consider what the sixth shall say about us? If we are logically anthropomorphic, yes. We and

> . . . all that tread
> The globe are but a handful to the tribes
> That slumber in its bosom. Take the wings
> Of morning; pierce the Barcan wilderness
> Or lose thyself in the continuous woods
> Where rolls the Oregon, and hears no sound
> Save his own dashings—yet the dead are there,
> And millions in those solitudes, since first

> The flight of years began, have laid them down
> In their last sleep.

And so, in time, shall we. And if there be, indeed, a special nobility inherent in the human race—a special cosmic value, distinctive from and superior to all other life—by what token shall it be manifest?

By a society decently respectful of its own and all other life, capable of inhabiting the earth without defiling it? Or by a society like that of John Burroughs' potato bug, which exterminated the potato, and thereby exterminated itself? As one or the other shall we be judged in "the derisive silence of eternity."

A Criticism of the Booster Spirit [1923]

Four years removed from his stint as secretary of the Albuquerque Chamber of Commerce but no less committed to civic betterment, Leopold distinguishes in this hard-hitting address to an Albuquerque civic society, Ten Dons, between growth from unearned increment and improvement resulting from the creation of value. The essay was found as a typescript with edits in Leopold's hand.

When the historians of the future write the story of our generation, the growth and spread of the Booster spirit will, I think, rank as one of the outstanding phenomena of the twentieth century.

Boosterism is not as yet firmly established in any country but our own. Perhaps it will not be. But it is an undeniable fact that it touches the daily lives of a hundred million Americans, and dominates the lives of some of them. This alone establishes it as one of the great political and economic forces of our time.

Boosterism is hard to gather up within the confines of a definition. From the viewpoint of a Booster, it might be called the application to civics of the truism that faith will move mountains. From the viewpoint of the critical observer, it might be defined by paraphrasing Decatur: "My ____! May she always be right! But right or wrong, my ____!" Directions: insert in the blank the name of any lodge, ward, corporation, luncheon club, city, county, or state which you own, or which owns you, and Boost!

Boosterism is new, at least in its abuses. Civic or group loyalty, in the more solid sense, is of course as old as morality and civilization, but the (to me) unholy wedlock between the moral principle of loyalty and the technique of billboard advertising is a recent and not altogether pleasing addition to the things that are under the sun.

The uses of Boosterism need no defense from me. The organizer of the Chamber of Commerce extols them adequately when he awakens the civic

A Criticism of the Booster Spirit

consciousness of our town for forty per cent of the gate receipts. The only thing about Boosterism that is not expounded to us daily is its abuses and fallacies. These I will attempt to describe.

The philosophy of boost is premised on certain tenets, which are proclaimed every little while by menu cards, convention badges, billboards, windshields and civic orators, but which do not appear to have yet been collected in a creed. To write such a creed is not an appropriate task for an unbeliever. Possibly I entirely misapprehend the matter, but this is as I understand it:

1. To be big and grow bigger is the end and aim of cities and citizens. To be small when young is excusable, but to stay small is failure.
2. The way to grow big is to advertise advantages and ignore defects, thereby abolishing them. Self-criticism is akin to treason.
3. Growth by labor, frugality, or natural increase is slow and old-fashioned. Growth in population is attained by decoying it from some other town or state. Growth in wealth is attained by attracting tourists or capital from elsewhere, or extracting appropriations from public treasuries.
4. Earned increment may indicate industry, but unearned increment proves vision and brains.
5. Unanimity is the only defensible attitude toward public questions. Minority opinions merely complicate the situation.
6. Taxes are crushing enterprise, and must be reduced, but our appropriations are entirely inadequate.
7. Bribing conventions and setting stool-pigeons for tourists are signs of friendly rivalry between cities.
8. The up-and-comingness of a town varies directly as the congestion of its billboards, luncheon clubs, and traffic, and inversely as its parking space.
9. Educational institutions, libraries, and parks are valuable business assets. They attract strangers.
10. Skilled craftsmen may move to our town if they want to, but we *must* have oil men and motor tourists, for they are the salt of the earth.

This may not be all of it, but it is enough. Let us examine in detail whether and why these propositions are true.

First, what, concretely, is our ambition as a city? "100,000 by 1930"—we have blazoned it forth like an army with banners. This is well and good—a city has as much "right" to resolve to attain a phenomenal growth in population as a citizen has to resolve to have 15 children, if not more so. And how are we going to get the 100,000? By advertising our climate, by craft and strategy in manipulating the location of institutions, and by increasing tourist traffic. There is nothing necessarily wrong about any of these things. But none of them are creative effort in any real sense. Supposing every city likewise resolved to live and grow by its wits instead of by creating values? (Most of them are, and maybe that is what is wrong with the country.)

Moreover, just why do we wish to grow by unearned increment instead of an earned increment derived from our own basic resources? Does it ever occur to the booster that we have fifty million acres of range in this state injured or ruined by overgrazing, that could be made into a source of wealth and prosperity beside which his tourists and sanatoria are mere bubbles? That we have a potential agriculture in this valley, crippled by seepage and threatened by silting, that is declining by neglect while he is playing with conventions and brass bands? That the lack of public interest in these real resources is causing them to deteriorate instead of develop?

Moreover, just why are we so much more intense about decoying newcomers to New Mexico than we are about securing better education, better recreational facilities, better public health service, and cleaner government for the citizens already here? Did anybody ever see a boosters' program that dealt with any of these internal betterments with even a fraction of the earnestness and ingenuity which it devotes to log-rolling for doubtful appropriations or entertaining motorists?

Can anyone deny that the vast fund of time, brains, and money now devoted to making our city big would actually make it better if diverted to betterment instead of bigness?

Moreover are we sure that if we effected these internal betterments for our own citizens, that we would have to bribe, threaten and cajole new people and new institutions to come here? I am afraid we could not keep them away.

In boosting, as in the Inquisition, the end justifies the means. Just now the boosters are lashing the latent patriotism of the Nation to build by public subscription a huge memorial sanatorium to the War Mothers of America—in Albuquerque. The sweep and daring of the idea is as splendid as its avowed motive is sordid and miserable. Would you want the Marble Manufacturer to conceive the splendid idea of passing the hat among your friends, in order that he might build, for cost plus ten per cent, a monument to *your* mother? What is the difference? Will this splendid, monstrous scheme find favor where stand the crosses, row on row, in Flanders fields? But, say the boosters, while the scheme is selfish for Albuquerque, it will give expression to unselfish and lofty motives throughout the Nation. Indeed! Was the Statue of Liberty thus conceived? Did the Westminster Chamber of Commerce boost the Abbey? And why not charge a fee of admission to the battleground at Gettysburg? Perhaps we shall accomplish this thing, and its greatness, like the grass, will grow and bury the inglorious memory of its inception.

> Pile the bodies high at Austerlitz and Waterloo
> Shovel them under and let me work—
> I am the grass, I cover all.

As Machiavelli, our preceptor since the Great War, once said: "Our experience has been that those who have done great things have held good faith of little account."

The booster is intensely provincial. A year ago he demanded a National Park for New Mexico. He did not know where or how, but he knew jolly well why: A National Park would be a tourist-getter of the first water, and tourists are to be desired above all things. They come, they see, they spend, and they are even known to come back.

Now a National Park is nothing more or less than the given word of the United States government that the place so designated is superlative among natural wonders. Had New Mexico places worthy of this high guarantee? If so, were they so situated as to comprise a unit that could be encompassed by a boundary which would say "Here it is"? If not, was there any danger of our government being induced to certify as superlative something that was not so? The booster should worry! Could the (his) government deny that any state outclassed New Mexico in natural wonders? Could the (his) government deny that the other children had been served pie and he had not? The government came pretty near giving its querulous child a bottle of water to hush it up.

As with the given word of the United States, so with its pocketbook. The boosters of Santa Fe and Las Vegas conceived a scenic highway across the Sangre de Cristo Range. To recreate and inspire their own citizens? No indeed, to fetch tourists. It was "the wonderland of the Americas," and like all wonderlands except that of the refreshing Alice, it far excelled Switzerland, with which the boosters are always entirely familiar. What would this scenic highway cost? "A mere detail, that—find out later—all we know is that it would cost too much for us to build. It's about time the government did something for northern New Mexico anyhow." Could the government, in justice to existing needs for roads elsewhere, afford to appropriate? "We are not representing 'elsewhere'—we are building up our city." Should the "wonderland" continue under the Agricultural Department as a National Forest? "It will continue under the department that helps us get the money. If your department won't, there are plenty of others that will."

Thus is the pork barrel filled—and emptied. Thus do we attain "less government in business, and more business in government." Thus are the burdens of taxation reduced. Thus do we build cities and attain prosperity.

Thus also does little Willie kick and squeal when his father denies his unseasonable demand for a new bicycle.

The booster's yardstick is the dollar, and if he recognizes any other standard of value, or any other agency of accomplishment, he makes it a point of pride not to admit it. Even works of charity are bought and sold, like cabbages or gasoline. Do we want to do something for the Boy Scouts? We levy a subscription, hire an architect, and build them a cabin in the mountains (which the Scouts ought to have built themselves), and proceed to forget the Scout movement. We can not see that what we should give toward such causes is usually not much money, but a little human interest.

A few months ago somebody discovered that the Bursum Bill raised the question of possible disintegration of the Pueblo Indian communes. A booster editor, commenting on the situation, cooly pointed out that the tourist-getting value of the Indians depended on their distinctive culture, which should therefore be preserved until our industrial development made it no longer possible to do so. This was, I hope, the ultimate impertinence of boosterism in the Southwest. That the Indian culture and ours should have been placed in competition for the possession of this country was inevitable, but the cool assumption that this last little fragment must necessarily disappear in order that an infinitesimal percentage of soot, bricks, and dollars may be added to our own, betrays a fundamental disrespect for the Creator, who made not only boosters, but mankind, in his image.

The booster seems almost proud of the ugliness and destruction that accompany industrialism. That some of this is inevitable and necessary I am the first to admit. That it can not be mitigated I emphatically deny. Is there any real economic necessity for the army of billboards that marches across the peaceful landscapes of the Rio Grande Valley, flaunting its ribald banners in the face of the eternal hills, and shouting at every turn of the road what is the best brand of chewing gum, tires, or tobacco? And to top off this indignity, there is even a billboard erected by a Business Woman's Club, proclaiming the virtues of our city! "Et tu, Brute?"

Whence cometh this noise? Is it a business necessity? On the contrary, it is one of those competitive business evils that merely cost money and benefit nobody. A zone for billboards at the entrance to the city, advising travellers of the kind and location of its services, would be justifiable. But a gauntlet of billboards fifty miles long is not only bad business, but miserably bad taste.

Is there any sound economic reason why our comely public buildings can not be grouped around a public plaza, instead of cramped into scattered lots where they will shortly be elbowed by such a maze of butcher shops and five-and-ten-cent stores as to be visible only from an airplane? No reason, except that boosters and politicians do not know and will not learn the modern devices of public finance like the "Excess Condemnation Plan," which would give us a plaza without additional cost. In all this needless ugliness the booster is not so much ruthless as clumsy. A hundred percenter in making the flag fly and the eagle scream, he is awkward in self-government. Worshipping commerce, he is slow to regulate its own abuses.

The typical booster is entirely out of contact with the most fundamental of his boasted resources, the soil. Ask the average one how many bushels an acre of corn produces in our valley and he doesn't know, but he will quote you yards of statistics on what the tourist spends in our town. Ask him what is wrong with the livestock industry and he will answer drouth, or foreign competition, or other accessories-after-the-fact. He doesn't know that the fundamental reason is lack of a stable land tenure to produce grass, and that his own unintelligent and irresponsible politics is in turn responsible for this. Knowing nothing of the soil, he does not have the confidence of those who till the soil, and accordingly his spasmodic efforts at drainage or other betterments come to naught. Happily, there are exceptions to this. Banks and boosters, in places, are doing a splendid and successful work in rebuilding the agriculture of the south, where the boll-weevil wrecked the cotton crop and the boosters had to do something or move out.

Growing away from the soil has spiritual as well as economic consequences which sometimes lead one to doubt whether the booster's hundred per cent Americanism attaches itself to the country, or only to the living which we by hook or crook extract from it. Recently our boosters "discovered" the Sandias. Since Coronado came, and before, they have offered cool shades and peace to the inhabitants of this valley, but suddenly we realize that they are there, and that they are beautiful. Do we rejoice that our citizens shall henceforth enjoy them? Not so. "I love thy rocks and rills, thy woods and templed hills"—as tourist bait.

The booster is covetous of trifles, blissfully devoid of public policy, intolerant of minorities, ruthless, unscrupulous, provincial, and extravagant of the government's purse as he is generous of his own, but he is not proud. Cleaning the boots of tourists, conventions, prospective investors, and other dispensers of "prosperity" is congenial labor, performed with an obsequiousness worthy of a head-waiter, and with as good an eye for gratuities. "Ich Dien, but please like our town." Conventions present their demands for civic hospitality in the same categorical imperative as an ultimatum to the Turks, and the boosters receive it with a polite humility that a Turk could never emulate. The ordinary relations of guest and host, premised on mutual consideration and self respect, are displaced by a brazen self-interest that survives nowhere else but in diplomacy and international relations. Auto tourists demand "service," under pain of blackballing the town, and they get it. New enterprises demand a "bonus," under pain of locating elsewhere. It is a hopeful sign that this "bonus" system is rapidly falling into disrepute.

It is characteristic of the "small boy" psychology of the booster that he recognizes no kind of civic service save his own. I once heard a subscription committee vehemently berate a dentist for his refusal to sign on the dotted line. They charged him with ignoring his civic obligations. As a matter of fact, the highly skilled and utterly conscientious professional services of that particular dentist had done more for the town, even measured by its own materialistic standards, than the new Chamber of Commerce which the committee purported to represent. He had lived the precept of Carlisle, who said "The latest gospel in this world is, Know thy work and do it."

Likewise characteristic of the small boy is the booster tendency toward mutual admiration societies. The solemn altruism and lofty ethical codes of our four-and-twenty varieties of luncheon clubs are not—as many outsiders aver—an hypocrisy, forgotten in the daily practice of their members. On the contrary, these codes are a uniform, like the plumes and swords of fraternal orders, and worn for a like purpose—the lifting up of the individual out of the treadmill of industry. Whether we know it or not, we all need and seek this lifting up.

And even boosters are lifted. The full measure of devotion often given to booster movements commands respect and admiration, regardless of the present fallacies of the cause. Look back on the birth of political and religious liberty, and the more recent birth of internationalism, and see how each is full of fallacies and extremes ranging from the tragic to the ludicrous. But sincerity is never ludicrous. Let us critics, therefore beware, in ridiculing the fallacies of this new thing "lest we laugh in the wrong place, and thus commit impiety when we think we are achieving wit."

One more admission of the possibility that true vitality and greatness underlies the booster idea. I once knew a doctor, who on the completion of his medical studies, returned to his home town to practice. He soon saw that the place was too small for him. "I realized," he says, "that I would either have to move to the kind of a town I needed, or else make over my home town into that kind of a place. I decided to make over my home town." And he did. He did it through a Chamber of Commerce.

The sweep and daring, the utter simplicity and directness of such occasional manifestations of the booster spirit give the lie to any easy assumption that it is all froth and noise. Somewhere, somehow, it contains the germ of a better order of things. Even in our town there are symptoms of it.

Every day on my way to my office I pass a booster billboard which exhorts me as follows: "Cities do not happen—BE A BUILDER—Support your Chamber of Commerce." Splendid truths, the first two. I detest billboards, but this one interests me. Be a builder! There is a real ring in those words. I look over at the towering beauty of "The Franciscan," and am proud.

> What vigor raised those spires; what joyful hand
> Put strength into those arches, gave the free
> Rock this immense and grotesque dignity,
> Making the structure greater than it planned!
> What laughter shook the builders as they scanned
> Those grinning gargoyles, and a jubilee
> Spirit enlarged the workers' energy;
> While, laid with love, each stone was made to stand!

The boosters built this lovely thing. I recall the travail of civic spirit, the fight with civic sloth and inertia, which converted the dream to reality. I recall the contest between the mediocrity which wanted "plain commerical architecture" and the vision which saw a building reflecting the history and traditions of the Southwest. The vision won. Out of this immense, vigorous, unlovely thing called Boosterism, provincial as a carpet-bagger, ruthless as the Juggernaut, intolerant as a Prussian, boisterous as Huckleberry Finn, but courageous as the Vikings grew "The Franciscan." Is it too much to hope that this force, harnessed to a finer ideal, may some day accomplish good as well as big things? That our future standard of civic values may even exclude quantity, obtained at the expense of quality, as not worth while? When this is accomplished shall we vindicate the truth that "the virtue of a living democracy consists not in its ability to avoid mistakes, but in its ability to profit by them."

Pioneers and Gullies [1924]

As suggested in "Standards of Conservation," Leopold particularly noted the condition of the mountain watersheds during his frequent inspection trips on the national forests of the Southwest. He began to write about the problem of soil erosion at least as early as 1921 in a series of essays addressed to foresters, scientists, and the general public. This 1924 article in *Sunset Magazine*, the popular journal of the Southern Pacific Railroad, is actually a recast version of a speech titled "Erosion as a Menace to the Social and Economic Future of the Southwest," which Leopold delivered to the New Mexico Association for Science in December 1922. In the article Leopold takes issue with contemporary notions of "development" and argues, for perhaps the first time in print, the need for a sense of obligation on the part of private landowners. The Blue River valley in the White Mountains of Arizona, which he selects as an example, was the site of his first assignment when he joined the Forest Service in 1909.

Pioneering a new country is hard labor. It has absorbed the best brawn and brains of the Nordic race since before the dawn of history. Anthropologists tell us that we, the Nordics, have a racial genius for pioneering, surpassing all other races in ability to reduce the wilderness to possession.

But if we saw a Nordic settler perspiring profusely to put a new field under irrigation while a flood was eating away his older field for lack of a few protective works, we should call that settler an inefficient pioneer. Yet that is exactly what we seem to be doing in trying to develop the Southwest. The only difference is that while one individual is putting the new field under irrigation, another individual is losing the older field from floods, and a third is causing the floods through misuse of his range. This scattering of cause and effect and of loss and gain among different owners or industries may give the individual his alibi, but it changes not one whit the inefficiency of our joint enterprise in "developing" the country. We, the community, are saving at the spigot and wasting at the bunghole, and it is time we realized it and mended our ways.

While our Government and our capitalists are laboring to bring new land under irrigation by the construction of huge and expensive works, floods are tearing away, in small parcels, here and there, an aggregate of old land, much of it already irrigated, which is comparable to the new land in area and value. The opening of these great reclamation projects we celebrate by oratory and monuments, but the loss of our existing farms we dismiss as an act of God—like the storm or the earthquake, inevitable. But it is not an act of God; on the contrary, it is the direct result of our own misuse of the country we are trying to improve.

Proof? A survey of 30 small agricultural valleys in the mountain sections of Arizona and New Mexico shows 12 wholly or partly ruined, 9 with erosion started, and 9 with little or no erosion. Roughly, we are losing nearly half of these mountain valleys.

The total irrigable acreage of U.S. Reclamation Projects in the two states is 430,000 acres. The loss to date in the 30 mountain valleys is 10,000 acres. Doubling this for the additional losses in small creeks not covered by the survey, but within the mountain area, would give 20,000 acres. The mountain area surveyed is one-seventh of the total area of the two States, and does not include such valleys as the Gila and San Juan where the really big losses have occurred. The Pueblo flood alone is known to have torn out 2500 acres. Olmstead* says the Gila destroyed 2500 acres in 1915 in Graham county alone, and is threatening 30,000 acres of farm land in this county. I feel safe in stating that erosion has destroyed agricultural land running into six figures in the two states.

Let us consider just one of these eroded valleys in detail. Blue river, in the White mountains, originally flowed through about 4000 acres of cultivated land. This land supported about 45 ranches and 300 people. Floods tore out the land, and today 400 acres remain cultivable, supporting about 20 ranches and 90 people. The land lost would now be worth $150 per acre, or $540,000. This loss, as cash, would pay for a tidy little reclamation project. It would have warranted an expenditure of $100 per acre for protective works. But after all, a cash value can not express the actual loss. Not only were 34 established homes destroyed, but the land carried away was a "key" resource, necessary for the proper utilization of the range, timber, and recreational values on half a million acres of adjacent mountains. There is no other land in the region suitable for homes, stock-ranches, mills, roads, and schools. Let us get the full significance of this by examining each item in detail.

Take, for instance, the adjacent range. On this lost farm land the stockmen lived and had their alfalfa, grain fields, gardens and orchards. With no fields, all feed for saddle and work horses and weak range stock must be

---

*"Flood Control on the Gila River," Frank H. Olmstead, U.S. Geological Survey, 1917.

either dispensed with or packed in 60 miles from the railroad at great cost. This may make the difference between a profitable and an unprofitable stock-raising operation. In one case where detailed figures were worked out, it was found that the loss of 60 acres of farming land to a stockman running 850 cattle caused a loss of 24% in gross income and increased his cost of production $6.50 per head. Moreover a stock ranch deprived of its garden patch, orchard, milk cows, and poultry is no fit place to establish a home and raise a family. Regardless of the profit of the business, it is an unsocial institution.

But this is not all. The destruction of the bottom lands destroyed the only feasible location for the road necessary to connect the ranches with each other, with schools and with the outside world, and to enable timber and minerals to be hauled out to market. Floods have left no place for a road. Children must now ride to school on horseback, and during floods they can not even do that. The Government and the counties are now actually spending half a million dollars on a road through this country, but it can not tap what remains of the Blue River community because it is unsafe to put a road on a sandbar. It must clamber high over the rocks and hills, at huge expense.

We, the community, have "developed" Blue River by overgrazing the range, washing out half a million in land, taking the profits out of the livestock industry, cutting the ranch homes by two-thirds, destroying conditions necessary for keeping families in the other third, leaving the timber without an outlet to the place where it is needed, and now we are spending half a million to build a road round this place of desolation which we have created. And to "replace" this smiling valley which Nature gave us free, we are spending another half a million to reclaim an equal acreage of desert some other place.

Just what is the nature of this process by which overgrazing of the range destroys the valleys necessary to make the range industry profitable? In past years most engineers and conservationists have believed that moderate grazing did not produce erosion. History and experience have shown, however, that this theory must be applied with caution. In scantily watered country, to graze the range at all often means to overgraze the water-courses and bottom lands. Some concentration of stock at these points is difficult to avoid, even under careful management. When a bad flood encountered a virgin watercourse full of vigorous trees, willows, vines, weeds and grass, it may have scoured it pretty severely, but the living roots remained to spring up and recover the land and cause the next more moderate flood to heal the scars instead of enlarging them. But when floods encounter a watercourse through bare fields, timber grazed clear of all undergrowth, and earth-scars like roads, trails, and ditches built parallel with the stream, the gouges left by one flood are liable to be enlarged by the next flood; an unprotected channel is excavated; the trees merely act as levers to pry off the undermined banks, the process of oxbowing cuts first one side of the bottom and then the other, eating into the very base of the hills; side-gullies running back from the deepened creek-channel cut at right angles into the remaining bottoms and benches, draining the natural *cienagas* and hay meadows and changing the grasses to a less resistant forage type, and in the long run our "improved" valley becomes a desolation of sandbars, rockpiles and driftwood, a sad monument to the unintelligence and misspent energy of us, the pioneers.

This is what I mean when I say that in the Southwest it is doubtful whether we are creating more useful land with the labor of our hands than we are unintentionally destroying with the trampling of our feet. But why is it that Nature is so quick to punish unintelligent "development" in the Southwest?

Every region seems to have a different resistance to every kind of use or abuse by man. The degree and nature of this resistance seem to be determined by climate. The more arid the climate, the less the resistance of the region to abuse. In Europe, with its wet climate, many centuries of use and abuse by man have altered but not destroyed the land, vegetation, and animal life. Britain, New England, Canada, the South, the Middle West and the Northwest are broadly in the same category.

Palestine and Asia Minor, on the other hand, have a semi-arid climate. Scores of centuries of use and abuse by man have to a large extent destroyed the vegetation, animal life and even the soil itself. The Bible is full of evidence that the mountains of the Holy Land in the time of the Prophets supported real forests; the range was abundant and excellent; many living streams found their source in the higher lands. Great forest fires swept the mountains unchecked. Grazing was the principal industry and doubtless the range was at least locally overgrazed, just as is happening today in the Southwest. The forests have long since disappeared, and the mountains are too bare to support a forest fire.

Vanished forests can be replaced by huge expenditures, but no country has ever replaced lost agricultural land on a big scale. The Incas came the nearest to it in the terraces which they built with soil packed in on their backs. We will not have the patience to do that. We have no way to restore the soil to lands that have washed away. Soil is the fundamental resource, and its loss the most serious of all losses.

Soil is not the only resource which shows a lower resistance to use and abuse in arid countries. Forests grow more slowly and are exceedingly difficult to reproduce after cutting. Desirable forage under excessive use loses its vitality and reverts to weeds. Game shows less resistance to hunting than in wet regions. All of these factors interact in a very complex manner with each other and with soil conditions, necessitating the fullest and most skilful coöperation among the various professions in charge of conservation work.

Coming back to soil: Erosion of soil is always accompanied by disturbance or damage to the usable water supply. In valleys like the Mimbres, the Sapello and the Blue, what were once trout streams have been entirely covered up by debris. While the water may be still there, it is buried. In streams like the Galisteo and Puerco, the channel has so deepened that water can not be diverted into the ditches which once supported orchards and farms. And finally let us not forget that most of the land destroyed on our watersheds is being dumped as silt into our great irrigation reservoirs, gradually reducing their storage capacity. After they are full of mud, then what?

In these early and hopeful days, the loss of land by erosion can always be temporarily made good by reclaiming new lands. The destruction of easy and cheap road routes can be met by rebuilding the road on the hills. The filling of dams can be staved off by building them higher or building new ones. But mark this well: the total possible acreage of tillable irrigable land, the total possible acre-feet of accessible water and the total storage capacity of dam-sites—these three things set the limits of the total possible future development of the Southwest. The virgin supply of each was limited; the subsequent losses, no matter with what energy we "replace" them, are steadily lowering the limits already set. To a degree we are facing the question of whether we are here to "skin" the Southwest and then get out, or whether we are here to found a permanent civilized community with room to grow and improve. We can not long continue to accept our losses without admitting that the former, rather than the latter, is by way of becoming the real result of our occupancy.

So far little has been said about remedies, which are, of course, the thing really worth talking about. It has been asserted that erosion is the result of overgrazing, and that some local overgrazing is difficult to avoid, even on ranges that are not overstocked. But nobody advocates that we cease grazing.

The situation does not call for a taboo upon grazing, but rather constitutes a challenge to the craftsmanship of our stockmen and the technical skill of grazing experts in devising controls that will work, and to the courage of our administrators in enforcing those controls in a manner fair both to the conflicting interests and to the community.

The stockmen must recognize that the privilege of grazing use carries with it the obligation to minimize and control its effects by more skilful and conservative methods. The day will come when the ownership of land will carry with it the obligation to so use and protect it with respect to erosion that it is not a menace to other landowners and the public. Just as it is illegal for one landowner to menace the public peace or health by maintaining disorderly or unsanitary conditions on his land, so will it become illegal for him to menace the public streams, reservoirs, irrigation projects, or the lands of his neighbors by allowing erosion to take place. But it is cheaper to prevent erosion than to cure it, and the cost of such prevention must some day be passed on uniformly by all landowners to all consumers of their products.

But enforced responsibility of landowners is of the future. What are the prevention methods that can be used *now* by those owners sufficiently progressive, or sufficiently menaced by impending loss, to do so?

A diagnosis of the process of destruction gives the most reliable pointers as to the best process of prevention and cure. First and foremost, a vigorous growth of grass on the watershed, and more especially on the watercourses, is essential. The science of range management has made great strides in the last decade in developing conservative methods of range use. Some of them were described in the October, 1923 issue of *Sunset*. It is coming to be generally admitted by stockmen that they are more profitable than the old destructive methods. Why, then, are they not in general use?

Because under present conditions it does not pay the average stockman to use them. In the National forests a genuine and frequently successful effort has been made to prevent overgrazing by careful regulation, but on the public range outside of the forests no control of any kind is exercised. First come, first served. This lack of regulation causes each stockman to try to get as much stock as possible on the range at the earliest possible moment, resulting in continuous and disastrous overgrazing. Further procrastination in effecting a public-domain policy is unthinkable. It matters a great deal more that some decisive policy be adopted than that such policy be ideally correct. The prospects are that the stockmen are about to agree on a Federal leasing bill. It will be the tenth principal attempt at legislation since 1900. Why not pass it?

So much for the first step in watershed conservation—the prevention of erosion by maintaining the grass cover on the range in general. It is by all odds the most important step. But there are other steps necessary. For instance, in the Southwest there has been a striking coincidence between the inception of erosion and the eating out of the native willows by stock. Willows formerly grew in most of our cañons and river bottoms. They are a palatable winter feed and soon disappear. There is no doubt that the grazing out of the willows has been the direct cause of streambank erosion in hundreds of cases.

It is very hard to maintain grass and willows on watercourses used by stock. The natural concentration of stock, especially in winter and spring, destroys them. Therefore many bottoms will have to be fenced, and merely lightly grazed as reserve pastures, and the willows artificially restored by planting cuttings on the banks of the stream. Experiments in such work indicate that banks can be willowed, cuttings two feet apart, for less than fifty cents per one hundred feet. The usefulness of such pastures can be regarded as offsetting the cost of the fencing. Of course the fences must be so located as to leave sufficient water-gaps and to provide for road routes.

Fencing will be practicable only in the wider cañons. Willowing will be practicable only on fenced lightly grazed pastures or agricultural fields. For watercourses that can not be fenced, some plant to replace the native willows but resistant to grazing either through having thorns or being non-palatable, will have to be found. The machinery of the Department of Agriculture has been set to work to find such a plant through its plant exploration service.

Farming lands can often be economically protected from bank-cutting by inexpensive works. Felling trees into the channel at strategic places and chaining the end of the trunk to its own stump is a method that makes the flood divert itself, rather than to divert it by the sheer strength of expensive walls and dams. Willows can often be advantageously planted under the protection of such felled tree-tops. Riprapping with woven wire fencing strung parallel with the bank on green spring-set cottonwood posts, which grow and form trees, is a cheap and good method.

Often it is necessary for landowners along a creek to work out a unified plan, else there is danger that the diligence of one owner will result merely in passing the trouble down the creek to his neighbors. Here is a fine opportunity for leadership and technical advice by county agents, Forest officers, and similar officials. It is unfortunate that our agricultural colleges have not seized their opportunity to develop erosion-control technique for the benefit of stockmen and farmers.

In addition to preventing erosion by conserving a grass cover on the range, and controlling it by protecting stream banks, there remains the huge problem of gully-control on the watershed as a whole. Many lands are being so cut up by gullies as to drain and dry the soil and thus change the type of forage. Many a fine glade, park, *valle, canada* or hay meadow has a gully gradually eating along its whole length, where a few minutes of throwing logs, stones, and brush into the head of the gully would prevent its further spread. There is a best way even in plugging a gully, and this best way deserves the careful study of engineers. Where an acre of grazing land worth $2.00 or of a hay meadow worth $40.00 can be saved by plugging a gully at a cost in time of 25¢, it is disregarding the public welfare and the principles of sound private business not to plug it.

Natural resources are interdependent, and in semi-arid countries are often set in a hair-trigger equilibrium which is quickly upset by uncontrolled use. As a consequence, uncontrolled use of one local resource may menace the economic system of whole regions. Therefore, to protect the public interest, certain resources must remain in public ownership, and ultimately the use of all resources will have to be put under public regulation, regardless of ownership. This is the fundamental reason why the Nation retains ownership of the mountain forests and one of the reasons why the Nation builds and regulates reclamation projects. But while partial provision has been made, through the Forest Service and Reclamation Service, to conserve the forests and the water supply, no provision has been made to conserve that fundamental resource, land.

The first step to remedy this omission is to reform the conditions of land tenure, especially on the unreserved public domain, so that the livestock industry can practice the conservation methods which the science of range management has already worked out.

The second step is for all agencies concerned, under the leadership of the agricultural colleges, to develop and demonstrate the cheapest and best methods of artificial erosion control, and urge all landowners to utilize them. This will enable owners to control some of the losses of land that will otherwise continue unchecked.

The third step, which must come later, is to put all land in the region threatened by erosion under Government inspection as to the adequacy of erosion control, and to force all owners to conserve their lands to the extent that is found reasonable and practicable. If they fail to do so, the Government must install the necessary controls and assess the landowner with the cost.

The greater the delay in the first two steps, the more urgent and drastic becomes the third.

Grass, Brush, Timber, and Fire in Southern Arizona [1924]

This classic of ecological analysis may have been Leopold's response to his colleagues' criticism of his treatment of the erosion problem in "Some Fundamentals of Conservation in the Southwest." Through careful observation and inferential reasoning, he develops an interpretation that integrates soils, vegetation, topography and climate, geologic and human history, fire, and livestock grazing into a single system of interactions. In this pathbreaking article published in the *Journal of Forestry*, he directly challenges a number of Forest Service dogmas, particularly those concerning the role of grazing and fire, and calls for a redirection of administrative policy.

One of the first things which a forester hears when he begins to travel among the cow-camps of the southern Arizona foothills is the story of how the brush has "taken the country." At first he is inclined to classify this with the legend, prevalent among the old timers of some of the northern states, about the hard winters that occurred years ago. The belief in the encroachment of brush, however, is often remarkably circumstantial. A cow-man will tell about how in the 1880's on a certain mesa he could see his cattle several miles, whereas now on the same mesa he can not even find them in a day's hunt. The legend of brush encroachment must be taken seriously.

Along with it goes an almost universal story about the great number of cattle which the southern Arizona foothills carried in the old days. The old timers say that there is not one cow now where there used to be 10, 20, 30, and so on. This again might be dismissed but for the figures cited as to the brandings of old cattle outfits, of which the location and area of range are readily determinable. This story likewise must be taken seriously.

In some quarters the forester will find a naive belief that the two stories represent cause and effect, that by putting more cattle on the range the old days of prosperity for the range industry might somehow be restored.

Grass, Brush, Timber, and Fire 115

The country in which the forester finds these prevalent beliefs consists of rough foothills corresponding in elevation to the woodland type. Above lie the forests of western yellow pine. Below lie the semi-desert ranges characteristic of the southern Arizona plains. The area we are dealing with is large, comprising the greater part of the Prescott, Tonto, Coronado, and Crook National Forests as well as much range outside the Forests. The brush that has "taken the country" comprises dozens of species, in which various oaks, manzanita, mountain mahogany and ceanothus predominate. Here and there alligator junipers of very large size occur. Along the creek bottoms the brush becomes a hardwood forest.

Five facts are so conspicuous in this foothill region as to immediately arrest the attention of a forester.

(1) Widespread abnormal erosion. This is universal along watercourses with sheet erosion in certain formations, especially granite.
(2) Universal fire scars on all the junipers, oaks, or other trees old enough to bear them.
(3) Old juniper stumps, often levelled to the ground, evidently by fire.
(4) Much juniper reproduction merging to pine reproduction in the upper limits of the type.
(5) Great thrift and size in the junipers or other woodland species which have survived fire.

A closer examination reveals the following additional facts:

First, the reproduction is remarkably even aged. A few ring counts immediately establish the significant fact that none of it is over 40 years old. It is therefore contemporaneous with settlement; this region having been settled and completely stocked with cattle in the 1880's.

Second, the reproduction is encroaching on the parks. These parks, in spite of heavy grazing, still contain some grass. It would appear, therefore, that this reproduction has something to do with grass.

Third, one frequently sees manzanita, young juniper or young pines growing within a foot or two of badly fire-scarred juniper trees. These growths being very susceptible to fire damage, they could obviously not have survived the fires which produced the scars. Ring counts show that these growths are less than 40 years old. One is forced to the conclusion that there have been no widespread fires during the last 40 years.

Fourth, a close examination of the erosion indicates that it, too, dates back about 40 years and is therefore contemporaneous with settlement, removal of grass, and cessation of fires.

These observations coordinate themselves in the following theory of what has happened: Previous to the settlement of the country, fires started by lightning and Indians kept the brush thin, kept the juniper and other woodland species decimated, and gave the grass the upper hand with respect to possession of the soil. In spite of the periodic fires, this grass prevented erosion. Then came the settlers with their great herds of livestock. These ranges had never been grazed and they grazed them to death, thus removing the grass and automatically checking the possibility of widespread fires. The removal of the grass relieved the brush species of root competition and of fire damage and thereby caused them to spread and "take the country." The removal of grass-root competition and of fire damage brought in the reproduction. In brief, the climax type is and always has been woodland. The thick grass and thin brush of pre-settlement days represented a temporary type. The substitution of grazing for fire brought on a transition of thin grass and thick brush. This transition type is now reverting to the climax type—woodland.

There may be other theories which would coordinate these observable phenomena, but if there are such theories nobody has propounded them, and I have been unable to formulate them.

One of the most interesting checks of the foregoing theory is the behavior of species like manzanita and piñon. These species are notoriously susceptible to fire damage at all ages. Take manzanita: One finds innumerable localities where manzanita thickets are being suppressed and obliterated by pine or juniper reproduction. The particular manzanita characteristic of the region (*Archtostaphylos pungens*) is propagated by brush fires, seedling (not coppice) reproduction taking the ground whenever a fire has killed the other brush species or reduced them to coppice. It is easy to think back to the days when these manzanita thickets, now being killed, were first established by a fire in what was then grass and brush. Cattle next removed the grass. Pine and juniper then reproduced due to the absence of grass and fire, and are now overtopping the manzanita. Take piñon: It is naturally a component of the climax woodland type but mature piñons are hardly to be found in the region; just a specimen here and there sufficient to perpetuate the species which has evidently been decimated through centuries of fires. Nevertheless today there is a large proportion of piñon in the woodland reproduction which is coming in under some of the Prescott brushfields.

Another interesting check is found in the present movement of type boundaries. Yellow pine is reproducing down hill into the woodland type. Juniper is reproducing down hill into the semi-desert type. This down-hill movement of type lines is so conspicuous and so universal as to establish beyond a doubt that the virgin condition previous to settlement represented a temporary type due to some kind of damage, and completely refutes the possible assumption that the virgin conditions were climax and the present tendency is away from rather than toward a climax.

A third interesting check is found in the parks. In general there are two alternative hypotheses for Southwestern parks—the one assuming chemical or physical soil conditions unfavorable to forests and the other assuming the exclusion of forests by damage. When the occasional forest tree found in any park is scrubby, it indicates in general defective soil conditions. When the occasional forest tree shows vigor and thrift, it indicates that the park was established by damage and that the soil is suitable. Nothing could be more conspicuous than the vigor and thrift of the ancient junipers scattered through the parks of the southern Arizona foothills. We may safely assume that these parks were not caused by defective soil conditions. That they were caused by grass fires is evidenced by the survival of grass species in spite of the extra heavy grazing which occurs in them and by the universal fire scars that prevail on the old junipers in them. The fact that they are now reproducing to juniper clinches the argument.

A fourth check bears on the hypothesis that the virgin grass was heavy enough to carry severe fires. The check consists in the occurrence of "islands" where topography has prevented grazing. One will find small benches high on the face of precipitous cliffs which, in spite of poor and dry soil, bear an amazing stand of grasses simply because they have never been grazed. One even finds huge blocks of stone at the base of cliffs where a little soil has gathered on the top of the block and a thrifty stand of grasses survives simply because livestock could not get at it.

The most impressive check of all is the occurrence of junipers evidently killed by a single fire from 50 years to many centuries ago, on areas where there is now neither brush nor grass and where the junipers were so scattered (as evidenced by their remains) that it is absolutely necessary to assume a connecting medium. If the connecting medium had been brush it could hardly have been totally wiped out because neither fire nor grazing exterminates a brushfield. It is necessary to assume that the connecting medium consisted of grass. It is significant that the above described phenomenon occurs mostly on granitic formations where it is easy to think that a heavy stand of grass might have been exterminated by even moderate grazing due to the loose nature of the soil.

Assuming that all the foregoing theory is correct, let us now consider what it teaches us about erosion. Why has erosion been enormously augmented during the last 40 years? Why has not the encroachment of brush checked the erosion which was induced by the removal of the grass? Why did not the fires of pre-settlement days cause as much erosion as the grazing of post-settlement days?

It is obvious at the start that these questions can not be answered without rejecting some of our traditional theories of erosion. The substance of these traditional theories and the extent to which they must be amended before they can be applied to the Southwest, I have discussed elsewhere.[1] It

---

1. "A Plea for Recognition of Artificial Works in Forest Erosion Control Policy," *Journal of Forestry* (March, 1921); "Pioneers and Gullies," *Sunset Magazine* (May, 1924); *Watershed Handbook,* Southwestern District, issued December, 1923.

will be well to repeat, however, that the acceptance of my theory as to the ecology of these brushfields carries with it the acceptance of the fact that at least in this region grass is a much more effective conserver of watersheds than foresters were at first willing to admit, and that grazing is the prime factor in destroying watershed values. In rough topography grazing always means some degree of localized overgrazing, and localized overgrazing means earth-scars. All recent experimentation indicates that earth-scars are the big causative agent of erosion. An excellent example is cited by Bates, who shows that the logging road built to denude Area B at Wagon Wheel Gap has caused more siltage than the denudation itself. Another conspicuous example is on the GOS cattle range in the Gila Forest, where earth-scars due to concentration of cattle along the water-courses have caused an entire trout stream to be buried by detritus, in spite of the fact that conservative range management has preserved the remainder of the watershed in an excellent condition.

Let us now consider the bearing of this theory on Forest administration. We have learned that during the pre-settlement period of no grazing and severe fires, erosion was not abnormally active. We have learned that during the post-settlement period of no fires and severe grazing, erosion became exceedingly active. Has our administrative policy applied these facts?

It has not. Until very recently we have administered the southern Arizona Forests on the assumption that while overgrazing was bad for erosion, fire was worse, and that therefore we must keep the brush hazard grazed down to the extent necessary to prevent serious fires.

In making this assumption we have accepted the traditional theory as to the place of fire and forests in erosion, and rejected the plain story written on the face of Nature. He who runs may read that it was not until fires ceased and grazing began that abnormal erosion occurred. We have likewise rejected the story written in our own fire statistics, which shows that on the Tonto Forest only about 1/3 of 1% of the hazard area burns over each year, and that it would therefore take 300 years for fire to cover the forest once. Even if the more conservative grazing policy which now prevails should largely enhance the present brush hazard by restoring a little grass, neither the potential danger of fire damage nor the potential cost of fire control could compare with the existing watershed damage. Moreover the reduction of the brush hazard by grazing is to a large degree impossible. This brush that has "taken the country" consists of many species, varying greatly in palatability. Heavy grazing of the palatable species would simply result in the unpalatable species closing in, and our hazard would still be there.

There is one point with respect to which both past policy and present policy are correct, and that is the paramount value of watersheds. The old policy simply erred in its diagnoses of how to conserve the watershed. The range industry on the Tonto Forest represents a present capital value of around three millions. Since this is about one third of the total Roosevelt Reservoir drainage we may assume roughly that the range industry affecting the Reservoir is worth nine millions. The Roosevelt Dam and the irrigation works of the Salt River Valley represent a cash expense by the Government of around twelve millions. The agricultural lands dependent upon this irrigation system are worth about fifty millions, not counting dependent industries. Grazing interests worth nine millions, therefore, must be balanced against agricultural interests worth sixty-two millions. To the extent that there is a conflict between the existence of the range industry and the permanence of reclamation, there can be no doubt that the range industry must give way.

In discussing administrative policy, I have tried to make three points clear: First, 15 years of Forest administration were based on an incorrect interpretation of ecological facts and were, therefore, in part misdirected. Second, this error of interpretation has now been recognized and administrative policy corrected accordingly. Third, while there can be no doubt about the enormous value of European traditions to American forestry, this error illustrates that there can also be no doubt about the great danger of European traditions to American forestry; this error also illustrates that there can be no doubt about the great danger of European traditions uncritically accepted and applied, especially in such complex fields as erosion.

The present situation in the southern Arizona brushfields may be summed up administratively as follows:

(1) There has been great damage to the watershed resources.
(2) There has been great benefit to the timber resources.
(3) There has been great damage to the range resources.

Whether the benefit to timber could have been obtained with lesser damage to watersheds and ranges is an academic question dealing with bygones and need not be discussed. Our present job is to conserve the benefit to timber and minimize the damage to watershed and range in so far as technical skill and good administration can do it. Wholesale exclusion of grazing is neither skill nor administration, and should be used only as a last resort. The problem which faces us constitutes a challenge to our technical competency as foresters—a challenge we have hardly as yet answered, much less actually attempted to meet. We are dealing right now with a fraction of a cycle involving centuries. We can not obstruct or reverse the cycle, but we can bend it; in what degree remains to be shown.

There are some interesting sidelights which enter into the foregoing discussions but which could not there be covered in detail. One of them is the extreme age of the junipers and juniper stumps. In one case I found a 36" alligator juniper with over half its basal cross-section eaten out by fire. On each edge of this huge scar were four overlapping healings. The last healing on each edge of the scar counted forty rings. Within 24" of the scar were two yellow pines of 20" diameter just emerging from the blackjack stage. Each must have been 130 years old. Neither showed any scars, but upon chopping into the side adjacent to the juniper, each was found to contain a buried fire-scald in the fortieth ring. It was perfectly evident that these 130-year pines had grown in the interval between the fires which consumed half the basal cross-section of the juniper, and the subsequent fires which resulted in the latest series of four healings. The fires which really ate into the juniper would most certainly have killed any pine standing only 24" distant. The conclusion is that the juniper attained its present diameter more than 130 years ago. The size of the main scar certainly indicates a long series of repetitions of scarring, drying and burning at the base of the juniper. The time necessary to attain a 36" diameter is in itself a matter of centuries. Consider now that other junipers killed by fire 40 years ago were found to still retain ¼" twigs, and then try to interpret in terms of centuries the meaning of the innumerable stumps of juniper (the wood is almost immune to decay) which dot the surface of the Arizona foothills. Who can doubt that we have in these junipers a graphic record of forest history extending back behind and beyond the Christian era? Who can doubt that this article discloses merely the main broad outlines of the story?

The following instance also tells us something about the intervals at which fires occurred. I mentioned a juniper with a big scar and four successive healings of which the last counted forty rings. The last was considerably the thickest. In a general way I would say that the previous fires probably occurred at intervals of approximately a decade. Ten years is plenty of time for a lusty growth of grass to come back and accumulate the fuel for another fire. This would reconcile my general theory with the known fact that fires injure most species of grass, it being entirely thinkable for the grass to recover from any such injury during a ten-year interval.

The foregoing likewise strengthens the supposition that root competition with grass rather than fire, was the salient factor in keeping down the brush during pre-settlement days. Brush species which coppice with as much vigor as those of the Arizona brushfields could stage quite a comeback during a ten-year surcease of fire if they were not inhibited by an additional competitor like grass roots.

Whether grass competition or fire was the principal deterrent of timber reproduction is hard to answer because the two factors were always paired, never isolated. Probably either one would have inhibited extensive reproduction. In northern Arizona there are great areas where removal of grass by grazing has caused spectacular encroachment of juniper on park areas. But here again both grass competition and fire evidently cause the original park, and both were removed before reproduction came in.

It is very interesting to compare what has happened in the woodland type with what has happened in the semi-desert type immediately below it. Here also old timers testify to a radical encroachment of brush species like mesquite and cat's-claw. They insist, however, that while this semi-desert type originally contained much grass, it never contained enough grass to carry fire. There are no signs of old fires. The encroachment of brush in this type can therefore be ascribed only to the removal of grass competition.

There are many loose masonry walls of Indian origin in the headwaters of drainages both in the woodland and semi-desert types. These have been fondly called "erosion-control works" by some enthusiastic forest officers, but it is perfectly evident that they were built as agricultural terraces, and that their function in erosion control was accidental. It is significant that any number of these terraces now contain heavy brush and even timber. Since they are prehistoric, the Indians could not have had metals, and therefore could not have easily cleared them of timber or brush. Therefore their sites must have been either barren or grassy when the Indians built them. This conforms with the belief that brush has encroached in both the woodland and semi-desert ranges.

In the brush fields of California the drift of administrative policy is toward heavy grazing as a means of reducing fire hazard. If the ecology of these California brushfields is similar to the ecology of the Arizona brushfields, it would appear obvious that either my Arizona theory or the California grazing policy is wrong. The point is that there is no similarity. The rainfall of the California brushfields is nearly twice that of the Arizona brushfields. Its seasonal distribution is different, and from what I can learn there is a great deal more duff and more herbs and other inflammable material under the California brush. It would appear, therefore, that the California tendency toward heavier grazing and the tendency in the Southwestern District toward much lighter grazing are not inconsistent because the two regions are not comparable.

The radical encroachment of brush in southern Arizona has had some interesting effects on game. There is one mountain range on the Tonto where the brush has become so thick as to almost prohibit travel, and where a thrifty stock of black bears have established themselves. The old hunters assure me that there were no black bears in these mountains when the country was first settled. It is likewise a significant fact that the wild turkey has been exterminated throughout most of the Arizona brushfields, whereas it has merely been decimated further north. It seems possible that turkeys require a certain proportion of open space in order to thrive. Plenty of open spaces originally existed, but the recent encroachment of brush has abolished them, and possibly thus made the birds fall an easier prey to predatory animals.

The cumulative abnormal erosion which has occurred coincident with the encroachment of brush and the decimation of grass naturally has its worst effect in the siltage of reservoirs. The data kept by Southwestern reclamation interests on siltage of reservoirs is regrettably inadequate, but it is sufficient to indicate one salient fact, viz., that the greater part of the loosened material is at the present time in transit toward the reservoir, rather than already dumped into it. Blockading this detritus in transit is therefore just as important as desilting the storage sites. The methods of blockading it will obviously be a combination of mechanical and vegetative obstructions, and with these foresters should be particularly qualified to deal. This fact further accentuates the responsibility of the Forest Service, and indicates that the watershed work of the future belongs quite as much to the forester as to the hydrographer and engineer.

The River of the Mother of God [1924]

The *Yale Review*, a literary magazine, turned down this most poignant of Leopold's wilderness essays. It remained in his desk as a yellowed, slightly edited typescript evoking the mystery of unknown places.

I am conscious of a considerable personal debt to the continent of South America.

It has given me, for instance, rubber for motor tires, which have carried me to lonely places on the face of Mother Earth where all her ways are pleasantness, and all her paths are peace.

It has given me coffee, and to brew it, many a memorable campfire with the dawn-wind rustling in autumnal trees.

It has given me rare woods, pleasant fruits, leather, medicines, nitrates to make my garden bloom, and books about strange beasts and ancient peoples. I am not unmindful of my obligation for these things. But more than all of these, it has given me the River of the Mother of God.

The river has been in my mind so long that I cannot recall just when or how I first heard of it. All that I remember is that long ago a Spanish Captain, wandering in some far Andean height, sent back word that he had found where a mighty river falls into the trackless Amazonian forest, and disappears. He had named it *el Rio Madre de Dios*. The Spanish Captain never came back. Like the river, he disappeared. But ever since some maps of South America have shown a short heavy line running eastward beyond the Andes, a river without beginning and without end, and labelled it the River of the Mother of God.

That short heavy line flung down upon the blank vastness of tropical wilderness has always seemed the perfect symbol of the Unknown Places of the earth. And its name, resonant of the clank of silver armor and the cruel progress of the Cross, yet carrying a hush of reverence and a murmur of the prows of galleons on the seven seas, has always seemed the symbol of Conquest, the Conquest that has reduced those Unknown Places, one by one, until now there are none left.

And when I read that MacMillan has planted the Radio among the Eskimos of the furthest polar seas, and that Everest is all but climbed, and that Russia is founding fisheries in Wrangel Land, I know the time is not far off when there will no more be a short line on the map, without beginning and without end, no mighty river to fall from far Andean heights into the Amazonian wilderness, and disappear. Motor boats will sputter through those trackless forests, the clank of steam hoists will be heard in the Mountain of the Sun, and there will be phonographs and chewing gum upon the River of the Mother of God.

No doubt it was "for this the earth lay preparing quintillions of years, for this the revolving centuries truly and steadily rolled." But it marks a new epoch in the history of mankind, an epoch in which Unknown Places disappear as a dominant fact in human life.

Ever since paleolithic man became conscious that his own home hunting ground was only part of a greater world, Unknown Places have been a seemingly fixed fact in human environment, and usually a major influence in human lives. Sumerian tribes, venturing the Unknown Places, found the valley of the Euphrates and an imperial destiny. Phoenician sailors, venturing the unknown seas, found Carthage and Cornwall and established commerce upon the earth. Hanno, Ulysses, Eric, Columbus—history is but a succession of adventures into the Unknown. For unnumbered centuries the test of men and nations has been whether they "chose rather to live miserably in this realm, pestered with inhabitants, or to venture forth, as becometh men, into those remote lands."

And now, speaking geographically, the end of the Unknown is at hand. This fact in our environment, seemingly as fixed as the wind and the sunset, has at last reached the vanishing point. Is it to be expected that it shall be lost from human experience without something likewise being lost from human character?

I think not. In fact, there is an instinctive human reaction against the loss of fundamental environmental influences, of which history records many examples. The chase, for instance, was a fundamental fact in the life of all nomadic tribes. Again and again, when these tribes conquered and took possession of agricultural regions, where they settled down and became civilized and had no further need of hunting, they nevertheless continued it as a sport, and as such it persists to this day, with ten million devotees in America alone.

It is this same reaction against the loss of adventure into the unknown which causes the hundreds of thousands to sally forth each year upon little expeditions, afoot, by pack train, or by canoe, into the odd bits of wilderness which commerce and "development" have regretfully and temporarily left us here and there. Modest adventurers to be sure, compared with Hanno, or Lewis and Clark. But so is the sportsman, with his setter dog in pursuit of partridges, a modest adventurer compared with his Neolithic ancestor in single combat with the Auroch bull. The point is that along with the necessity for expression of racial instincts there happily goes that capacity for illusion which enables little boys to fish happily in wash-tubs. That capacity is a precious thing, if not overworked.

But there is a basic difference between the adventures of the chase and the adventures of wilderness travel. Production of game for the chase can, with proper skill, be superimposed upon agriculture and forestry and can thus be indefinitely perpetuated. But the wilderness cannot be superimposed upon anything. The wilderness and economics are, in every ordinary sense, mutually exclusive. If the wilderness is to be perpetuated at all, it must be in areas exclusively dedicated to that purpose.

We come now to the question: Is it possible to preserve the element of Unknown Places in our national life? Is it practicable to do so, without undue loss in economic values? I say "yes" to both questions. But we must act vigorously and quickly, before the remaining bits of wilderness have disappeared.

Like parks and playgrounds and other "useless" things, any system of wilderness areas would have to be owned and held for public use by the Government. The fortunate thing is that the Government already owns enough of them, scattered here and there in the poorer and rougher parts of the National Forests and National Parks, to make a very good start. The one thing needful is for the Government to draw a line around each one and say: "This is wilderness, and wilderness it shall remain." A place where Americans may "venture forth, as becometh men, into remote lands."

Such a policy would not subtract even a fraction of one per cent from our economic wealth, but would preserve a fraction of what has, since first the flight of years began, been wealth to the human spirit.

There is a current advertisement of Wells' Outline of History which says "The unforgivable sin is standing still. In all Nature, to cease to grow is to perish." I suppose this pretty accurately summarizes the rebuttal which the Economic American would make to the proposal of a national system of wilderness playgrounds. But what is standing still? And what constitutes growth? The Economic American has shown very plainly that he thinks growth is the number of ciphers added yearly to the national population and the national bank-roll. But the Gigantosaurus tried out that definition of growth for several million years. He was a quantitative economist of the first water. He added two ciphers to his stature, and a staggering row of them to his numbers. But he perished, the blind victim of natural and "economic" laws. They made him, and they destroyed him.

There has been just one really new thing since the Gigantosaurus. That new thing is Man, the first creature in all the immensities of time and space whose evolution is self-directed. The first creature, in any spiritual sense, to create his own environment. Is it not in that fact, rather than in mere ciphers of dollars or population, that we have grown?

The question of wilderness playgrounds is a question in self-control of environment. If we had not exercised that control in other ways, we would already be in process of destruction by our own ciphers. Wilderness playgrounds simply represent a new need for exercising it in a new direction. Have we grown enough to realize that before it is too late?

I say "too late" because wilderness is the one thing we can not build to order. When our ciphers result in slums, we can tear down enough of them to re-establish parks and playgrounds. When they choke traffic, we can tear down enough of them to build highways and subways. But when our ciphers have choked out the last vestige of the Unknown Places, we cannot build new ones. To artificially create wilderness areas would overwork the capacity for illusion of even little boys with wash-tubs.

Just what is it that is choking out our last vestiges of wilderness? Is it real economic need for farmlands? Go out and see them—they contain no farmlands worthy of the name. Is it real economic need for timber? They contain timber to be sure, much of it better to look at than to saw, but until we start growing timber on the eighty million acres of fire-gutted wastes created by our "economic" system we have small call to begrudge what timber they contain. The thing that is choking out the wilderness is not true economics at all, but rather that Frankenstein which our boosters have builded, the "Good Roads Movement."

This movement, entirely sound and beneficial in its inception, has been boosted until it resembles a gold-rush, with about the same regard for ethics and good craftsmanship. The spilled treasures of Nature and of the Government seem to incite about the same kind of stampede in the human mind.

In this case the yellow lure is the Motor Tourist. Like Mammon, he must now be spelled with a capital, and as with Mammon, we grovel at his feet, and he rules us with the insolence characteristic of a new god. We offer up our groves and our greenswards for him to camp upon, and he litters them with cans and with rubbish. We hand him our wild life and our wild flowers, and humbly continue the gesture after there are none left to hand. But of all offerings foolish roads are to him the most pleasing of sacrifice.

(Since they are mostly to be paid for by a distant treasury or by a distant posterity, they are likewise pleasing to us.)

And of all foolish roads, the most pleasing is the one that "opens up" some last little vestige of virgin wilderness. With the unholy zeal of fanatics we hunt them out and pile them upon his altar, while from the throats of a thousand luncheon clubs and Chambers of Commerce and Greater Gopher Prairie Associations rises the solemn chant "There is No God but Gasoline and Motor is his Prophet!"

The more benignant aspects of the Great God Motor and the really sound elements of the Good Roads Movement need no defense from me. They are cried from every housetop, and we all know them. What I am trying to picture is the tragic absurdity of trying to whip the March of Empire into a gallop.

Very specifically, I am pointing out that in this headlong stampede for speed and ciphers we are crushing the last remnants of something that ought to be preserved for the spiritual and physical welfare of future Americans, even at the cost of acquiring a few less millions of wealth or population in the long run. Something that has helped build the race for such innumerable centuries that we may logically suppose it will help preserve it in the centuries to come.

Failing this, it seems to me we fail in the ultimate test of our vaunted superiority—the self-control of environment. We fall back into the biological category of the potato bug which exterminated the potato, and thereby exterminated itself.

Conserving the Covered Wagon [1925]

In this *Sunset Magazine* companion to "Pioneers and Gullies," Leopold once again evokes the image of the pioneer, this time to argue for preservation of remnants of "covered wagon wilderness," free of roads and motorized vehicles, as a link to the virtues of America's pioneer past.

One evening I was talking to a settler in one of those irrigated valleys that stretch like a green ribbon across the colorful wastes of southern Arizona. He was showing me his farm, and he was proud of it. Broad acres of alfalfa bloom, fields of ripening grain, and a dip and a sweep of laden orchards redolent of milk and honey, all created with the labor of his own hands. Over in one corner I noticed a little patch of the original desert, an island of sandy hillocks, sprawling mesquite trees, with a giant cactus stark against the sky, and musical with the sunset whistle of quail. "Why don't you clear and level that too, and complete your farm?" I asked, secretly fearing he intended to do so.

"Oh, that's for my boys—a sample of what I made the farm out of," he replied quietly. There was no further explanation. I might comprehend his idea, or think him a fool, as I chose.

I chose to think him a very wise man—wise beyond his kind and his generation. That little patch of untamed desert enormously increased the significance of his achievement, and conversely, his achievement enormously increased the significance of the little patch. He was handing down to his sons not only a piece of real estate, but a Romance written upon the oldest of all books, the land. The Romance of The March of Empire.

It set me to thinking. Our fathers set great store by this Winning of the West, but what do we know about it? Many of us have never seen what it was won from. And how much less will the next generation know? If we think we are going to learn by cruising round the mountains in a Ford, we are largely deceiving ourselves. There is a vast difference between the days of the "Free

Tourist Campground—Wood and Water Furnished," and the Covered Wagon Days.

> We pitched our tents where the buffalo feed,
> Unheard of streams were our flagons;
> And we sowed our sons like the apple seed
> In the trail of the prairie wagons.

Yes—sowed them so thick that tens of thousands are killed each year trying to keep out of the way of each other's motors. Is this thickness necessarily a blessing to the sons? Perhaps. But not an unmixed blessing. For those who are so inclined, we might at least preserve a sample of the Covered Wagon Life. For after all, the measure of civilization is in its contrasts. A modern city is a national asset, not because the citizen has planted his iron heel on the breast of nature, but because of the different kinds of man his control over nature has enabled him to be. Saturday morning he stands like a god, directing the wheels of industry that have dominion over the earth. Saturday afternoon he is playing golf on a kindly greensward. Saturday evening he may till a homely garden or he may turn a button and direct the mysteries of the firmament to bring him the words and songs and deeds of all the nations. And if, once in a while, he has the opportunity to flee the city, throw a diamond hitch upon a packmule, and disappear into the wilderness of the Covered Wagon Days, he is just that much more civilized than he would be without the opportunity. It makes him one more kind of a man—a pioneer.

We do not realize how many Americans have an instinctive craving for the wilderness life, or how valuable to the nation has been their opportunity of exercising that instinct, because up to this time the opportunity has been automatically supplied. Little patches of Covered Wagon wilderness have persisted at the very doors of our cities. But now these little patches are being wiped out at a rate which takes one's breath away. And the thing that is wiping them out is the motor car and the motor highway. It is of these, their uses and their abuses, that I would speak.

Motor cars and highways are of course the very instruments which have restored to millions of city dwellers their contact with the land and with nature. For this reason and to this extent they are a benefaction to mankind. But even a benefaction can be carried too far. It was one of Shakespeare's characters who said:

> For Virtue, grown into a pleurisy
> Dies of its own too-much.

To my mind the Good Roads Movement has become a Good Roads Mania; it has grown into a pleurisy. We are building good roads to give the rancher access to the city, which is good, and to give the city dweller access to recreation in the forests and mountains, which is good, but we now, out of sheer momentum, are thrusting more and ever more roads into every little remaining patch of wilderness, which in many cases is sheer stupidity. For by so doing we are cutting off, irrevocably and forever, our national contact with the Covered Wagon days.

Pick up any outdoor magazine and the chances are that on the first page you will find an article describing the adventure of some well-to-do sportsman who has been to Alaska, or British Columbia, or Africa, or Siberia in search of wilderness and the life and hardy sports that go with it. He has pushed to the Back of Beyond, and he tells of it with infinite zest. It has been his Big Adventure. Why? Because he brought home the tusk of an elephant or the hide of a brown bear? No, fundamentally no. Rather because he has proved himself to be still another kind of man than his friends gave him credit for. He has been, if only for one fleeting month, a pioneer, and met the test. He has justified the Blood of the Conquerors.

But have these well-to-do travelers in foreign wilds a monopoly on the Covered Wagon blood? Here is the point of the whole matter. They have not. In every village and in every city suburb and in every skyscraper are dozens of the self-same blood. But they lack the opportunity. It is the opportunity, not the desire, on which the well-to-do are coming to have a monopoly. And the reason is the gradually increasing destruction of the nearby wilderness by good roads. The American of moderate means can not go to Alaska, or Africa, or British Columbia. He must seek his big adventure in the nearby wilderness, or go without it.

Ten years ago, for instance, there were five big regions in the National Forests of Arizona and New Mexico where the Covered Wagon blood could disport at will. In any one of them a man could pack up a mule and disappear into the tall uncut for a month without ever crossing his back track. Today there is just one of the five left. The Forest Service, the largest custodian of land in either State, has naturally and rightly joined with the good roads movement, and today has built or is helping to build good roads right through the vitals of four of these five big regions. As wilderness, they are gone, and gone forever. So far so good. But shall the Forest Service now do the same with the fifth and the last?

Round this last little remnant of the original Southwest lies an economic empire without any wilderness playground or the faintest chance of acquiring one. Texas, Oklahoma, and the rich valleys and mines of Arizona and New Mexico already support millions of Americans. The high mountains of the National Forests are their natural and necessary recreation grounds. The greater part of these mountain areas is already irrevocably dedicated to the motorized forms of recreation. Is it unreasonable or visionary to ask the Forest Service to preserve the one remaining portion of unmotorized wilderness for those who prefer that sort of place?

Would it be unreasonable or visionary to ask the Government to set aside similar remnants of wilderness here and there throughout the National Forests and National Parks? Say one such area, if possible, in each State?

As a matter of fact, the officials of the Forest Service are already seriously considering doing just that. Colonel William B. Greeley, Chief Forester, in addressing a meeting of the American Game Protective Association, put it this way:

> We all recognize what the forest background of the United States has meant to this country—how it has given stamina and resourcefulness and mental and physical vigor to every oncoming generation of Americans. We must preserve something of that forest background for the future. It seems to me that in the National Forests, while we are building roads, as we must; while we are developing areas for the utilization of timber, as we must; while we are opening up extensive regions for the camper, the summer vacationist, and the masses of people who have the God-given right to enjoy these areas—we should keep here and there as part of the picture some bit of wilderness frontier, some hinterland of mountain and upland lake that the roads and automobiles will have to pass by.
>
> The law laid down for the guidance of the Forest Service was that these public properties must be administered for the greatest good of the greatest number in the long run. When Secretary Wilson laid down that rule, probably he was thinking more of timber and water and forage than anything else, but today the same rule applies just as clearly as it did in the time of Roosevelt in 1905. I think we can all agree that the greatest good of the greatest number of American citizens in the long run does require that in their own National Forests there should be preserved some bits of unspoiled wilderness where the young America of the future can take to the outdoors in the right way.

But let no man think that because a few foresters have tentatively formulated a wilderness policy, that the preservation of a system of wilderness remnants is assured in the National Forests. Do not forget that the good roads mania, and all forms of unthinking Boosterism that go with it, constitute a steam roller the like of which has seldom been seen in the history of mankind. No steam roller can overwhelm a good idea or a righteous policy, but it might very readily flatten out, one by one, the remaining opportunities for applying this particular policy. After these remnants are gone, a correct wilderness policy would be useless.

What I mean is this: The Forest Service will naturally select for wilderness playgrounds the roughest areas and those poorest from the economic standpoint. But it will be physically impossible to find any area which does not embrace some economic values. Sooner or later some private interest will wish to develop these values, at which time those who are thinking in terms of the national development in the broad sense and those who are thinking of local development in the narrow sense will come to grips. And forthwith the private interests will invoke the aid of the steam roller. They always do. And unless the wilderness idea represents the mandate of an organized, fighting and voting body of far-seeing Americans, the steam roller will win.

At the present moment, the most needed move is to secure recognition of the need for a Wilderness Area Policy from the National Conference on Outdoor Recreation, set up by President Coolidge for the express purpose of coördinating the many conflicting recreational interests which have arisen in recent years. If the spirit of the Covered Wagon really persists, as I firmly believe it does, its devotees must speak now, or forever hold their peace.

The Pig in the Parlor [1925]

In some ways the most recalcitrant audience Leopold tried to reach with his wilderness message was his colleagues in the Forest Service. Here he addresses the key question of roads in a pithy note in the mimeographed in-house *Service Bulletin* published by the Washington office.

In the May 11 *Service Bulletin* there is an item from D-6 which says in effect that the "wild-life enthusiasts" need not fret about the invasion of wilderness areas by roads, because in Germany there is a mile of dirt road for every 105 acres of forest, and a mile of hard road for every 220 acres of forest. Germany, it says, spends up to 35 cents per acre per year for forest roads, and because we have not attained such beatitude we need not worry yet about overdoing the road game.

In short, the wilderness area idea is assumed to be an anti-road idea. The assumption is incorrect. It is just exactly as incorrect as Editor Abbott's assumption that recreational development is anti-forestry. My plea is that the wilderness idea be not condemned, especially by foresters, without first acquiring at least a rudimentary understanding of what it is all about.

I do not know of a single "enthusiast" for wilderness areas who denies the need for more forest roads. It is not a question of how many roads, but a question of distribution of roads. The wilderness idea simply affirms that a well-balanced plan for the highest use of National Forests will exclude roads from certain areas so that the unmotorized forms of public recreation will not be left high and dry, just as summer homes are excluded from certain areas so that the camper will not be left high and dry. The only difference is that where a public camp ground requires a forty, a public wilderness area requires a few townships.

Roads and wilderness are merely a case of the pig in the parlor. We now recognize that the pig is all right—for bacon, which we all eat. But there no doubt was a time, soon after the discovery that many pigs meant much bacon, when our ancestors assumed that because the pig was so useful an institution he should be welcomed at all times and places. And I suppose that the first "enthusiast" who raised the question of limiting his distribution was construed to be uneconomic, visionary, and anti-pig.

Wilderness as a Form of Land Use [1925]

In the scholarly *Journal of Land and Public Utility Economics*, Leopold published his most sustained, comprehensive statement of the wilderness idea and the rationale for developing a policy of wilderness preservation as a component of land use in America. The influence of the leading historian of the American frontier, Frederick Jackson Turner, who moved to a house just two doors down the street from Leopold the same year that Leopold moved to Madison, is unmistakable in Leopold's argument for the cultural value of wilderness.

From the earliest times one of the principal criteria of civilization has been the ability to conquer the wilderness and convert it to economic use. To deny the validity of this criterion would be to deny history. But because the conquest of wilderness has produced beneficial reactions on social, political, and economic development, we have set up, more or less unconsciously, the converse assumption that the ultimate social, political, and economic development will be produced by conquering the wilderness entirely—that is, by eliminating it from our environment.

My purpose is to challenge the validity of such an assumption and to show how it is inconsistent with certain cultural ideas which we regard as most distinctly American.

Our system of land use is full of phenomena which are sound as tendencies but become unsound as ultimates. It is sound for a city to grow but unsound for it to cover its entire site with buildings. It was sound to cut down our forests but unsound to run out of wood. It was sound to expand our agriculture, but unsound to allow the momentum of that expansion to result in the present overproduction. To multiply examples of an obvious truth would be tedious. The question, in brief, is whether the benefits of wilderness-conquest will extend to ultimate wilderness-elimination.

The question is new because in America the point of elimination has only recently appeared upon the horizon of foreseeable events. During our four centuries of wilderness-conquest the possibility of disappearance has

Wilderness as a Form of Land Use

been too remote to register in the national consciousness. Hence we have no mental language in which to discuss the matter. We must first set up some ideas and definitions.

What Is a Wilderness Area?

The term wilderness, as here used, means a wild, roadless area where those who are so inclined may enjoy primitive modes of travel and subsistence, such as exploration trips by pack-train or canoe.

The first idea is that wilderness is a resource, not only in the physical sense of the raw materials it contains, but also in the sense of a distinctive environment which may, if rightly used, yield certain social values. Such a conception ought not to be difficult, because we have lately learned to think of other forms of land use in the same way. We no longer think of a municipal golf links, for instance, as merely soil and grass.

The second idea is that the value of wilderness varies enormously with location. As with other resources, it is impossible to dissociate value from location. There are wilderness areas in Siberia which are probably very similar in character to parts of our Lake states, but their value to us is negligible, compared with what the value of a similar area in the Lake states would be, just as the value of a golf links would be negligible if located so as to be out of reach of golfers.

The third idea is that wilderness, in the sense of an environment as distinguished from a quantity of physical materials, lies somewhere between the class of non-reproducible resources like minerals, and the reproducible resources like forests. It does not disappear proportionately to use, as minerals do, because we can conceive of a wild area which, if properly administered, could be traveled indefinitely and still be as good as ever. On the other hand, wilderness certainly cannot be built at will, like a city park or a tennis court. If we should tear down improvements already made in order to build a wilderness, not only would the cost be prohibitive, but the result would probably be highly dissatisfying. Neither can a wilderness be grown like timber, because it is something more than trees. The practical point is that if we want wilderness, we must foresee our want and preserve the proper areas against the encroachment of inimical uses.

Fourth, wilderness exists in all degrees, from the little accidental wild spot at the head of a ravine in a Corn Belt woodlot to vast expanses of virgin country—

> Where nameless men by nameless rivers wander
> And in strange valleys die strange deaths alone.

What degree of wilderness, then, are we discussing? The answer is, *all degrees*. Wilderness is a relative condition. As a form of land use it cannot be a rigid entity of unchanging content, exclusive of all other forms. On the contrary, it must be a flexible thing, accommodating itself to other forms and blending with them in that highly localized give-and-take scheme of land-planning which employs the criterion of "highest use." By skilfully adjusting one use to another, the land planner builds a balanced whole without undue sacrifice of any function, and thus attains a maximum net utility of land.

Just as the application of the park idea in civic planning varies in degree from the provision of a public bench on a street corner to the establishment of a municipal forest playground as large as the city itself, so should the application of the wilderness idea vary in degree from the wild, roadless spot of a few acres left in the rougher parts of public forest devoted to timber-growing, to wild, roadless regions approaching in size a whole national forest or a whole national park. For it is not to be supposed that a public wilderness area is a new kind of public land reservation, distinct from public forests and public parks. It is rather a new kind of land-dedication within our system of public forests and parks, to be duly correlated with dedications to the other uses which that system is already obligated to accommodate.

Lastly, to round out our definitions, let us exclude from practical consideration any degree of wilderness so absolute as to forbid reasonable protection. It would be idle to discuss wilderness areas if they are to be left subject to destruction by forest fires, or wide open to abuse. Experience has demonstrated, however, that a very modest and unobtrusive framework of trails, telephone line and lookout stations will suffice for protective purposes. Such improvements do not destroy the wild flavor of the area, and are necessary if it is to be kept in usable condition.

Wilderness Areas in a Balanced Land System

What kind of case, then, can be made for wilderness as a form of land use?

To preserve any land in a wild condition is, of course, a reversal of economic tendency, but that fact alone should not condemn the proposal. A study of the history of land utilization shows that good use is largely a matter of good balance—of wise adjustment between opposing tendencies. The modern movements toward diversified crops and live stock on the farm, conservation of eroding soils, forestry, range management, game management, public parks—all these are attempts to balance opposing tendencies that have swung out of counterpoise.

One noteworthy thing about good balance is the nature of the opposing tendencies. In its more utilitarian aspect, as seen in modern agriculture, the needed adjustment is between economic uses. But in the public park movement the adjustment is between an economic use, on the one hand, and a purely social use on the other. Yet, after a century of actual experience, even the most rigid economic determinists have ceased to challenge the wisdom of a reasonable reversal of economic tendency in favor of public parks.

I submit that the wilderness is a parallel case. The parallelism is not yet generally recognized because we do not yet conceive of the wilderness environment as a resource. The accessible supply has heretofore been unlimited, like the supply of air-power, or tide-power, or sunsets, and we do not recognize anything as a resource until the demand becomes commensurable with the supply.

Now after three centuries of overabundance, and before we have even realized that we are dealing with a non-reproducible resource, we have come to the end of our pioneer environment and are about to push its remnants into the Pacific. For three centuries that environment has determined the character of our development; it may, in fact, be said that, coupled with the character of our racial stocks, it is the very stuff America is made of. Shall we now exterminate this thing that made us American?

Ouspensky says that, biologically speaking, the determining characteristic of rational beings is that their evolution is self-directed. John Burroughs cites the opposite example of the potato bug, which, blindly obedient to the law of increase, exterminates the potato and thereby exterminates itself. Which are we?

What the Wilderness Has Contributed to American Culture

Our wilderness environment cannot, of course, be preserved on any considerable scale as an economic fact. But, like many other receding economic facts, it can be preserved for the ends of sport. But what is the justification of sport, as the word is here used?

Physical combat between men, for instance, for unnumbered centuries was an economic fact. When it disappeared as such, a sound instinct led us to preserve it in the form of athletic sports and games. Physical combat between men and beasts since first the flight of years began was an economic fact, but when it disappeared as such, the instinct of the race led us to hunt and fish for sport. The transition of these tests of skill from an economic to a social basis has in no way destroyed their efficacy as human experiences—in fact, the change may be regarded in some respects as an improvement.

Football requires the same kind of back-bone as battle but avoids its moral and physical retrogressions. Hunting for sport in its highest form is an improvement on hunting for food in that there has been added, to the test of skill, an ethical code which the hunter formulates for himself and must often execute without the moral support of bystanders.

In these cases the surviving sport is actually an improvement on the receding economic fact. Public wilderness areas are essentially a means for allowing the more virile and primitive forms of outdoor recreation to survive the receding economic fact of pioneering. These forms should survive because they likewise are an improvement on pioneering itself.

There is little question that many of the attributes most distinctive of America and Americans are the impress of the wilderness and the life that accompanied it. If we have any such thing as an American culture (and I think we have), its distinguishing marks are a certain vigorous individualism combined with ability to organize, a certain intellectual curiosity bent to practical ends, a lack of subservience to stiff social forms, and an intolerance of drones, all of which are the distinctive characteristics of successful pioneers. These, if anything, are the indigenous part of our Americanism, the qualities that set it apart as a new rather than an imitative contribution to civilization. Many observers see these qualities not only bred into our people, but built into our institutions. Is it not a bit beside the point for us to be so solicitous about preserving those institutions without giving so much as a thought to preserving the environment which produced them and which may now be one of our effective means of keeping them alive?

Wilderness Locations

But the proposal to establish wilderness areas is idle unless acted on before the wilderness has disappeared. Just what is the present status of wilderness remnants in the United States?

Large areas of half a million acres and upward are disappearing very rapidly, not so much by reason of economic need, as by extension of motor roads. Smaller areas are still relatively abundant in the mountainous parts of the country, and will so continue for a long time.

The disappearance of large areas is illustrated by the following instance: In 1910 there were six roadless regions in Arizona and New Mexico, ranging in size from half a million to a million acres, where the finest type of mountain wilderness pack trips could be enjoyed. Today roads have eliminated all but one area of about half a million acres.

In California there were seven large areas ten years ago, but today there are only two left unmotorized.

In the Lake states no large unmotorized playgrounds remain. The motor launch, as well as the motor road, is rapidly wiping out the remnants of canoe country.

In the Northwest large roadless areas are still relatively numerous. The land plans of the Forest Service call for exclusion of roads from several areas of moderate size.

Unless the present attempts to preserve such areas are greatly strengthened and extended, however, it may be predicted with certainty that, except in the Northwest, all of the large areas already in public ownership will be invaded by motors in another decade.

In selecting areas for retention as wilderness, the vital factor of location must be more decisively recognized. A few areas in the national forests of Idaho or Montana are better than none, but, after all, they will be of limited usefulness to the citizen of Chicago or New Orleans who has a great desire but a small purse and a short vacation. Wild areas in the poor lands of the Ozarks and the Lake states would be within his reach. For the great urban populations concentrated on the Atlantic seaboards, wild areas in both ends of the Appalachians would be especially valuable.

Are the remaining large wilderness areas disappearing so rapidly because they contain agricultural lands suitable for settlement? No; most of them are entirely devoid of either existing or potential agriculture. Is it because they contain timber which should be cut? It is true that some of them do contain valuable timber, and in a few cases this fact is leading to a legitimate extension of logging operations; but in most of the remaining wilderness the timber is either too thin and scattered for exploitation, or else the topography is too difficult for the timber alone to carry the cost of roads or railroads. In view of the general belief that lumber is being overproduced in relation to the growing scarcity of stumpage, and will probably so continue for several decades, the sacrifice of wilderness for timber can hardly be justified on grounds of necessity.

Generally speaking, it is not timber, and certainly not agriculture, which is causing the decimation of wilderness areas, but rather the desire to attract tourists. The accumulated momentum of the good-roads movement constitutes a mighty force, which, skilfully manipulated by every little mountain village possessed of a chamber of commerce and a desire to become a metropolis, is bringing about the extension of motor roads into every remaining bit of wild country, whether or not there is economic justification for the extension.

Our remaining wild lands are wild because they are poor. But this poverty does not deter the booster from building expensive roads through them as bait for motor tourists.

I am not without admiration for this spirit of enterprise in backwoods villages, nor am I attempting a censorious pose toward the subsidization of their ambitions from the public treasuries; nor yet am I asserting that the resulting roads are devoid of any economic utility. I do maintain, (1) that such extensions of our road systems into the wilderness are seldom yielding a return sufficient to amortize the public investment; (2) that even where they do yield such a return, their construction is not necessarily in the public interest, any more than obtaining an economic return from the last vacant lot in a parkless city would be in the public interest. On the contrary, the public interest demands the careful planning of a system of wilderness areas and the permanent reversal of the ordinary economic process within their borders.

To be sure, to the extent that the motor-tourist business is the cause of invasion of these wilderness playgrounds, one kind of recreational use is merely substituted for another. But this substitution is a vitally serious matter from the point of view of good balance. It is just as unwise to devote 100% of the recreational resources of our public parks and forests to motorists as it would be to devote 100% of our city parks to merry-go-rounds. It would be just as unreasonable to ask the aged to indorse a park with only swings and trapezes, or the children a park with only benches, or the motorists a park with only bridlepaths, as to ask the wilderness recreationist to indorse a universal priority for motor roads. Yet that is what our land plans—or rather lack of them—are now doing; and so sacred is our dogma of "development" that there is no effective protest. The inexorable molding of the individual American to a standardized pattern in his economic activities makes all the more undesirable this unnecessary standardization of his recreational tastes.

Practical Aspects of Establishing Wilderness Areas

Public wilderness playgrounds differ from all other public areas in that both their establishment and maintenance would entail very low costs. The wilderness is the one kind of public land that requires no improvements. To be sure, a simple system of fire protection and administrative patrol would be required, but the cost would not exceed two or three cents per acre per year. Even that would not usually be a new cost, since the greater part of the needed areas are already under administration in the rougher parts of the national forests and parks. The action needed is the permanent differentiation of a suitable system of wild areas within our national park and forest system.

In regions such as the Lake states, where the public domain has largely disappeared, lands would have to be purchased; but that will have to be done, in any event, to round out our park and forest system. In such cases a lesser degree of wilderness may have to suffice, the only ordinary utilities practicable to exclude being cottages, hotels, roads, and motor boats.

The retention of certain wild areas in both national forests and national parks will introduce a healthy variety into the wilderness idea itself, the forest areas serving as public hunting grounds, the park areas as public wildlife sanctuaries, and both kinds as public playgrounds in which the wilderness environments and modes of travel may be preserved and enjoyed.

The Cultural Value of Wilderness

Are these things worth preserving? This is the vital question. I cannot give an unbiased answer. I can only picture the day that is almost upon us when canoe travel will consist in paddling in the noisy wake of a motor launch and portaging through the back yard of a summer cottage. When that day comes, canoe travel will be dead, and dead, too, will be a part of our Americanism. Joliet and LaSalle will be words in a book, Champlain will be a blue spot on a map, and canoes will be merely things of wood and canvas, with a connotation of white duck pants and bathing "beauties."

The day is almost upon us when a pack-train must wind its way up a graveled highway and turn out its bell-mare in the pasture of a summer hotel. When that day comes the pack-train will be dead, the diamond hitch will be merely rope, and Kit Carson and Jim Bridger will be names in a history lesson. Rendezvous will be French for "date," and Forty-Nine will be the number preceding fifty. And thenceforth the march of empire will be a matter of gasoline and four-wheel brakes.

European outdoor recreation is largely devoid of the thing that wilderness areas would be the means of preserving in this country. Europeans do not camp, cook, or pack in the woods for pleasure. They hunt and fish when they can afford it, but their hunting and fishing is merely hunting and fishing, staged in a setting of ready-made hunting lodges, elaborate fare, and hired beaters. The whole thing carries the atmosphere of a picnic rather than that of a pack trip. The test of skill is confined almost entirely to the act of killing, itself. Its value as a human experience is reduced accordingly.

There is a strong movement in this country to preserve the distinctive democracy of our field sports by preserving free hunting and fishing, as distinguished from the European condition of commercialized hunting and fishing privileges. Public shooting grounds and organized cooperative relations between sportsmen and landowners are the means proposed for keeping these sports within reach of the American of moderate means. Free hunting and fishing is a most worthy objective, but it deals with only one of the two distinctive characteristics of American sport. The other characteristic is that our test of skill is primarily the act of living in the open, and only secondarily the act of killing game. It is to preserve this primary characteristic that public wilderness playgrounds are necessary.

Herbert Hoover aptly says that there is no point in increasing the average American's leisure by perfecting the organization of industry, if the expansion of industry is allowed to destroy the recreational resources on which leisure may be beneficially employed. Surely the wilderness is one of the most valuable of these resources, and surely the building of unproductive roads in the wrong places at public expense is one of the least valuable of industries. If we are unable to steer the Juggernaut of our own prosperity, then surely there is an impotence in our vaunted Americanism that augurs ill for our future. The self-directed evolution of rational beings does not apply to us until we become collectively, as well as individually, rational and self-directing.

Wilderness as a form of land-use is, of course, premised on a qualitative conception of progress. It is premised on the assumption that enlarging the range of individual experience is as important as enlarging the number of individuals; that the expansion of commerce is a means, not an end; that the environment of the American pioneers had values of its own, and was not merely a punishment which they endured in order that we might ride in motors. It is premised on the assumption that the rocks and rills and templed hills of this America are something more than economic materials, and should not be dedicated exclusively to economic use.

The vanguard of American thought on the use of land has already recognized all this, in theory. Are we too poor in spirit, in pocket, or in idle acres to recognize it likewise in fact?

The Home Builder Conserves [1928]

From 1924 to 1928 Leopold served as associate director of the Forest Products Laboratory in Madison, Wisconsin, the principal research arm of the Forest Service at the time. It was a position in which he was not entirely happy, since the laboratory was concerned almost exclusively with research on forest products rather than with the growing trees, and he published few articles related to its work. Those few, however, are models of how to reach a targeted audience with a message. In this piece published in *American Forests and Forest Life*, he attempts to reach the thinking conservationist with the laboratory's concern about proper utilization of wood and ends with a characteristically Leopoldian definition of citizenship.

It has been known for years that our processes for converting forest trees into houses, furniture, implements, newspapers and a thousand other necessary wooden products were wasteful, but nobody knew, for the country as a whole, just how wasteful or just why. But the Forest Products Laboratory of the United States Forest Service, which has been working since 1910 on the more economical utilization of forests, recently concluded that after two decades of study and research it was prepared not only to estimate pretty closely the total amount of wood lost under present practices, but to tell the nation how a substantial fraction of the loss might be averted at a profit.

The Laboratory's estimate is that two-thirds of the total wood cut each year from our forests is lost in the process of converting trees into wooden products. It further estimates that at least one-third of this loss is unnecessary and could be avoided if the forest-using industries would adopt the methods of waste prevention which the most progressive firms and individuals have tried out and found commercially feasible.

Good wood costs good money. Why then should a competitive industry need to be urged to adopt methods of preventing wood waste already found to be feasible and often profitable? Why do they not seize upon these improved methods of their own accord? Every American conservationist should note and ponder well the answer to this question.

The improved methods are not automatically adopted and used because they are not known, or they require more skill and equipment than the old methods. Also they require better organization of the producers and markets before the individual firm can take advantage of them. Another point is that the prejudices and customs of the consuming public prevent the sale of the resulting product.

The American public for many years has been abusing the wasteful lumberman. A public which lives in wooden houses should be careful about throwing stones at lumbermen, even wasteful ones, until it has learned how its own arbitrary demands as to kinds and qualities of lumber, help cause the waste which it decries. If all the housewives of the United States should suddenly decline to buy anything but wish-bone cuts of chicken, the poultrymen of the country would be in the same boat as many lumbermen are today.

A householder, for instance, is adding a room to his dwelling. He is buying the joists for the floor. He wants a good job and good material, so he may insist on practically clear 2" x 8" pine, even though he could get knotty pine for twenty-five to thirty-five per cent less. Now, pine trees produce clear and knotty material in the ratio of about one to three, whereas the demand, as measured by price, is in inverse ratio. Consequently large quantities of knotty material are left without an adequate market, and thus tend to be directly or indirectly wasted. In some uses practically clear material is of course necessary, but in the case of the householder and his floor joists for a small room, clear material is unnecessary, and therefore wasteful. The reason why clear material is unnecessary is that the number and kind of joists put under a house floor is nearly always in excess of what is required for *strength*. Enough additional joists are put in to give a high *stiffness*, or ability to resist deflection, with its accompanying vibrations and squeaks. Research has shown that a knotty joist, while not as strong, is practically as stiff as a clear one. Therefore, clear joists in this case represent an unnecessary excess of strength, and help to induce a waste of forest.

It may be unreasonable to hope that every householder in the United States can be taught that knotty joists are as stiff as clear ones, and therefore just as good for small houses. But it ought to be possible to teach it to the architects, contractors and engineers who deal daily in such matters and who write the specifications for the bulk of lumber purchases.

Let us now assume that the same householder is buying the lumber for his interior trim, which is to be painted. The dealer offers him his choice of white pine or white fir. He has heard somewhere, somehow, that white fir is poor lumber. So he cheerfully pays extra money for the pine.

Is his choice a wise one? Not from the standpoint of conservation. White fir is perfectly good for interior trim. In strength qualities it is equal to sugar pine, and except in toughness it is equal to spruce. By refusing to buy white fir for the purposes for which it is suited, the average consumer unconsciously contributes to the huge waste of so-called "inferior species," and hastens the day when lumber will soar out of reach of the average citizen's pocketbook.

Again, suppose that our householder is buying siding for the exterior walls of his building. He declines to accept certain material because it contains blue sap-stain. He is afraid this blue stain represents the initial stages of decay. As a matter of fact, research has shown that blue-stain is not decay, and that it does not affect the strength or paint-holding qualities of the board. Nevertheless, millions of feet of perfectly usable lumber are "degraded" because of prejudice against blue-stain. At the same time our householder may be unknowingly accepting sills infected with real decay, and thus, because of the greater exposure of the sills to dampness, be shortening the life of his building by many years.

The long and short of the matter is that forest conservation depends in part on intelligent consumption, as well as intelligent production of lumber. Intelligent lumber consumption depends on overcoming misinformation and prejudice, of which the foregoing are examples, and supplanting it with scientific information and a real understanding of the properties of wood. The ultimate consumer probably has too many other things to think about to acquire more than a smattering of such matters, but he is entitled to better advice than he usually gets from the retailer who sells him lumber. Our industrial functions are becoming highly specialized. There are any number of retailers who have never seen a sawmill, who do not know the forest trees from which their lumber is cut, and who are far from posted on the utilization problems bearing on forest conservation. This may not be their fault, but to acquire a real understanding of such matters would redound to the advantage of a retail lumber dealer. There is still another category of wood waste to which the ultimate consumer indirectly contributes, but which could be more quickly eliminated by an intelligent interest on his part.

Take, for instance, our universal insistence on clear hardwoods for furniture and interior woodwork. A sound knot is today absolutely taboo on the face of a drawer or a baseboard or a window casing. Contrast this with the attitude of the craftsmen who built the handmade furniture of colonial times. A sound, tight knot on a drawer face worried them not at all—in fact, I have seen wonderful old walnut dressers where they actually seemed to prefer knotty pieces because of their ornamental effect. Consider this, and the fact that the greater part of our enormous hardwood waste occurs in the process of trimming out knots. Is it too much to hope that fashion may some day lift the ban against sound knots in places where they enhance the beauty of the wood and do not injure strength?

One of the biggest leaks in National timber supply occurs in connection with the manufacture of articles requiring small pieces of wood, which the manufacturer cuts at his factory from wide long boards, whereas they could just as well be cut at the sawmill from low-grade boards, slabs, edgings, and woods waste. Cutting these small pieces from mill and woods waste would release the wide, long boards for uses where they are really necessary, and thus help to do away with that curse of conservation, the waste-burner. It would be financially profitable to all concerned, and would materially lengthen the life of our remaining forests. These small pieces cut from waste are called "small dimension stock." The principal reason why the manufacture and use of small dimension stock is not more prevalent is that the sizes and kinds are not sufficiently standardized—an obstacle which could be removed by better organized trade practices. The ultimate consumer interested in forest conservation can aid the cause materially by patronizing the firms which sell goods made from such stock, and which are doing their part, through trade associations and otherwise, in extending its use.

Another big leak in the National timber supply is in the still widespread failure to apply preservative treatments to wood used in contact with soil or dampness. Railroads, mines, and telephone companies, for instance, use enormous quantities of cross-ties, poles, and other timbers exposed to decay. Treating such timbers with creosote, zinc chloride, or other good preservatives doubles or trebles their durability for a comparatively slight additional cost. If the life of the ties in a mile of railroad track is trebled by preservatives, it follows that the area of forest necessary to grow ties for replacements is reduced to a third of that necessary for unpreserved ties. It means that a given area of forest will grow ties for three times as great a mileage of track where preservatives are used as where they are not used. The thinking conservationist can encourage the practice of preservative treatments by voicing his approval of the railroads that use them. Mines, by the way, are much less progressive than railroads with respect to this matter.

Large wastes result from unscientific drying of wood, through the necessity for discarding or degrading boards which have developed checks, cracks, cupping, or warping during the drying process. These defects are often needlessly increased by reason of careless piling and improper operation of dry kilns. Research has devised methods of reducing these losses. The ultimate consumer has, however, little opportunity to directly encourage the adoption of scientific drying methods by the industries.

So much for the second category of wood wastes, on which the intelligent consumer can exert an indirect but nevertheless valuable personal influence. There still remains a third category, which, in so far as the general public is concerned, can only be influenced by those who happen to be bankers or financiers and who thus control the purse strings of capital by which forest-using industries are established, expanded, contracted, or abandoned. Just as the far-western bankers had to learn by bitter experience that an adequate area of productive range was just as important to the safety of a livestock loan as the number of cattle or sheep offered as security, so do the bankers of forest regions have to learn that an assured supply of raw material is just as important to the safety of an investment in a wood-using plant as is the plant itself, its product, and its market.

There is one thing about a wood-using plant which even bankers have not generally realized, namely, that a stable supply of raw material often depends on a proper grouping of diversified industries in each forest community. Thus in a certain "woods" town three sawmills might be a bad risk, whereas one sawmill, one paper mill, and one box factory might each be good risks. Why? Because their raw material needs are complementary instead of competitive; because what one wastes the other uses. It is the same problem as diversification in agriculture, and holds out the same possibilities for industrial permanence and prosperity, and for conservation of natural resources.

The foregoing does not attempt to give even an outline of the task which the Nation is facing in the prevention of forest wastes. It merely sketches some of the ways in which the thinking citizen can aid. Even the thinking citizen is too apt to assume that his only power as a conservationist lies in his vote. Such an assumption is wrong. At least an equal power lies in his daily thought, speech, and action, and especially in his habits as a buyer and user of wood.

I admit that the effective exercise of his power as a purchaser and user of forest products depends on his being well posted. But most problems of good citizenship in these days seem to resolve themselves into just that. Good citizenship is the only effective patriotism, and patriotism requires less and less of making the eagle scream, but more and more of making him think.

Ho! Compadres Piñoneros! [1929]

This paean to the piñon jay appeared in *Forest Fire and Other Verse*, a second volume of Forest Service doggerel edited by Leopold's close friend, John D. Guthrie. Inspired perhaps by one of Leopold's bow-and-arrow hunting trips to the Gila Wilderness in the late 1920s, the poem reveals his longing for the Southwest, which he felt especially during his years of desk-bound exile in the Forest Products Laboratory.

> Ho for piney lanes of sunshine
> On the tops of basking mesas!—
> Ho for singing groves of piñon
> On the foothills in the fall.
> Ho for lanes of silky gramma,
> Tang of sage and scent of cedar
> On the pleasant hills of autumn
> Where the Piñoneros call.
>
> Bosky shades of cedar thickets
> On the tops of basking mesas—
> Pale blue flocks of Piñoneros,
> Wheeling gaily from the crest
> Down the yellow glades of pingue
> Soaked in gold and drenched in sunshine,
> Pitchy—nutty—tasting sunshine
> Of the foothills of the west!
>
> Ho Compadres! Piñoneros!
> Jolly band of shameless loafers,
> Full of fun and full of business!
> Ho Compadres! Vagrants all—
> Ho for singing groves of piñons!

Ho! Compadres Piñoneros!

Ho for singing autumn weather!
Piñoneros! My Compadres
Of the foothills in the fall!—

Pale blue flocks of Piñoneros
Wheeling off across the mesas—
Calling . . . Calling. Piñoneros,
Each one softly to the rest
Calling dimly from the distance . . .
Piñoneros faintly calling
Calling faint . . . but ever calling
To the foothills of the west.

Report to the American Game Conference on an American Game Policy [1930]

Leopold served on dozens of professional committees during his career, almost always taking the lead in drafting the reports. Perhaps most noteworthy was his service during 1929–1932 as chairman of the Committee on Game Policy of the American Game Association, forerunner to the Wildlife Management Institute and the National Wildlife Federation. The game policy proposed by Leopold's committee was adopted by resolution at the seventeenth American Game Conference in December 1930. It became the basis for wildlife administration policies in many states and set the general direction of the wildlife profession in America for decades. The report included an explanatory appendix not reprinted here.

Introduction

Demand for hunting is outstripping supply. If hunting as a recreation is to continue, game production must be increased. Where? How? By whom? For whom? These are the questions with which a game policy must deal.

In the case of ordinary economic products, the free play of economic forces automatically adjusts supply to demand.

Game production, however, is not so simple. Irreplaceable species may be destroyed before these forces become operative. Moreover, game is not a primary crop, but a secondary by-product of farm and forest lands, obtainable only when the farming and forestry cropping methods are suitably modified in favor of the game. Economic forces must act through these primary land uses, rather than directly.

It is axiomatic that timber and farm crops must be bought and sold, otherwise they would not be produced at all. Is this also true of game? Some say yes, but the majority adhere to the deep-rooted American pioneer tradition that hunting is a free privilege, and insist that it can be kept so, in spite of the contrary pressure of economic law.

The two opposing schools of thought have so far nullified each other, because the proponents of each have insisted that the two ideas cannot coexist; that one must prevail to the exclusion of the other.

This Committee contends that they can and should coexist, each on its appropriate kind of land, and often in close proximity to each other.

We submit that public hunting under the license system is workable for game species inhabiting cheap land which the public can afford to own (or lease) and operate, but that compensation to the landowner in some form or other is the only workable system for producing game on expensive private farm land.

We submit that recognition of this principle, and a spirit of mutual cooperation in acting upon it, will bend the two hitherto opposing schools of thought to a new and common direction.

We do not pretend to foresee or prescribe all of the detailed actions necessary to accomplish this. This report, however, segregates certain fundamental moves which have this new and common direction. We urge all factions to co-operate in executing them, and to let experience dictate succeeding steps.

We believe, in short, that experiment, not doctrine or prophecy, is the key to an American Game Policy.

Seven fundamental actions are recommended (Part A) for adoption by the American Game Conference as an American Game Policy.

An Appendix (Part B) presents in additional detail how the seven fundamental suggestions were arrived at by the Committee, and describes such ways and means as are known to it for carrying them out. These particular ways and means are not offered as final. Better ones may be developed by experimentation.

The proposed policy offers no panacea. We urge frank recognition of the fact that there is no panacea; that game conservation faces a crisis in many states; that it is only a question of time before it does so in all states; that the present order is radically unsatisfactory; and that mild modifications of it will not do. We are convinced that only bold action, guided by as much wisdom as we can muster from time to time, can restore America's game resources. Timidity, optimism, or unbending insistence on old grooves of thought and action will surely either destroy the remaining resources, or force the adoption of policies which will limit their use to a few.

COMMITTEE ON GAME POLICY
Aldo Leopold, *Chairman*

S. F. Rathbun	I. Zellerbach	P. S. Lovejoy
Wm. J. Tucker	Seth E. Gordon	Paul G. Redington
John C. Phillips	George A. Lawyer	H. C. Bryant
J. W. Titcomb	John B. Burnham	A. Willis Robertson
	R. Fred Pettit	

(A) American Game Policy

Need of Game Management

Game can be safely hunted only when the stock on each parcel of land is protected against overkilling and provided with cover, food, and some protection from natural enemies. These provisions constitute game management.

The present system of restrictive legislation cannot prevent overkilling without prohibition of all shooting and never provides cover or food, except by accident. Continual restocking of range not provided with protection, cover, and food is no remedy. Hence in the long run the present system holds out no hope of conserving game, unless it is supplemented by game management on a large scale.

Inducements for Landowners

Only the landholder can practice management efficiently, because he is the only person who resides on the land and has complete authority over it. All others are absentees. Absentees can provide the essentials: protection, cover, and food, but only with the landholder's co-operation, and at a higher cost.

With rare exceptions, the landholder is not yet practicing management. There are three ways to induce him to do so:

1. Buy him out, and become the landowner.
2. Compensate him directly or indirectly for producing a game crop and for the privilege of harvesting it.
3. Cede him the title to the game, so that he will own it and can buy and sell it just as he owns, buys, and sells his poultry.

The first way is feasible on cheap lands, but prohibitive elsewhere.
The second is feasible anywhere.
The third way is the English system, and incompatible with American tradition and thought. It is not considered in this report.

There Are No Other Alternatives

Even if the system still prevalent in most states were effective in producing a game crop, it is increasingly ineffective in maintaining free public hunting on farms, because as hunters increase, trespass becomes a nuisance, and posting follows. Closed seasons, posting, or both, are the inevitable result on farm lands.

The attempt to stave off posting by exchanging free public hunting for free public restocking is insufficient, because it gives the landowner no stake in the welfare of the game. The less the game thrives, the less will be the trespass nuisance he has to endure. Moreover, it is applicable only to species which can be restocked by artificial propagation or by buying the excess wild stock of other states or countries. The end of purchasable wild stock is in sight.

Moreover public restocking of private lands is prohibitive in cost. One license will usually plant just about one bird.

Kinds of Land and Classes of Game

Game land is of two kinds: (A) that which is cheap enough for the public to buy and manage, and (B) that which is too expensive for the public to buy in quantity, and which therefore must be managed by the present owners, or not at all.

Game is of four classes:
I. Farm game, which inhabits Class B land. It thrives best on farms with suitable cover.
II. Forest and range game, which inhabits Class A lands. It thrives best on land partially farmed.
III. Wilderness game, which inhabits very cheap Class A land. It is excluded by farming, or other economic uses.
IV. Migratory game which inhabits both classes of land. It thrives on farms if marshlands are left undrained.

Need of Facts, Skill, and Funds

Cover, food, and protection (i.e., management) do not increase game unless they are of the right kind. Game management may be unduly expensive unless skillfully dovetailed with the management of the primary crop.

To select the right kind of management and to apply it skillfully requires biological facts and men who can advise the landowner how to apply them. The facts must be discovered and the men trained. In short, game management must be recognized as a distinct profession and developed accordingly.

All these actions will require large additional funds, both public and private.

Need of Co-operation

The public, not the sportsman, owns the game.

The public is (and the sportsman ought to be) just as much interested in conserving non-game species, forests, fish, and other wild life as in conserving game.

In the long run lop-sided programs dealing with game only, songbirds only, forests only, or fish only, will fail because they cost too much, use up too much energy in friction, and lack sufficient volume of support.

No game program can command the good will or funds necessary to success, without harmonious co-operation between sportsmen and other conservationists.

To this end sportsmen must recognize conservation as one integral whole, of which game restoration is only a part. In predator-control and other activities where game management conflicts in part with other wild life, sportsmen must join with nature lovers in seeking and accepting the findings of impartial research.

Program

How can all the foregoing characteristics of the land, the game, the landowner, the sportsman, and the public be knit together into a feasible and effective program of game restoration?

A detailed program cannot be predicted far in advance. The Committee is convinced, however, that any program must begin with seven basic moves or actions. If these are adequately started, experience may be trusted to guide the more distant future.

The seven basic actions now needed are:

1. *Extend public ownership and management* of game lands just as far and as fast as land prices and available funds permit. Such extensions must often be for forestry, watershed, and recreation, as well as for game purposes.
2. *Recognize the landowner as the custodian of public game on all other land*, protect him from the irresponsible shooter, and compensate him for putting his land in productive condition. Compensate him either publicly or privately, with either cash, service, or protection, for the use of his land and for his labor, on condition that he preserves the game seed and otherwise safeguards the public interest. In short, make game management a partnership enterprise to which the landholder, the sportsman, and the public each contributes appropriate services, and from which each derives appropriate rewards.
3. *Experiment* to determine in each state the merits and demerits of various ways of bringing the three parties into productive relationship with each other. Encourage the adoption of *all ways which promise to result in game management*. Let the alternative ways

compete for the use of the land, subjecting them to public regulation if this becomes necessary.
4. *Train men* for skillful game administration, management, and fact-finding. Make game a profession like forestry, agriculture, and other forms of applied biology.
5. *Find facts* on what to do on the land to make game abundant.
6. *Recognize the non-shooting protectionist and the scientist* as sharing with sportsmen and landowners the responsibility for conservation of wild life as a whole. Insist on a joint conservation program, *jointly formulated and jointly financed*.
7. *Provide funds*. Insist on public funds from general taxation for all betterments serving wild life as a whole. Let the sportsmen pay for all betterments serving game alone. Seek private funds to help carry the cost of education and research.

It is imperative that these seven basic actions be no further delayed by debates among sportsmen as to which of the alternative forms of relationship with landowners should be adopted *to the exclusion of the others*, or by futile attempts to manage game without the landowner's co-operation, or to hunt it without his consent.

Relations with landowners must of course be adapted to local customs and conditions before they can be put into local operation. This is the task of local agencies, and it is a bigger and more important task than writing this policy.

Game Methods: The American Way [1931]

In this article published in *American Game* in 1931, Leopold tackles one of the most contentious issues underlying the American Game Policy, the policy's preference for the American tradition of public ownership of wildlife with compensation to the landowner for use of his land, in contrast with the European system of individual ownership. An influential group of American sportsmen had just organized the More Game Birds in America Foundation to propagate game through private ownership and captive breeding, thus posing a direct challenge to the system Leopold advocated. While the game policy resolved the issue in concise, diplomatic language, Leopold here presents a more extended, personal, and artful defense of his position.

The game policy adopted by the 1930 American Game Conference begins with this assertion:

> With rare exceptions, the landholder is not yet practicing management. There are three ways to induce him to do so:
> 1. Buy him out, and become the landowner.
> 2. Compensate him directly or indirectly for producing a game crop and for the privilege of harvesting it.
> 3. Cede him the title to the game, so that he will own it and can buy and sell it just as he owns, buys, and sells his poultry.
>
> The first way is feasible on cheap lands, but prohibitive elsewhere.
> The second is feasible anywhere.
> *The third is the English system, and incompatible with American tradition and thought. It is not considered in this report.*

The *Game Breeder*, in a recent editorial asks the committee which drafted the policy to explain what it means by "incompatible with American tradition."

Captain Percy R. Creed, in his "Rambling Thoughts of a Perverted Britisher," also takes humorous exception to some recent discussion of "the abuses of the European Game System."

This paper is to express a personal view of what the policy means in its references to the European practices. In no degree does it commit the other members of the committee, or any person or group.

No comparison of American and European practices is worth while unless it first contrasts the biological and economic circumstances obtaining in the two continents. Just how do economics and biology condition the problem of game conservation in America and Europe?

Our definition must begin with present and prospective human population densities. Europe has many people but little land. America has much land but comparatively few people. *To supply any given proportion of the population with any given amount of game, Europe must raise a denser stand of game per acre, and hence practice a more intensive form of game management, than America.* This is the first theorem which conditions our problem.

To illustrate: The Scottish moors support about one grouse per three acres on the average, and one per acre as a maximum. This is a very dense stand, obtainable only through intensive management. The Wisconsin sand plains support about one grouse per 40 acres. This is a very thin stand, occurring "naturally" without any management at all. Section (b) of the chart illustrates the contrast.

A crude or extensive system of game management would raise the Wisconsin grouse density to (let us say) one per eight acres, or five times the present stand. On the other hand, a complete or intensive system of game management would doubtless raise the Wisconsin grouse density to that of Scotland, or 20–40 times the present stand.

As nearly as we now know, disease would frustrate any attempt to raise the density higher than this in either place.

We are working, therefore, between an upper and a lower limit set by biological nature and economic accident. Art cannot raise the upper limit. Delay can depress the lower and exterminate the species. The two limits constitute the upper and lower edge of our game policy "slate." The two limits are far apart. Between them lie a wide range of choices.

The denser the stand, the larger the proportion of it which may be safely killed. In fact, in Scotch grouse stands nearing the upper limit of density, it is imperative to kill two-thirds. If our present stand permits of a forty-fold increase, our present kill could be raised much more than forty-fold. Section (c) of the graph would indicate 160-fold.

158   Game Methods

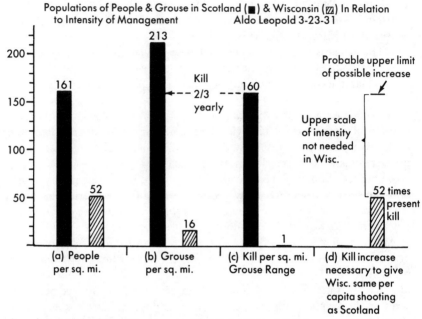

A rendering of the figure that accompanied "Game Methods" and appeared in *American Game*, March–April, 1931.

The second theorem is aesthetic and therefore debatable. It is this: *The recreational value of a head of game is inverse to the artificiality of its origin, and hence in a broad way to the intensiveness of the system of game management which produced it.* I may be presumptuous in calling this a theorem. It is my personal feeling as a sportsman. That does not prove it is true. I suspect, though, that many feel as I do. Possibly more people in America feel that way than in Europe. I have never been to Europe and hence do not know. If so, maybe this is where "Americanism" begins to enter our definition. Even if it be so, I do not imply any "superiority" for my idea. Wild nature is only one generation behind us here—in Europe centuries have elapsed. Aesthetic standards of living are, I suppose, more or less delimited in their range by the actualities of experience.

I will not try to illustrate this theorem, because it does not lie within the field of logic. It is either true or not true, according to one's personal views. I have often wondered whether the moral fervor of the "one gallus" element in this country may not be explained as a blind groping to reconcile this feeling, of which they are sure, even though they have not formulated it in words, with the practical limitations of the American game problem of which they are not sure, because these limitations have only recently been isolated and defined.

The third theorem follows from the first two: *A game policy should seek a happy medium* between the evident necessity of *some* management, and the aesthetic desideratum of *not too much*. We would be foolish not to take advantage of a relatively low human population density. On the assumption that about one-third of both Wisconsin and Scotland constitute grouse range, and that on this range the Scotch kill 160 grouse per square mile (see Section c) could, *with intensive management*, be duplicated, to what degree would it *need to be duplicated* to furnish Wisconsin citizens with the same per capita shooting as Scotland? Section (c) indicates a fifty-fold increase in kill would suffice. This, as best I can see, would necessitate about a six- or ten-fold increase in the stand. (It should be remembered that our present stand is so low as to support virtually no kill, as evidenced by most years being closed.)

Our total kill could doubtless be increased 160-fold before reaching the biological upper limit. In short, we need to use only one-third the possible scale of intensity of management (see Section d). The third theorem would indicate this is better aesthetics than to blindly copy the Scotch stand through complete control of all factors. Again, absolutely no "superiority" is here implied. If we had as many people per square mile as Britain, we too would be forced to practice intensive methods, or do without our shooting. (If there be any question of "superiority" involved at all, it is whether we will prove capable of regulating our own future human population density by some qualitative standard, or whether, like the grouse, we will automatically fill up the large biological niche which Columbus found for us, and which Mr. Edison and Mr. Ford, through "management" of our human environment, are constantly making larger. I fear we will. The boosters fear we will not, or else they fear there will be some needless delay about it.)

Some, but not too much, management is good aesthetics, but what may interest some readers more is that it appears to be also good business. Game management is a form of agriculture, and like the other forms it presumably obeys the Malthusian law of diminishing returns. All other land crops do. That is to say, a dollar or an hour spent in quintupling the present accidental stand of grouse would be expected to go farther than a dollar spent to quintuple it again. Why? Because we are then approaching the biological upper limit of density, where it takes more work to add a bird than lower down on the bio-economic "slate." I say "presumably" because I have not the figures to prove it. We have had so little management in this country that comparable figures do not exist, and European figures are hard to convert. *These cheaper costs for the lower scale of game populations constitute the fourth theorem.* Whether this would also be true of yields or kill may be doubtful, because yield increases more rapidly than density.

This fourth theorem has an important corollary which is true within certain limits: a dollar spread over a large area will raise more game than the same dollar spread over a small area. This is overlooked by those optimists who would shortcut the farmer because "the state can raise game for all" on small public shooting grounds, leased or owned for the purpose. It is also overlooked by those game farmers whose operations have been confined to private estates which did not care about costs, or whose owners want lots of shooting on their own small area, regardless of cost.

The fifth theorem is stated in the game policy: *Only the landholder can practice game management cheaply.* This is because it normally consists of many small jobs scattered through the whole gamut of the seasons. The farmer or forester can perform these jobs "on the side," often without any separate cash cost.

So much for the bio-economic laws that limit the slate on which our picture is to be drawn. It is always easier to cite laws than to identify the conditions to which they apply. What next?

The present game policy sketches the picture, at least in some of its broader present outlines. It proposes low-intensity management on all lands (rather than high intensity management on a few lands) in order to take advantage of the lower costs and lesser artificiality thus obtainable. It proposes that the public be its own manager wherever it can own the land, but admits the necessity of working through the present owner where it cannot. The signers of the policy do not deceive themselves by believing that it can map an exact route to these objectives, or that it can be fully applied over the whole country. They do believe that this, or some other policy, can be applied to the extent that the public understands and supports it, and that the details of the route must be worked out a step at a time by some "try and see" procedure.

Those who would have us adopt the European idea scoff at the notion, prevalent in our one-gallus circles, that it would be undemocratic. "Produce the game as Europeans do," they say, "and there will be shooting for all; certainly there will be a lot more than there is now." That this rejoinder is mainly true must be admitted without cavil or qualification. It carries one implication, however, which needs to be brought out in the open for a clearer view. All European countries, as nearly as I can find out, *market the meat* as well as the shooting privilege. When the European sportsman leases (or builds) his shooting ground, he intends to recover a part of his expense by selling the meat which he does not need for his own use. *Consequently he leases more land, spends more money, and shoots a larger bag, than he needs for his own use.* Consequently there is less land, and less game, left for the other fellow of lighter purse. In short, open markets, in Europe, make for a *narrow*, rather than a wide, distribution of the shooting privilege. It seems pretty clear, likewise, that they make for *intensive* rather than extensive management. There seems no reason to believe that open markets would make for anything different here.

Many defenders of the one-gallus interest condemn these large bags on ethical grounds. They do not realize that large bags are a necessary cog in European game economics. They also forget that in spite of the large bags, European nonmigratory game is not overkilled, as it is here. It seems to me footless to argue which system is the more ethical. They are simply *different*, due to the accident of historical evolution. Which do we want? The American idea was born on the homestead, and is based on the basic idea of a "mess of game for the family." I suppose the European system harks back to the idea of a mess of game for the feudal community, to secure which the lord or baron did all the hunting for his whole demesne. Regardless of ethics, our system is an indigenous growth, or as Doyle says, "an American folkway, not to be discarded with impunity." I prefer it for this reason. There is nothing to prevent us from adopting the European technique for *producing* a game crop, and at the same time rejecting the European customs governing the intensity of the operation and the European system for its *harvesting and distribution*. The game policy, by and large, proposes just this. To this extent, and no farther, does it "advocate the European system," and it does this only for farm lands too expensive for public ownership and operation.

The most radical prophet of the one-gallus camp has never publicly urged so huge a program of socialized public recreation.

One thorn has festered in the heart of European naturalists and nature-lovers for years: the ruthless suppression of predators which goes with game management in most European countries. W. T. Hudson has voiced his protest over the disappearance of one predatory species after another, and his resulting contempt for the aesthetic horizon of sportsmen and sportsmanship. He who runs may read that American protectionists mortally hate and fear the impending (?) American counterpart of this sacrifice. Are their fears justified?

I am no prophet. I would point out, however, that stringent predator control is usually unnecessary save in the upper scale of intensive game management. If the third theorem is true, then we do not need that kind of management.

When I say that stringent predator control is unnecessary for the lower scale of management, I am voicing, not a personal opinion or a personal hope, but what seems to be the visible trend of biological research. One after another, the analysts of the life-equation of game birds are finding that food, cover, disease, or some other environmental factor needs attention first, and predators afterward. A dollar spent on these other factors will go farthest at the outset. This is not to say that no predator control is needed. It does mean that extensive or low-grade management—enough, let us say, to quintuple our crop—can best be achieved by light, local, seasonal and *selective* handling of the predator-factor. In game mammals this trend is not so clear as in birds, and it is clearer in the northern than in the southern states.

Furthermore it should not be overlooked that European predator-policy is empirical, not scientific. Its standards were set before biological science was born. Our standards can be better. This is not Europe's fault, but our own good luck, in that our biology preceded our management.

Is it too much to hope, then, that the group-cooperative wild life enterprise advocated by the game policy may ultimately evolve an *American* attitude toward predators, based on the new biology, and recognizing the nature-lover and farmer, as well as the sportsman, as joint partners? Utopian? Perhaps, but what is the other alternative? The loss of an income-producing by-product for the farmer, the spiritual impoverishment of the farm as the place to raise a boy, the end of hunting as a sport, and the loss of millions of additional songbirds, which would, like game, increase to fill the coverts and eat the food created by game management. These, or the European system with another Hudson. The aspirants for this job are already whetting their pens. The outcome in this country will depend on whether all parties spend the next ten years thinking and learning and trying things *on land*, or whether they spend it bandying phrases. The whole thing is relatively simple and real on the land. It is abstruse or fearsome mainly on paper. The basic trouble is that the great majority of both our sportsmen and protectionists are people who push a pen, not a plow. To him who reads fencerows, thickets, and fields as the letters of a great and tragic history, the direction to go is clear.

Ten years ago I would have said that there is still another distinctively American idea in game management. The voluntary limitation of too-effective equipments and too-destructive practices seemed to be making greater headway here than in Europe. We made great parade of our "code" concerning closed markets, small gauges, non-potting, "rather-watch-the-dog-than-shoot," et cetera, and we cast in a snoofy remark about European driven game, to clinch the proof of our superior ethics. It would take a braver man than I to press this point today, in the face of our pell-mell mechanization of equipments, our wholesale organization of lures and baits, and the wellnigh universal attitude that the bag limit is the minimum proof of prowess, rather than the maximum limit of respectability (for which it was originally intended). On this matter I can only express the personal hope that this has been a temporary stampede. In American camping, in fishing on both sides of the water, and in the new American sport of hunting with a bow and arrow, the idea of voluntarily making sport difficult still lives, even though it cannot be said to prevail.

Perhaps it can with time win back its moral trenches in the field of sport with gun and dog.

Game management in both America and Europe has one obligation in common: it must be the art which conceals art. Our chance of concealment is the better, because we have lots of land, and no need of resorting to intensive forms. There are of course those who scoff at any degree of management whatsoever. To them I would only say: "Look to the beam in your own eye." Most of our atavistic instincts, including hunting, find their exercise only through the frank acceptance of illusion. It isn't really necessary to see the lady home—in most communities she is quite safe anyhow. To keep a dog to guard the "castle" expresses our love for dogs, not our solicitude for the family. To kill a mess of game "by strength of hound" or quickness of trigger, and bring it home to the family, is just about as necessary to most grown Americans as for their very young sons to go fishing in the family washtub. And that, in my opinion, is very necessary indeed.

Game and Wild Life Conservation [1932]

The controversy over European versus American systems of game management was still simmering when a writer in the pages of the *Condor*, an ornithological journal that appealed mainly to preservationists, lumped Leopold together with his principal opponents, the advocates of artificial propagation, and castigated both for their "pernicious" emphasis on shooting rather than preserving the wildlife resource. Leopold, who was himself an ornithologist who had frequently contributed to the *Condor* and whose sympathies were as much preservationist as commodity-oriented, was deeply aggrieved by the attack and responded in a pointed defense of what he considered his more-realistic approach to conservation. This article in the *Condor* thus provides not only a revealing glimpse of the tensions within the conservation fraternity but also Leopold's most focused statement of the position he occupied along the spectrum of conservation concerns.

This is a reply to Mr. T. T. McCabe's well written and persuasive *exposé* of two recent manifestations of the sportsman's movement: my *Game Survey of the North Central States*, and the several publications issued by More Game Birds in America. Both are, I take it, inclusively condemned as "a framework of pernicious doctrines, too often speciously glossed over."

Mr. McCabe's attitude raises what seems to me a fundamental issue. I hope that it may provoke some badly needed cerebration among both protectionists and sportsmen, and especially among those intergrades like myself, who share the aspirations of both.

There are many sportsmen who laugh at any attempt to embody the protectionist point-of-view in any game program. "Whatever you do the protectionists will be against it." Mr. McCabe's paper furnishes scant comfort to those of us who have been holding out against this attitude, because we see in it the indefinite continuation of the present deadlock, from which the sharpest pens gain much glory, but the game gains nothing except a further chance to disappear.

More Game Birds, on the one hand, and the *Game Survey* (as further developed in the "American Game Policy") on the other, represent the opposite wings of the sportsman's camp. From their very inception they agreed to disagree on the very issues with respect to which Mr. McCabe presumably finds them both "pernicious," namely: predator control, exotics, degree of commercialization, and artificial propagation. This divergence, great enough to seem fundamental to two groups of hardened sportsmen, would, I had hoped, be perceptible to readers of the *Condor*.

I do not imply that Mr. McCabe should agree with either More Game Birds or myself on these moot questions. I ask, though, whether it is good for conservation for him to dismiss both, with one breath, as equally subversive of what he considers sound policy. (I think this is not too strong a statement, since Mr. McCabe says "these proposals are an offer . . . to the nation, for its game birds," to which he would reply, "Not for sale.")

Of course, no disagreement is ever as simple as it looks on paper. A partial explanation of this one lies, I think, in the fact that Mr. McCabe's game policy, whether he realizes it or not, consists of a system of *personal wishes* which might be realized if America consisted of 120 million ornithologists, whereas mine is a system of *proposed public actions* designed to fit the unpleasant fact that America consists largely of business men, farmers, and Rotarians, busily playing the national game of economic expansion. Most of them admit that birds, trees, and flowers are nice to have around, but few of them would admit that the present "depression" in waterfowl is more important than the one in banks, or that the status of the blue goose has more bearing on the cultural future of America than the price of U.S. Steel.

Now if Mr. McCabe and I had the courage to challenge this universal priority for things material and things economic, we might consistently hoist the banner "Not For Sale" and die heroically under the heels of the mob. But have we not already compromised ourselves? I realize that every time I turn on an electric light, or ride on a Pullman, or pocket the unearned increment on a stock, or a bond, or a piece of real estate, I am "selling out" to the enemies of conservation. When I submit these thoughts to a printing press, I am helping cut down the woods. When I pour cream in my coffee, I am helping to drain a marsh for cows to graze, and to exterminate the birds of Brazil. When I go birding or hunting in my Ford, I am devastating an oil field, and re-electing an imperialist to get me rubber. Nay more: when I father more than two children I am creating an insatiable need for more printing presses, more cows, more coffee, more oil, and more rubber, to supply which more birds, more trees, and more flowers will either be killed, or what is just as destructive, evicted from their several environments.

What to do? I see only two courses open to the likes of us. One is to go live on locusts in the wilderness, if there is any wilderness left. The other is surreptitiously to set up within the economic Juggernaut certain new cogs and wheels whereby the residual love of nature, inherent even in Rotarians, may be made to recreate at least a fraction of those values which their love of "progress" is destroying. A briefer way to put it is: if we want Mr. Babbitt to rebuild outdoor America, we must let him use the same tools wherewith he destroyed it. He knows no other.

I by no means imply that Mr. McCabe should agree with this view. I do imply that to accept the economic order which is destroying wild life disqualifies us from rejecting any and all economic tools for its restoration, on the grounds that such tools are impure and unholy.

With what other than economic tools, for instance, can we cope with progressive eviction of game (and most other wild life) from our rich agricultural lands by clean farming and drainage? Does anyone still believe that restrictive game laws alone will halt the wave of destruction which sweeps majestically across the continent, regardless of closed seasons, paper refuges, bird-books-for-school-children, game farms, Izaak Walton Leagues, Audubon Societies, or the other feeble palliatives which we protectionists and sportsmen, jointly or separately, have so far erected as barriers in its path? Does Mr. McCabe know a way to induce the average farmer to leave the birds some food and cover without paying him for it? To raise the fund for such payment without in some way taxing sportsmen?

I have tried to build a mechanism whereby the sportsmen and the Ammunition Industry could contribute financially to the solution of this problem, without dictating the answer themselves. The mechanism consists of a series of game fellowships, set up in the agricultural colleges, to examine the question of whether slick-and-clean agriculture is really economic, and if not, to advise farmers how they can, by leaving a little cover and food, raise a game crop, and market the surplus by sale of shooting privileges to sportsmen. This mechanism is, I take it, specious. Have the protectionists a better one to offer?

Another mechanism which I have tried to build is the committee of sportsmen and protectionists charged with setting forth a new wild life policy. Has Mr. McCabe read it?

These things I have done, and I make no apology for them. Even if they should ultimately succeed, they will not restore the good old days of free hunting of wholly natural wild life (which I loved as well as Mr. McCabe), but they may restore something. That something will be more native to America, and available on more democratic terms, than More Game Birds pheasants, even though it be less so than Mr. McCabe's dreams of days gone by.

Let me admit that my cogs and wheels are designed to perpetuate wild life to shoot, as well as wild life to look at. This is because I believe that hunting takes rank with agriculture and nature study as one of three fundamentally valuable human contacts with the soil. Secondly, because hunting revenue offers the only available "coin of the realm" for buying from Mr. Babbitt the environmental modifications necessary to offset the inroads of industry.

I admit the possibility that I am wrong about hunting. The total cessation of it would certainly conserve some forms of wild life in some places. Any ecologist must, however, admit that the resulting distribution and assortment of species would be very irregular and arbitrary, and quite unrelated to human needs. The richest lands would be totally devoid of game because of the lack of cover, and the poorer lands nearly so because of the lack of food. The intermediate zones might have a great deal of game. Each species would shrink to those localities where economic accident offered the requisite assortment of environmental requirements. That same condition—namely the fortuitous (as distinguished from purposeful) make-up of wild life environments—shares, with overshooting, the credit for our present deplorable situation.

The protectionists will, at this point, remind me of the possibilities of inviolate sanctuaries, publicly owned, in which habitable environments are perpetuated at public expense. Let us by all means have as many as possible. But will Mr. Babbitt vote the necessary funds for the huge expansion in sanctuaries which we need? He hasn't so far. It is "blood money" which has bought a large part of what we have. Moreover, sanctuaries propose to salvage only a few samples of wild life. I, for one, demand more. I demand of Mr. Babbitt that game and wild life be one of the normal products of every farm, and the enjoyment of it a part of the normal environment of every boy, whether he live next door to a public sanctuary or elsewhere.

Mr. McCabe taxes me with omitting any mention of game production on public lands, where the one-gallus hunter will have free access to it. I can only infer that he has not read the American Game Policy. Has any group ever proposed a larger public land program, and called for more wild life production thereon? The Policy admits, to be sure, the unpleasant fact that lands must be cheap in order to be public. It advocates the paid-hunting system only for those lands too expensive for the public to own.

Finally Mr. McCabe taxes me with too much interest in exotics. Modesty forbids me to refute this charge in detail. I have persuaded two states to go out of the pheasant business, and several others to limit it to half their area. I devised the "glaciation hypothesis" which seems to exclude pheasants from about a third of the United States. On the other hand, I have recommended the continuation of pheasants and Hungarians in certain regions where economic changes have so radically altered the environment as to make the restoration of native game prohibitive in cost. Just what native species would Mr. McCabe recommend for east-central Wisconsin, or for northern Iowa, or for farm land in Massachusetts?

Let it by no chance be inferred that because I speak as a sportsman I defend the whole history of the sportman's movement. Hindsight shows that history contains any number of blunders, much bad ecology, and not a few actions which must be construed as either stubbornness or hypocrisy. For every one of these, one could point out a counterpart in the history of the protectionists, only there has been no "Emergency Committee" with either the means or the desire to compile and advertise them. Fifteen years ago, for instance, the protectionists closed the prairie chicken in Iowa, and then sat calmly by while plow and cow pushed the species almost to the brink of oblivion. Was this a blunder? Yes—but what of it? Is there any human aspiration which ever scored a victory without losing to some extent its capacity for self-criticism? The worthiness of any cause is not measured by its clean record, but by its readiness to see the blots when they are pointed out, and to change its mind. Is there not some way in which our two factions can point out each other's sophistries and blunders without losing sight of our common love for what Mr. Babbitt is trampling under foot? Must the past mistakes of each group automatically condemn every future effort of either to correct them?

To me, the most hopeful sign in the sportsman's movement is that several little groups have publicly avowed that the old program is a failure. Each is struggling to devise a new formula. I am conceited enough to believe that the formula my little group is trying to put together comes as near meeting the ugly realities of economics on the one hand, and the ideals of the protectionists on the other, as any yet devised. Mr. McCabe's paper will neither help nor hinder its future acceptance or rejection among sportsmen, but it may hinder its thoughtful consideration by the protectionist camp, and thus prevent what I had devoutly hoped for: their active participation in its development, modification and growth.

Lest this be construed as an idle boast, let me point out that as chairman of the Game Policy Committee, I asked the A.O.U. to appoint a representative to sit on or with the Committee, and to pull the reins whenever the Committee got into proposals subversive of the protectionists' point of view. He has not yet pulled. I hereby invite Mr. McCabe to sit with him.

In short, I beg for a little selectivity in weighing the new departures proposed by the other fellow. I also pray for the day when some little group of protectionists will publicly avow that their old formula of restriction is not the whole Alpha-to-Omega of conservation. With both sides in doubt as to the infallibility of their own past dogmas, we might actually hang together long enough to save some wild life. At present, we are getting good and ready to hang separately.

Grand-Opera Game [1932]

Leopold had his personal preferences in game species just as in management techniques. Here, in a popular essay submitted to the *Sportsman*, a magazine that he noted had not run any previous articles dealing with management of the sort proposed in the American Game Policy, Leopold reveals his preference for wild quail over pen-raised pheasants. In rejecting the piece, the editor gave as his reason that "it is impossible even for you, with your extensive knowledge and experience, to deal with such an intricate and diversified subject as game management within the compass at our disposal." One suspects instead a marked difference in predilections.

It has been aptly said that pheasant shooting is a good show, but quail and prairie chicken are grand opera.

This implies that there is a big spread in the price of tickets. Sometimes there is, but with the right management on the right land, and enough of it, a quail or chicken crop may be quite as inexpensive as one of lowly pheasants.

Our thinking on these questions has been muddled by the automatic assumption that to produce game we must buy stock and confine it in chicken-wire slums for breeding and subsequent release. Pheasants tolerate slums, but quail barely, and the lordly chicken not at all.

We Americans are just awakening to the fact that there is another and intrinsically better way, long practiced on the Scottish moors and English manors. That way is to so modify the range that the game produces itself. Herbert L. Stoddard, the first American to practice it successfully on a large scale, calls it "game management."

Game management lubricates the engine we call "Nature," rather than building a substitute engine in the form of a propagating plant. The motive power is that natural force implied in the biblical injunction, "Go forth and replenish the earth," and which the professors define impersonally as "the tendency of any species to increase to the capacity of its environment."

The game manager simply enlarges the capacity of the environment by improving cover and food, and by protecting the game against natural enemies and overshooting. The increase follows.

Artificial rearing is, of course, one form of game management, and a very useful one, but it does not produce "grand opera."

Of the two alternative forms of production, we mechanized Americans usually choose artificial rearing, probably because the propagating plant, like our factories, is composed of wire, troughs, bins, hoppers, and other things we can touch and see, whereas a piece of land improved for game is composed of—well, just land. Somehow, the game-manufacturing plant seems the more likely to work. Our faith in factories, however, is somewhat sobered of late. Possibly we are now ready to appreciate that an improved environment is the more conservative of the two alternative investments, provided (as in all investments) the enterprise be skillfully directed.

Directing the "wild management" of game is an enterprise in botanical and zoological engineering. To produce cheaply, the game manager, like other engineers, must take advantage of what is already on the ground, and build only that which is necessary. Here, for instance, is a typical midwestern country estate of less than a hundred acres, located on some river bluff or lakeshore. The estate carries a covey or two of quail, but the owner desires half a dozen coveys. How does he go about getting them?

The property, like as not, consists of wooded and ungrazed bluffs abutting on grazed or cultivated upland farms. In such event the existing covey in winter is almost invariably located on the boundary between the woods and fields, as at "1." Why? Because this is the only spot now offering feeding, hiding, and sleeping quarters all within a short radius. The covey feeds in the corn at 1a, roosts in the grassy draw at 1b, and on cold days finds a safe solarium in the warm, brushy, south slope at 1c.

Let me emphasize again that these three daily requirements—board, room, and club—must all lie within a short radius. Bobwhite, let it be remembered, does not drive a car, and though sound of wind and limb (as your setter knows, and mayhap his owner also), his legs are short.

This "raison d'être" for covey No. 1 is also our ways and means for bringing Coveys 2, 3, 4, 5, and 6 into existence. Up in the hayfield, for example, is another grassy, brushy draw, but no corn to go with it. Plant some at 2a. The new covey will be there next winter, because you have expanded the "capacity of environment" by creating one new covey range.

At "3" is some grazed timber offering neither cover nor food. Fence the south slope of the draw, let it grow up to haws, hazel, and sumac, and then plant a little patch of grain.

At 4 and 5 cover abounds, but food is absent. Supply it, and the immutable laws of biology will supply the quail.

At 6 is food, but no cover. Plant it at 6a, with the advice of your

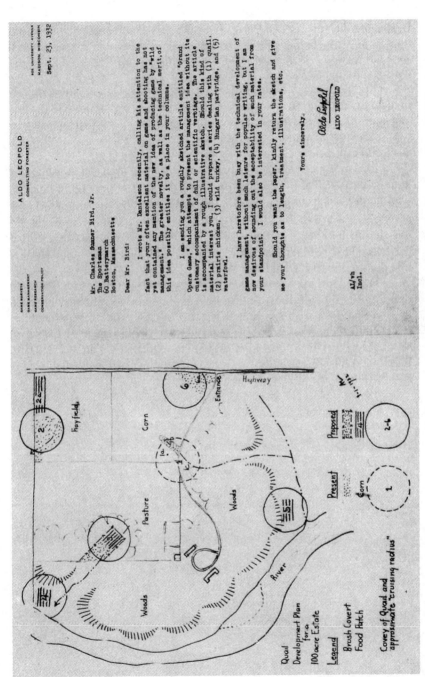

Leopold's cover letter and illustrative sketch for "Grand Opera Game," submitted to but rejected by *The Sportsman*. (Leopold Collection X25 2222, UW Archives)

landscaper, for this is "The Entrance." Some day landscapers will learn not only to greet the home-coming eye with pretty bushes, but bushes of such sort and quantity as to greet the home-coming ear with that even prettier whistle: "Bobwhite."

This, then is the technic of the biological engineer. His engine is the law of increase, his tools the axe, plow, fence, fieldglass, and, if desired, the dog and gun. What more delightful avocation than to take a piece of land and, by cautious experimentation, to prove how it works? What more substantial service to conservation than to practice it on one's own land?

Like any other engineer, it behooves the game manager to know the limitations as well as the possibilities of his materials. One cannot build up wild quail beyond an average of a bird per acre. Why? I said before that grand-opera game doesn't tolerate slums, but that is merely answering one question by another. The real answer is that nobody knows. We scientists partially conceal our ignorance by labelling the bird-per-acre limit a "saturation point," and let it go at that.

Having built up a good stand of birds, the owner may shoot, but not over a third or a half. In its results, light shooting is indistinguishable from no shooting. Next year's crop is determined by the number of covey locations, and will be what it will be, regardless of whether the owner elects to shoot a few birds in season, or prefers to pin a "sanctuary" sign on every tree. But, shot or unshot, the whir of a rising covey in one's own field or covert is a goodly thing to hear. Quail are grand-opera game.

The Virgin Southwest [1933]

In 1933, while on a temporary assignment supervising erosion control work in the national forests of Arizona and New Mexico, Leopold substantially revised a manuscript he had written six years earlier about the erosion problem in the Southwest, apparently for presentation at the Laboratory of Anthropology in Santa Fe. Originally intended as a chapter in a book on southwestern game that Leopold never completed, it had been titled "The Virgin Southwest and What the White Man Has Done to It." The essay reveals Leopold's skill at reading both the landscape and historical descriptions of it in order to discover processes of environmental change. The typescript, heavily edited in Leopold's hand, was reviewed by the editor of a southwestern periodical but remained unpublished. Leopold's reference to the journal of Alexander Ohio Pattie is probably a slip—it should be the "personal narrative" of James Ohio Pattie.

The major premise of civilization is that the attainments of one generation shall be available to the next.

Some social sciences, by their very nature, cast doubt upon the validity of this premise. Archeology, for example, describes an endless caravan of societies, now dead, for whom it did not hold true. On the other hand the art of government is concerned with bolstering such doubts, not wholly, I take it, out of altruistic regard for the unborn, but rather out of an imperative need for confidence among the living. The changing "tempo" of the generations, so convincingly described by Ortega in *The Revolt of the Masses*, consists, perhaps, of fluctuations in their social confidence. Be that as it may, any matter which challenges the validity of the major premise is, ipso facto, a matter of concern to all thoughtful men.

It is only recently that the biological sciences have had new occasion to challenge it, in their discovery of an abnormal erosion-rate in some of our best soils. The rate is rapid elsewhere, but in the Southwest and the adjoining semi-arid regions it is nothing short of alarming. At the mouth of one Utah canyon, for example, erosive deposits display seasonal color-layers, from which a chronology similar to that of tree-rings has been built. It shows more movement of soil since the introduction of livestock to the watershed fifty years ago than had previously occurred since the recession of the glacial epoch.

"Discovery" is a slow process. It is almost a generation since certain ecologists, range-managers, foresters, and engineers saw and described the present southwestern situation, but it is only a year or two since the social consequences of its continuance were given credence by the lay public. Even statesmen now show signs of being aware that the best soils are slipping, sliding, toward the sea, and that the basic cause of this abnormal movement is the devegetation of the range through overgrazing by domestic livestock.

I think, though, that the thoughtful citizen still entertains a mental reservation—he regards this thing as important, *if true*. This is only natural, since he is unable to weigh personally the technical evidence; he must take the ecologist's word for it. The fact of abnormal erosion, however, can be established on historical as well as ecological evidence. This paper aims to present such evidence as gathered from a single document: the journal of Alexander Ohio Pattie, who trapped beaver in this region almost a generation before the Santa Fe Trail opened it to wholesale economic exploitation. Certain other material of miscellaneous origin, and certain personal observations, are interjected to give relief to the Pattie narrative.

Pattie was a young Missourian of the Boone and Kenton tradition, with an eye for game, grass, and timber. He travelled down the Rio Grande and the Gila, trapping beavers, in 1824. In 1825 (?) he came back up the Grand Canyon of the Colorado.

On the journey from Santa Fe to San Felipe, Pattie speaks of "a handsome plain, covered with herds of domestic animals." Continuing down the Upper Rio Grande Valley to Socorro, he "traversed the same beautiful plain country," on which grazed "the same multitude of domestic animals." There must have been heavy grass, not only along the river but *on the mesas* adjoining the valley floor. There is little grass on these adjoining mesas today; many of them have become bare sand dunes.

Pattie remarks that the valley floor was not cultivated except at San Felipe and above Socorro. At these points the valley is narrow and the river has a steep gradient. They would be the easiest places to divert irrigation water from an *unsilted river channel*, because the flatter the gradient of a stream, the more work is required to build intake ditches up its banks.

While the main valley was used for grazing, the farming, according to Pattie, was mostly conducted in side streams like the Puerco. Today the reverse is true. The main valley is all farmed, except where seepage (due to silting) makes it too wet, while the side streams are fit only for grazing, because erosion has gutted all the irrigable land.

Couzens says that in 1859 (?) the channel of the Puerco was only 12 to 15 feet deep where it crosses the road from Isleta to Acoma and Zuni. Abert says that in 1846 it was only 10 or 12 feet deep at a point a few miles higher up. Today, at these same spots, the channel of the Puerco is a miniature Grand Canyon carved in clay. I recollect it as over a hundred yards wide and thirty feet deep.

Pattie remarks that at Socorro the valley was thinly timbered, but covered with willow and cottonwood brush in which "great numbers of bear, deer, and turkey" found refuge. One infers there was little large timber anywhere along the upper river. Today the ancient cottonwoods that line its irrigation ditches are its principal ornament. Most of these cottonwoods *are rooted in the ridges of silt* that have resulted from the annual cleaning of the ditch channels; in fact the older ditches *have raised themselves from five to ten feet above the valley floor* by gradual siltage.

What do these seemingly disjointed facts tell us about the virgin Southwest?

They tell us that in Pattie's day the Rio Grande drained a stable watershed, devoid of abnormal erosion. Even the sand dunes adjoining the river carried a heavy growth of grass. By reason of this grass, prairie fires swept across the valley and kept it devoid of large timber. The river channel, now so filled with silt that it is actually higher than the valley floor, was then so far below the floor that irrigation was difficult, except at the points where steep gradients facilitated the building of intakes. In short we now have scant grass, much erosion, and a river so choked with silt that it bogs its own bottoms with seepage and poisons their fertility with alkali. In Pattie's day there was grass everywhere, little erosion, a normal river, and bottomlands of sweet well-drained soil.

Pattie's testimony is really superfluous; there is hardly an acre that does not tell its own story to those who understand the speech of hills and rivers. The Galisteo which winds across Pattie's "handsome plain" has since been lived upon. We see the skeletons of ancient fruit trees, toppling one by one into the parched arroyo, which year by year gnaws away the loam of what was once a farm.

That farm was irrigated once—one can trace the old ditches winding across the remnants of bottomland. If irrigated, there must have been a stream. There is no stream now, only a trickle in the sand.

The stream banks must have been shallow and gentle, else the water could not have been led upon the land. They are not shallow now. The channel is a flood-torn chasm.

If there were ditches, there must have been wide stretches of level, friendly soil to irrigate. That soil has been dumped as silt into the main river; one farm washed away to curse another in the making, somewhere below.

Pattie's handsome plain is still green, at times, but it is the kind of green which could deceive only a tourist. It is not the greenness of grass, it is the greenness of tumbleweeds and snakeweed and pinque—worthless substitutes which a denuded nature has invoked to cover her nakedness. On this same Galisteo, Doniphan, in 1846, found "grass and water abundant and of good quality."

It might be argued by some that the farming which once occupied valleys like the Galisteo started the erosion which has since destroyed them. That this is not true is attested by valleys which were farmed before Coronado came, but did not erode until livestock was introduced. The San Jose is a case in point.

Everybody observes the lava cliffs which line its valley, black from centuries of oxidation in the sun. At their base runs a horizontal band of gray and red. It is like the tell-tale whiteness of a school-boy's face which shows, by contrast, where he has been washed. These cliffs have been washed—by erosion. The gray and red band is the measure of the soil which floods have torn away from the as yet unblackened rock—more farms gutted out to clog rivers and fill reservoirs, somewhere below.

Nothing has changed in the watershed of the San Jose except *grass*. Coronado and those who came after him brought sheep and goats and cattle to the Indians, and the subsequent over-grazing of the whole watershed is what upset its equilibrium. Throughout the Southwest the worst erosion is in the regions of the oldest settlements, because it is there that over-grazing has been most severe through the longest time.

On the north flank of Mt. Taylor is dramatic proof of the when and why of erosion. Winding up and down across the treeless foothills, threading in and out of crest and hollow, is an old earth-scar, overgrown with chemise and snakeweed, but nevertheless clearly traceable for miles, like the track of some great serpent. It is the Santa Fe Trail.

Wherever it crosses a hump the old roadbed is worn two and three foot deep into the soil. The grades followed, and the manner of winding around rocks and obstacles, are exactly the same as one would select in driving a heavy wagon over the same route today, with this significant exception: at the bottom of every hollow is now a steep-banked arroyo which no wagon could possibly cross. There are more of these arroyos than an army of engineers could bridge in months. There can be no two ways to read the history written on these hills. The story is as plain as a street blocked by many deep ditches. If you know the street had been used in 1849, you would know the ditches had been dug since.

How long since? In the bottoms of the arroyos which cross the old trail are chemise bushes and sometimes scraggly junipers, sprung from the juniper forests higher up on the mountain. Cut off the oldest and count back the annual rings of wood. They too will tell you 1850. Overgrazing started these arroyos, and in all probability it was overgrazing by the livestock which accompanied the immigrant trains, which thus destroyed this, their own wagon road. This probability is strengthened by the fact that the erosion is oldest and worst near the water-holes at which the Forty-Niners camped and grazed their thousands of animals.

Let us now rejoin Pattie on his trip down the Rio Grande. He left the river near what is now San Marcial and crossed southwesterly to the copper mines at Santa Rita del Cobre. From Santa Rita he went to the head of the Gila River to trap beavers. He caught "trout" where the river emerges from the mountains, probably near the present settlement of Cliffs. (If these were really trout, rather than 'bony-tails,' then the trout extended fifteen miles farther downstream than they do now.) The first night of trapping at this point yielded 30 beavers. But the important thing is not so much the abundance of beaver, as the fact that these hardy trappers *"were much fatigued by the difficulty of getting through the high grass which covered the heavily timbered bottom."*

Today, at this spot, and for miles above and below, the river is flanked by naked bars of sand and cobblestones, and the bottoms, except where fenced, are as bare of grass, as naked of timber, as the top of a billiard table.

Ascending the box of the Gila, Pattie describes *"a thick tangle of grapevines and underbrush"* through which he crawled, sometimes on hands and knees. At the forks of the river (now the XSX cattle ranch) the banks were still "very brushy, and frequented by numbers of bears." Here too, there is now little brush and many cobble bars.

Chop down the oldest of the young sycamores and alders that have found rootage on the cobble bars which have replaced Pattie's bottoms, and you will find that few are older than the cow-business, which invaded these hills in the early eighties.

A year later Pattie followed down the Gila to its junction with the Colorado. There on February 26, 1825, he tells in a single sentence more of what has befallen the Southwest than could be compressed into a volume today. This is the sentence: "At twelve we started up the Red River (Rio Colorado), which is between two and three hundred yards wide, a deep bold stream, *and the water at this point is entirely clear."*

The Colorado today discharges around 12,000 acre feet of silt per year, enough to cover half a township with a foot of mud. Clear water in the Colorado is unknown.

Again Pattie, this time at the mouth of the Little Colorado above the Grand Canyon: "On the fifteenth (April) we returned to the banks of the Red River, *which is here a clear beautiful stream."*

Pattie's finding clear water in February might be explained away on the grounds that headwater snows were still intact, but his report of clear water in both February and April, with no intimation of any opposite conditions during the interim journey along its course, indicates to me that our present Colorado resembles Pattie's only in location and name. The present river is never clear; not only that but it is carrying the richest soils of Arizona into the Gulf of California, which has become in very fact "the vermillion sea."

The foregoing comparisons of what Pattie saw and what we see today are merely random examples of what has happened, in some degree, to almost every watershed in the Southwest. On many of the National Forests and on a few well-managed private ranches the damage is partial and confined mainly to loss of bottomlands. Near many old settlements the damage is complete, erosion having exposed enough rocks to substitute what might be called a mechanical equilibrium for the vegetative one which once existed. In most places the damage is still in process, and the process is cumulative.

It has been necessary to offer proof of these changes because most people do not know that any change has taken place, and some who do know deny that overgrazing is the primary cause. They persist in believing either that abnormal erosion was always there, or that it is somehow an act of God instead of an act of goats, sheep, and cows.

In trying to picture the meaning of the term overgrazing, it is important that the reader divest his mind of the assumption that overgrazing constitutes a uniformly distributed excess of consumption over growth. More often than not the excessive utilization of one plant or type of ground is accompanied by the underutilization of another. For this reason the very diversity of the country has contributed to its undoing. If a mountain cow on a cold winter day has the choice of basking in the warm sun of a hardwood bottom, or of climbing upon the wind-swept mesa, or scrambling among the rocky slopes between the two, she will choose the bottom. In fact, she may browse the last bottomland willow to death before the bunch grass on the slopes is even touched. It seems as if the greater the diversity of types, the less uniform their utilization and the quicker the inception of damage.

The reader must grasp the fact that overgrazing is more than mere lack of visible forage. It is rather a lack of vigorous roots of desirable forage plants. An area is overgrazed to the extent its palatable plants are thinned out or weakened in growing power. It takes more than a few good rains, or a temporary removal of livestock, to cure this thinning or weakening of palatable plants. In some cases it may take years of skillful range management to effect a cure; in others erosion has so drained and leached the soil that restoration is a matter of decades; again it has removed the soil entirely. In the latter event restoration involves geological periods of time, and thus for human purposes must be dismissed as impossible.

There was once a widespread impression that forest fires, as well as overgrazing, were an important cause of watershed damage. Recent evidence in other regions supports this belief, but not here. On the contrary, observations on their sequence and relative importance in the Arizona brushfields, indicates that when the cattle came the grass went, the fires diminished, and erosion began.

The rivers on which we have built storage reservoirs or power dams deposit their deltas not only in the sea, but behind the dams. We build these to store water, and mortgage our irrigated valleys and our industries to pay for them, but every year they store a little less water and a little more mud. Reclamation, which should be for all time, thus becomes in part the source of a merely temporary prosperity.

The rivers which get their silt from the hills use it to scour the valleys which they once fertilized. Thus the Gila Valley, a garden since the Indian days, stretching almost across the breadth of Arizona, has lost 6000 acres of natural agricultural land since Pattie trapped beaver there, and may lose 33,000 acres more before the damage is complete. During the same period artificial reclamation has watered only 46,000 acres in the Gila drainage. Yet we think of reclamation as a net addition to the wealth of the arid west. In the Southwest it is more accurate to regard it, in part, as a mere offset to our own clumsy destruction of the natural bottoms which required no expensive dams and reservoirs, and which the Indians cultivated before irrigation bonds had a name and before the voice of the booster was heard in the land.

How now shall we sum up the degree of doubt as to the future—the possible damage to the tempo of our time, which inheres in this question?

We are, of course, in the position of a biographer who cannot evaluate a contemporary because he knows too much and understands too little. The forces at work are still in operation; it is too early to foresee their final outcome.

But we can say this with assurance: If erosion proceeds unchecked the ranges, irrigation reservoirs, and wild life will be gone.

If we do check it, we will lose the mountain valleys, and eventually the reservoirs will be much impaired, but the ranges can come back.

We can say this: That what we call "development" is not a uni-directional process, especially in a semi-arid country. To develop this land we have used engines that we could not control, and have started actions and reactions far different from those intended. Some of these are proving beneficial; most of them harmful. This land is too complex for the simple processes of "the mass-mind" armed with modern tools. To live in real harmony with such a country seems to require either a degree of public regulation we will not tolerate, or a degree of private enlightenment we do not possess.

But of course we must continue to live with it according to our lights. Two things hold promise of improving those lights. One is to apply science to land-use. The other is to cultivate a love of country a little less spangled with stars, and a little more imbued with that respect for mother-earth—the lack of which is, to me, the outstanding attribute of the machine-age.

The Conservation Ethic [1933]

While he was advising on soil erosion in the Southwest, Leopold delivered the fourth annual John Wesley Powell Lecture to the Southwestern Division of the American Association for the Advancement of Science in Las Cruces, New Mexico. It was the most important address of his career and his most comprehensive statement to date of the ethical aspects of conservation. Fifteen years later, Leopold reworked portions of the address, which had been published in the *Journal of Forestry*, for incorporation into "The Land Ethic," the capstone essay of *Sand County Almanac*. But "The Conservation Ethic," which was widely read and frequently cited, remains a vitally significant milepost in Leopold's intellectual development.

When god-like Odysseus returned from the wars in Troy, he hanged all on one rope some dozen slave-girls of his household whom he suspected of misbehavior during his absence.

This hanging involved no question of propriety, much less of justice. The girls were property. The disposal of property was then, as now, a matter of expediency, not of right and wrong.

Criteria of right and wrong were not lacking from Odysseus' Greece: witness the fidelity of his wife through the long years before at last his black-prowed galleys clove the wine-dark seas for home. The ethical structure of that day covered wives, but had not yet been extended to human chattels. During the three thousand years which have since elapsed, ethical criteria have been extended to many fields of conduct, with corresponding shrinkages in those judged by expediency only.

This extension of ethics, so far studied only by philosophers, is actually a process in ecological evolution. Its sequences may be described in biological as well as philosophical terms. An ethic, biologically, is a limitation on freedom of action in the struggle for existence. An ethic, philosophically, is a differentiation of social from anti-social conduct. These are two definitions of one thing. The thing has its origin in the tendency of interdependent individuals or societies to evolve modes of coöperation. The biologist calls these symbioses. Man elaborated certain advanced symbioses called politics and economics. Like their simpler biological antecedents, they enable individuals or groups to exploit each other in an orderly way. Their first yardstick was expediency.

The complexity of coöperative mechanisms increased with population density, and with the efficiency of tools. It was simpler, for example, to define the antisocial uses of sticks and stones in the days of the mastodons than of bullets and billboards in the age of motors.

At a certain stage of complexity, the human community found expediency-yardsticks no longer sufficient. One by one it has evolved and superimposed upon them a set of ethical yardsticks. The first ethics dealt with the relationship between individuals. The Mosaic Decalogue is an example. Later accretions dealt with the relationship between the individual and society. Christianity tries to integrate the individual to society, Democracy to integrate social organization to the individual.

There is as yet no ethic dealing with man's relationship to land and to the non-human animals and plants which grow upon it. Land, like Odysseus' slave-girls, is still property. The land-relation is still strictly economic, entailing privileges but not obligations.

The extension of ethics to this third element in human environment is, if we read evolution correctly, an ecological possibility. It is the third step in a sequence. The first two have already been taken. Civilized man exhibits in his own mind evidence that the third is needed. For example, his sense of right and wrong may be aroused quite as strongly by the desecration of a nearby woodlot as by a famine in China, a near-pogrom in Germany, or the murder of the slave-girls in ancient Greece. Individual thinkers since the days of Ezekial and Isaiah have asserted that the despoliation of land is not only inexpedient but wrong. Society, however, has not yet affirmed their belief. I regard the present conservation movement as the embryo of such an affirmation. I here discuss why this is, or should be, so.

Some scientists will dismiss this matter forthwith, on the ground that ecology has no relation to right and wrong. To such I reply that science, if not philosophy, should by now have made us cautious about dismissals. An ethic may be regarded as a mode of guidance for meeting ecological situations so new or intricate, or involving such deferred reactions, that the path of social expediency is not discernible to the average individual. Animal instincts are just this. Ethics are possibly a kind of advanced social instinct in-the-making.

Whatever the merits of this analogy, no ecologist can deny that our land-relation involves penalties and rewards which the individual does not see, and needs modes of guidance which do not yet exist. Call these what you will, science cannot escape its part in forming them.

Ecology—Its Role in History

A harmonious relation to land is more intricate, and of more consequence to civilization, than the historians of its progress seem to realize. Civilization is not, as they often assume, the enslavement of a stable and constant earth. It is a state of *mutual and interdependent coöperation* between human animals, other animals, plants, and soils, which may be disrupted at any moment by the failure of any of them. Land-despoliation has evicted nations, and can on occasion do it again. As long as six virgin continents awaited the plow, this was perhaps no tragic matter—eviction from one piece of soil could be recouped by despoiling another. But there are now wars and rumors of wars which foretell the impending saturation of the earth's best soils and climates. It thus becomes a matter of some importance, at least to ourselves, that our dominion, once gained, be self-perpetuating rather than self-destructive.

This instability of our land-relation calls for example. I will sketch a single aspect of it: the plant succession as a factor in history.

In the years following the Revolution, three groups were contending for control of the Mississippi valley: the native Indians, the French and English traders, and American settlers. Historians wonder what would have happened if the English at Detroit had thrown a little more weight into the Indian side of those tipsy scales which decided the outcome of the Colonial migration into the cane-lands of Kentucky. Yet who ever wondered why the cane-lands, when subjected to the particular mixture of forces represented by the cow, plow, fire, and axe of the pioneer, became bluegrass? What if the plant succession inherent in this "dark and bloody ground" had, under the impact of these forces, given us some worthless sedge, shrub, or weed? Would Boone and Kenton have held out? Would there have been any overflow into Ohio? Any Louisiana Purchase? Any trans-continental union of new states? Any Civil War? Any machine age? Any depression? The subsequent drama of American history, here and elsewhere, hung in large degree on the reaction of particular soils to the impact of particular forces exerted by a particular kind and degree of human occupation. No statesman-biologist selected those forces, nor foresaw their effects. That chain of events which on the Fourth of July we call our National Destiny hung on a "fortuitous concourse of elements," the interplay of which we now dimly decipher *by hindsight only*.

Contrast Kentucky with what hindsight tells us about the Southwest. The impact of occupancy here brought no bluegrass, nor other plant fitted to withstand the bumps and buffetings of misuse. Most of these soils, when grazed, reverted through a successive series of more and more worthless grasses, shrubs, and weeds to a condition of unstable equilibrium. Each recession of plant types bred erosion; each increment to erosion bred a further recession of plants. The result today is a progressive and mutual deterioration, not only of plants and soils, but of the animal community subsisting thereon. The early settlers did not expect this, on the cienegas of central New Mexico some even cut artificial gullies to hasten it. So subtle has been its progress that few people know anything about it. It is not discussed at polite tea-tables or go-getting luncheon clubs, but only in the arid halls of science.

All civilization seem to have been conditioned upon whether the plant succession, under the impact of occupancy, gave a stable and habitable assortment of vegetative types, or an unstable and uninhabitable assortment. The swampy forests of Caesar's Gaul were utterly changed by human use—for the better. Moses' land of milk and honey was utterly changed—for the worse. Both changes are the unpremeditated resultant of the impact between ecological and economic forces. We now decipher these reactions retrospectively. What could possibly be more important than to foresee and control them?

We of the machine age admire ourselves for our mechanical ingenuity; we harness cars to the solar energy impounded in carboniferous forests; we fly in mechanical birds; we make the ether carry our words or even our pictures. But are these not in one sense mere parlor tricks compared with our utter ineptitude in keeping land fit to live upon? Our engineering has attained the pearly gates of a near-millennium, but our applied biology still lives in nomad's tents of the stone age. If our system of land-use happens to be self-perpetuating, we stay. If it happens to be self-destructive we move, like Abraham, to pastures new.

Do I overdraw this paradox? I think not. Consider the transcontinental airmail which plies the skyways of the Southwest—a symbol of its final conquest. What does it see? A score of mountain valleys which were green gems of fertility when first described by Coronado, Espejo, Pattie, Abert, Sitgreaves, and Couzens. What are they now? Sandbars, wastes of cobbles and burroweed, a path for torrents. Rivers which Pattie says were clear, now muddy sewers for the wasting fertility of an empire. A "Public Domain," once a velvet carpet of rich buffalo-grass and grama, now an illimitable waste of rattlesnake-bush and tumbleweed, too impoverished to be accepted as a gift by the states within which it lies. Why? Because the ecology of this Southwest happened to be set on a hair-trigger. Because cows eat brush when the grass is gone, and thus postpone the penalties of over-utilization. Because certain grasses, when grazed too closely to bear seed-stalks, are weakened and give way to inferior grasses, and these to inferior shrubs, and these to weeds, and these to naked earth. Because rain which spatters upon vegetated soil stays clear and sinks, while rain which spatters upon devegetated soil seals its interstices with colloidal mud and hence must run away as floods, cutting the heart out of country as it goes. Are these phenomena any more difficult to foresee than the paths of stars which science deciphers without the error of a single second? Which is the more important to the permanence and welfare of civilization?

I do not here berate the astronomer for his precocity, but rather the ecologist for his lack of it. The days of his cloistered sequestration are over:

> Whether you will or not,
> You are a king, Tristram, for you are one
> Of the time-tested few that leave the world,
> When they are gone, not the same place it was.
> Mark what you leave.

Unforeseen ecological reactions not only make or break history in a few exceptional enterprises—they condition, circumscribe, delimit, and warp all enterprises, both economic and cultural, that pertain to land. In the cornbelt, after grazing and plowing out all the cover in the interests of "clean farming," we grew tearful about wild-life, and spent several decades passing laws for its restoration. We were like Canute commanding the tide. Only recently has research made it clear that the implements for restoration lie not in the legislature, but in the farmer's toolshed. Barbed wire and brains are doing what laws alone failed to do.

In other instances we take credit for shaking down apples which were, in all probability, ecological windfalls. In the Lake States and the Northeast lumbering, pulping, and fire accidentally created some scores of millions of acres of new second-growth. At the proper stage we find these thickets full of deer. For this we naively thank the wisdom of our game laws.

In short, the reaction of land to occupancy determines the nature and duration of civilization. In arid climates the land may be destroyed. In all climates the plant succession determines what economic activities can be supported. Their nature and intensity in turn determine not only the domestic but also the wild plant and animal life, the scenery, and the whole face of nature. We inherit the earth, but within the limits of the soil and the plant succession we also *rebuild* the earth—without plan, without knowledge of its properties, and without understanding of the increasingly coarse and powerful tools which science has placed at our disposal. We are remodelling the Alhambra with a steam-shovel.

Ecology and Economics

The conservation movement is, at the very least, an assertion that these interactions between man and land are too important to be left to chance, even that sacred variety of chance known as economic law.

We have three possible controls: Legislation, self-interest, and ethics. Before we can know where and how they will work, we must first understand the reactions. Such understanding arises only from research. At the present moment research, inadequate as it is, has nevertheless piled up a large store of facts which our land using industries are unwilling, or (they claim) unable, to apply. Why? A review of three sample fields will be attempted.

Soil science has so far relied on self-interest as the motive for conservation. The landholder is told that it pays to conserve his soil and its fertility. On good farms this economic formula has improved land-practice, but on poorer soils vast abuses still proceed unchecked. Public acquisition of submarginal soils is being urged as a remedy for their misuse. It has been applied to some extent, but it often comes too late to check erosion, and can hardly hope more than to ameliorate a phenomenon involving in some degree *every square foot* on the continent. Legislative compulsion might work on the best soils where it is least needed, but it seems hopeless on poor soils where the existing economic set-up hardly permits even uncontrolled private enterprise to make a profit. We must face the fact that, by and large, no defensible relationship between man and the soil of his nativity is as yet in sight.

Forestry exhibits another tragedy—or comedy—of *Homo sapiens*, astride the runaway Juggernaut of his own building, trying to be decent to his environment. A new profession was trained in the confident expectation that the shrinkage in virgin timber would, as a matter of self-interest, bring an expansion of timber-cropping. Foresters are cropping timber on certain parcels of poor land which happen to be public, but on the great bulk of private holdings they have accomplished little. Economics won't let them. Why? He would be bold indeed who claimed to know the whole answer, but these parts of it seem agreed upon: modern transport prevents profitable tree-cropping in cut-out regions until virgin stands in all others are first exhausted; substitutes for lumber have undermined confidence in the future need for it; carrying charges on stumpage reserves are so high as to force perennial liquidation, overproduction, depressed prices, and an appalling wastage of unmarketable grades which must be cut to get the higher grades; the mind of the forest owner lacks the point-of-view underlying sustained yield; the low wage-standards on which European forestry rests do not obtain in America.

A few tentative gropings toward industrial forestry were visible before 1929, but these have been mostly swept away by the depression, with the net result that forty years of "campaigning" have left us only such actual tree-cropping as is under-written by public treasuries. Only a blind man could see in this the beginnings of an orderly and harmonious use of the forest resource.

There are those who would remedy this failure by legislative compulsion of private owners. Can a landholder be successfully compelled to raise any crop, let alone a complex long-time crop like a forest, on land the private possession of which is, for the moment at least, a liability? Compulsion would merely hasten that avalanche of tax-delinquent land-titles now being dumped into the public lap.

Another and larger group seeks a remedy in more public ownership. Doubtless we need it—we are getting it whether we need it or not—but how far can it go? We cannot dodge the fact that the forest problem, like the soil problem, *is coextensive with the map of the United States.* How far can we tax other lands and industries to maintain forest lands and industries artificially? How confidently can we set out to run a hundred-yard dash with a twenty foot rope tying our ankle to the starting point? Well, we are bravely "getting set," anyhow.

The trend in wild-life conservation is possibly more encouraging than in either soils or forests. It has suddenly become apparent that farmers, out of self-interest, can be induced to crop game. Game crops are in demand, staple crops are not. For farm-species, therefore, the immediate future is relatively bright. Forest game has profited to some extent by the accidental establishment of new habitat following the decline of forest industries. Migratory game, on the other hand, has lost heavily through drainage and over-shooting; its future is black because motives of self-interest do not apply to the private cropping of birds so mobile that they "belong" to everybody, and hence to nobody. Only governments have interests coextensive with their annual movements, and the divided counsels of conservationists give governments ample alibi for doing little. Governments could crop migratory birds because their marshy habitat is cheap and concentrated, but we get only an annual crop of new hearings on how to divide the fast-dwindling remnant.

These three fields of conservation, while but fractions of the whole, suffice to illustrate the welter of conflicting forces, facts, and opinions which so far comprise the result of the effort to harmonize our machine civilization with the land whence comes its sustenance. We have accomplished little, but we should have learned much. What?

I can see clearly only two things:

First, that the economic cards are stacked against some of the most important reforms in land-use.

Second, that the scheme to circumvent this obstacle by public ownership, while highly desirable and good as far as it goes, can never go far enough. Many will take issue on this, but the issue is between two conflicting conceptions of the end towards which we are working.

One regards conservation as a kind of sacrificial offering, made for us vicariously by bureaus, on lands nobody wants for other purposes, in propitiation for the atrocities which still prevail everywhere else. We have made a real start on this kind of conservation, and we can carry it as far as the tax-string on our leg will reach. Obviously, though it conserves our self-respect better than our land. Many excellent people accept it, either because they despair of anything better, or because they fail to see the *universality of the reactions needing control*. That is to say their ecological education is not yet sufficient.

The other concept supports the public program, but regards it as merely extension, teaching, demonstration, an initial nucleus, a means to an end, but not the end itself. The real end is a *universal symbiosis with land*, economic and esthetic, public and private. To this school of thought public ownership is a patch but not a program.

Are we, then, limited to patchwork until such time as Mr. Babbitt has taken his Ph.D. in ecology and esthetics? Or do the new economic formulae offer a short-cut to harmony with our environment?

The Economic Isms

As nearly as I can see, all the new isms—Socialism, Communism, Fascism, and especially the late but not lamented Technocracy—outdo even Capitalism itself in their preoccupation with one thing: The distribution of more machine-made commodities to more people. They all proceed on the theory that if we can all keep warm and full, and all own a Ford and a radio, the good life will follow. Their programs differ only in ways to mobilize machines to this end. Though they despise each other, they are all, in respect of this objective, as identically alike as peas in a pod. They are competitive apostles of a single creed: *salvation by machinery*.

We are here concerned, not with their proposals for adjusting men and machinery to goods, but rather with their lack of any vital proposal for adjusting men and machines to land. To conservationists they offer only the old familiar palliatives: Public ownership and private compulsion. If these are insufficient now, by what magic are they to become sufficient after we change our collective label?

Let us apply economic reasoning to a sample problem and see where it takes us. As already pointed out, there is a huge area which the economist calls submarginal, because it has a minus value for exploitation. In its once-virgin condition, however, it could be "skinned" at a profit. It has been, and as a result erosion is washing it away. What shall we do about it?

By all the accepted tenets of current economics and science we ought to say "let her wash." Why? Because staple land-crops are overproduced, our population curve is flattening out, science is still raising the yields from better lands, we are spending millions from the public treasury to retire unneeded acreage, and here is nature offering to do the same thing free of charge; why not let her do it? This, I say, is economic reasoning. *Yet no man has so spoken.* I cannot help reading a meaning into this fact. To me it means that the average citizen shares in some degree the intuitive and instantaneous contempt with which the conservationist would regard such an attitude. We can, it seems, stomach the burning or plowing-under of over-produced cotton, coffee, or corn, but the destruction of mother-earth, however "sub-marginal," touches something deeper, some sub-economic stratum of the human intelligence wherein lies that something—perhaps the essence of civilization—which Wilson called "the decent opinion of mankind."

The Conservation Movement

We are confronted, then, by a contradiction. To build a better motor we tap the uttermost powers of the human brain; to build a better countryside we throw dice. Political systems take no cognizance of this disparity, offer no sufficient remedy. There is, however, a dormant but widespread consciousness that the destruction of land, and of the living things upon it, is wrong. A new minority have espoused an idea called conservation which tends to assert this as a positive principle. Does it contain seeds which are likely to grow?

Its own devotees, I confess, often give apparent grounds for skepticism. We have, as an extreme example, the cult of the barbless hook, which acquires self-esteem by a self-imposed limitation of armaments in catching fish. The limitation is commendable, but the illusion that it has something to do with salvation is as naive as some of the primitive taboos and mortifications which still adhere to religious sects. Such excrescences seem to indicate the whereabouts of a moral problem, however irrelevant they be in either defining or solving it.

Then there is the conservation-booster, who of late has been rewriting the conservation ticket in terms of "tourist-bait." He exhorts us to "conserve outdoor Wisconsin" because if we don't the motorist-on-vacation will streak through to Michigan, leaving us only a cloud of dust. Is Mr. Babbitt trumping up hard-boiled reasons to serve as a screen for doing what he thinks is right? His tenacity suggests that he is after something more than tourists. Have he and other thousands of "conservation workers" labored through all these barren decades fired by a dream of augmenting the sales of sandwiches and gasoline? I think not. Some of these people have hitched their wagon to a star—and that is something.

Any wagon so hitched offers the discerning politician a quick ride to glory. His agility in hopping up and seizing the reins adds little dignity to the cause, but it does add the testimony of his political nose to an important question: is this conservation something people really want? The political objective, to be sure, is often some trivial tinkering with the laws, some useless appropriation, or some pasting of pretty labels on ugly realities. How often, though, does any political action portray the real depth of the idea behind it? For political consumption a new thought must always be reduced to a posture or a phrase. It has happened before that great ideas were heralded by growing-pains in the body politic, semi-comic to those onlookers not yet infected by them. The insignificance of what we conservationists, in our political capacity, say and do, does not detract from the significance of our persistent desire to do something. To turn this desire into productive channels is the task of time, and ecology.

The recent trend in wild life conservation shows the direction in which ideas are evolving. At the inception of the movement fifty years ago, its underlying thesis was to save species from extermination. The means to this end were a series of restrictive enactments. The duty of the individual was to cherish and extend these enactments, and to see that his neighbor obeyed them. The whole structure was negative and prohibitory. It assumed land to be a constant in the ecological equation. Gun-powder and blood-lust were the variables needing control.

There is now being superimposed on this a positive and affirmatory ideology, the thesis of which is to prevent the deterioration of environment. The means to this end is research. The duty of the individual is to apply its findings to land, and to encourage his neighbor to do likewise. The soil and the plant succession are recognized as the basic variables which determine plant and animal life, both wild and domesticated, and likewise the quality and quantity of human satisfactions to be derived. Gun-powder is relegated to the status of a tool for harvesting one of these satisfactions. Blood-lust is a source of motive-power, like sex in social organization. Only one constant is assumed, and that is common to both equations: the love of nature.

This new idea is so far regarded as merely a new and promising means to better hunting and fishing, but its potential uses are much larger. To explain this, let us go back to the basic thesis—the preservation of fauna and flora.

Why do species become extinct? Because they first become rare. Why do they become rare? Because of shrinkage in the particular environments which their particular adaptations enable them to inhabit. Can such shrinkage be controlled? Yes, once the specifications are known. How known? Through ecological research. How controlled? By modifying the environment with those same tools and skills already used in agriculture and forestry.

Given, then, the knowledge and the desire, this idea of controlled wild culture or "management" can be applied not only to quail and trout, but to *any living thing* from bloodroots to Bell's vireos. Within the limits imposed by the plant succession, the soil, the size of the property, and the gamut of the seasons, the landholder can "raise" any wild plant, fish, bird, or mammal he wants to. A rare bird or flower need remain no rarer than the people willing to venture their skill in *building it a habitat*. Nor need we visualize this as a new diversion for the idle rich. The average dolled-up estate merely proves what we will some day learn to acknowledge: that bread and beauty grow best together. Their harmonious integration can make farming not only a business but an art; the land not only a food-factory but an instrument for self-expression, on which each can play music of his own choosing.

It is well to ponder the sweep of this thing. It offers us nothing less than a renaissance—a new creative stage—in the oldest, and potentially the most universal, of all the fine arts. "Landscaping," for ages dissociated from economic land-use, has suffered that dwarfing and distortion which always attends the relegation of esthetic or spiritual functions to parks and parlors. Hence it is hard for us to visualize a creative art of land-beauty which is the prerogative, not of esthetic priests but of dirt farmers, which deals not with plants but with biota, and which wields not only spade and pruning shears, but also draws rein on those invisible forces which determine the presence or absence of plants and animals. Yet such is this thing which lies to hand, if we want it. In it are the seeds of change, including, perhaps, a rebirth of that social dignity which ought to inhere in land-ownership, but which, for the moment, has passed to inferior professions, and which the current processes of land-skinning hardly deserve. In it, too, are perhaps the seeds of a new fellowship in land, a new solidarity in all men privileged to plow, a realization of Whitman's dream to *"plant companionship as thick as trees along all the rivers of America."* What bitter parody of such companionship, and trees, and rivers, is offered to this our generation!

I will not belabor the pipe-dream. It is no prediction, but merely an assertion that the idea of controlled environment contains colors and brushes wherewith society may some day paint a new and possibly a better picture of itself. Granted a community in which the combined beauty and utility of land determines the social status of its owner, and we will see a speedy dissolution of the economic obstacles which now beset conservation. Economic laws may be permanent, but their impact reflects what people want, which in turn reflects what they know and what they are. The economic set-up at any one moment is in some measure the result, as well as the cause, of the then prevailing standard of living. Such standards change. For example: some people discriminate against manufactured goods produced by child-labor or other anti-social processes. They have learned some of the abuses of machinery, and are willing to use their custom as a leverage for betterment.

Social pressures have also been exerted to modify ecological processes which happened to be simple enough for people to understand—witness the very effective boycott of birdskins for millinery ornament. We need postulate only a little further advance in ecological education to visualize the application of like pressures to other conservation problems.

For example: the lumberman who is now unable to practice forestry because the public is turning to synthetic boards may then be able to sell man-grown lumber "to keep the mountains green." Again: certain wools are produced by gutting the public domain; couldn't their competitors, who lead their sheep in greener pastures, so label their product? Must we view forever the irony of educating our sons with paper, the offal of which pollutes the rivers which they need quite as badly as books? Would not many people pay an extra penny for a "clean" newspaper? Government may some day busy itself with the legitimacy of labels used by land-industries to distinguish conservation products, rather than with the attempt to operate their lands for them.

I neither predict nor advocate these particular pressures—their wisdom or unwisdom is beyond my knowledge. I do assert that these abuses are just as real, and their correction every whit as urgent, as was the killing of egrets for hats. *They differ only in the number of links composing the ecological chain of cause and effect.* In egrets there were one or two links, which the mass-mind saw, believed, and acted upon. In these others there are many links; people do not see them, nor believe us who do. The ultimate issue, in conservation as in other social problems, is whether the mass-mind *wants to* extend its powers of comprehending the world in which it lives, or, granted the desire, *has the capacity to do so*. Ortega, in his *Revolt of the Masses*, has pointed the first question with devastating lucidity. The geneticists are gradually, with trepidations, coming to grips with the second. I do not know the answer to either. I simply affirm that a sufficiently enlightened society, by changing its wants and tolerances, can change the economic factors bearing on land. It can be said of nations, as of individuals: "as a man thinketh, so is he."

It may seem idle to project such imaginary elaborations of culture at a time when millions lack even the means of physical existence. Some may feel for it the same honest horror as the Senator from Michigan who lately arraigned Congress for protecting migratory birds at a time when fellow-humans lacked bread. The trouble with such deadly parallels is we can never be sure which is cause and which is effect. It is not inconceivable that the wave phenomena which have lately upset everything from banks to crime-rates might be less troublesome if the human medium in which they run *readjusted its tensions*. The stampede is an attribute of animals interested solely in grass.

Conservation Economics [1934]

A companion to "The Conservation Ethic," this essay critiques the limits of government-sponsored conservation, especially as practiced by New Deal agencies. It calls for the development of institutional incentives to induce private landowners to manage their land in the public interest. Delivered as an address to the Taylor-Hibbard Economics Club at the University of Wisconsin shortly after Leopold joined the Department of Agricultural Economics there, the text was subsequently published in the *Journal of Forestry*.

The moon, they say, was born when some mighty planet, zooming aimlessly through the firmament, happened to pass so near the earth as to lift off a piece of its substance and hurl it forth into space as a new and separate entity in the galaxy of heavenly bodies.

Conservation, I think, was "born" in somewhat this same manner in the year A.D. 1933. A mighty force, consisting of the pent-up desires and frustrated dreams of two generations of conservationists, passed near the national money-bags whilst opened wide for post-depression relief. Something large and heavy was lifted off and hurled forth into the galaxy of the alphabets. It is still moving too fast for us to be sure how big it is, or what cosmic forces draw rein on its career. My purpose is to discuss the new arrival and his prospects in life.

We must first of all understand the sequence of events which generated the lifting force. For the last half-century there has grown up a widespread conviction that our whip-hand over nature is no unmixed blessing. We have gained an easier living, but in the process of getting it we are losing two things of possibly equal value: (1) The permanence of the resources whence comes our bread and butter; (2) the opportunity of personal contact with natural beauty.

Conservation is the effort to so use the whip that these two losses will be minimized.

Its history in America may be compressed into two sentences: We tried to get conservation by buying land, by subsidizing desirable changes in land use, and by passing restrictive laws. The last method largely failed; the other two have produced some small samples of success.

The "New Deal" expenditures are the natural consequence of this experience. Public ownership or subsidy having given us the only taste of conservation we have ever enjoyed, the public money-bags being open, and private land being a drug on the market, we have suddenly decided to buy us a real mouthful, if not indeed, a square meal.

Is this good logic? Will we get a square meal? These are the questions of the hour.

Geography

The monumental Copeland Report on forestry, and some lesser labors in other fields, have recently shed much light on these questions, but it seems to me that we can further illuminate them by considering the simple geography of the phenomena which conservation seeks to control. Forests, erosion, and game each have certain characteristics and certain limitations affecting their dispersion over the land. Can these be made to fit the geographical peculiarities of public ownership? For instance:
1. Public lands are necessarily of limited dispersion.
2. The ratio of public to private land cannot exceed what the private tax-base plus operating revenues if any, will carry.
3. The minimum unit of public land must be large enough to carry a custodian.

Let us examine the geography of game in the light of these limitations. Wild game has an inherent intolerance of concentration. Few enthusiasts are aware of this simple but important fact. The most skillful culture cannot build a wild stand heavier than a bird per acre, or a deer per 20 acres. Take upland birds as an example. The safe limit of annual kill is one-third the population, hence under ideal conditions it takes 3 acres to put a bird in the hunting coat. Under the non-Utopian realities of actual practice it will likely take at least 6 acres. Perhaps half of Wisconsin is suitable to be cropped for birds. On this half the state could bag 3 million birds yearly, or 15 for each hunter now licensed. This is ample, but it assumes *all suitable land* to be cropped. By no stretch of the imagination could the public own all suitable land. Moreover, if it did, the land would no longer be farmed, whereupon its productive capacity for game would sink to a much lower figure. If the public owned and cropped a tenth of the state—3 million acres—it could produce only 5 birds for each hunter now licensed. What, now, is left for the unlicensed thousands who have leisure but no place to spend it, and for the non-residents who are the answer to the booster's prayer?

We can, to be sure, get heavier yields by artificial propagation, but the cost would be prohibitive and the esthetic quality of the product distinctly lower. It also happens that waterfowl differ from other game. They have no intolerance of concentration. Large-scale public ownership of marshlands, therefore, is feasible. It is also necessary, because the interstate movements of waterfowl render the incentive for their private production partially inoperative. Hence waterfowl stand is an exception to the rule.

It is clear, however, that the inherent dispersion of the phenomena dealt with in game management makes public game production a mere supplement to production on private lands. Game must grow as a by-product of other land uses. "Sport for all" is obtainable only by using all the land. Public game cropping as a sole dependence is excluded by the very nature of the game itself.

Consider, now, the geography of forests. Forestry is unique in that timber products can be grown in one place and used in another. This is not true of game or fish or erosion control, or scenery, or wildflowers, or birdsongs. Forestry is unique also in this respect: Consumption of timber products is not increasing. Hence it is probably feasible to relegate the timber-growing function to public lands. It is not, to be sure, a desirable solution of the forest problem, because the secondary functions of erosion control, wild life production, and recreation decline as dispersion decreases. Wood waste goes up as dispersion goes down. The social disciplines which private landowners might derive from timber-growing certainly are partly lost when the job is done vicariously by public agents. Until 1933, both foresters and lumbermen clung tenaciously to the theory that there must be both private and public forestry, despite the near-failure of all efforts to bring private practice into existence. Since 1933, however, there has been a virtual stampede for public ownership. Even Article X of the Lumber Code seems to be bending in the direction of a preparation for public acquisition of cutovers.

What, now, is the geography of soil erosion and floods? What is the dispersion of the phenomena which determines the regimen of the Mississippi?—which determine whether the topsoil on farms shall stay where it is, or be dumped into the Gulf of Mexico? Unless science has utterly deluded itself, the answer is at variance with the recent trend of land policy. With as much certainty as we know whether swallows hibernate in mud, and whether the elements are fire, water, and air, we know that the dispersion of potential erosion is as universal as the dispersion of cultivation, grazing, slope, and rain. How, then, shall we control it by purchasing a few headwaters and riverbanks and converting them into public forests? These spots are, I admit, usually the most vulnerable, and their public afforestation will, I admit, retard the degeneration of our soil and water resources, but will it assure the physical integrity of America in A.D. 2000, or even A.D. 1950?

Most assuredly not. It is a geographic axiom that there is no such assurance except in the *conservative use of every acre on every watershed in America*, whether it be farm or forest, public or private. In the West are dozens of irrigation projects "protected" by a headwater-patch of national forest, each subjoined by a watershed on which overgrazing, fire, and dry-farming have run riot. Most of these "protected" reservoirs began to choke with silt before the ink was dry on their bonds. This disease of erosion is a leprosy of the land, hardly to be cured by slapping a mustard plaster on the first sore. The only cure is the universal reformation of land-use, and the longer we dabble with palliatives, the more gigantic grows the job of restoration.

Let us now examine the geography of that subtle, complex, and (barring agriculture) most important of all the uses of land: recreation. Recreation is a perpetual battlefield because it is a single word denoting as many diverse things as there are diverse people. One can discuss it only in personal terms. A sawlog can be scaled, and a covey of quail is 15 birds, but there is no unit of either volume or value wherewith diverse persons can impersonally measure or compare recreational use. Those who have opinions about it must admit, like Whitman, that

> Whatever the sounding, whatever the sea or the sail,
> Man brings all things to the test of himself.

The salient geographic character of outdoor recreation, to my mind, is that recreational use is self-destructive. The more people are concentrated on a given area, the less is the chance of their finding what they seek. This is not true of the uncritical mob, but I see no more reason for running a national or state park to please the mob than a public art gallery or a public university. A slum is a slum, whether in the Bowery or on the Yellowstone. Dispersion, then, is the first principle of recreational planning. Dispersion of outdoor playgrounds has the equally important attribute of enhancing their accessibility.

It is inconceivable to me that the "leisure for all" revealed to us in Mr. Hoover's dream can be spent mainly, or even in large part, on public recreation grounds. Already the public grounds are so congested that the solitary recreationist must either invade such of their roadless hinterlands as may have temporarily escaped the CWA, or avoid them altogether. The expanding demand for recreation must in some way be spread over both public and private lands, or else, like Shakespeare's virtue, it will "die of its own toomuch."

Let it be clear that I do not challenge the purchase of public lands for conservation. For the first time in history we are buying on a scale commensurate with the size of the problem. I do challenge the growing assumption that bigger buying is a substitute for private conservation practice. Bigger buying, I fear, is serving as an escape-mechanism—it masks our failure to solve the harder problem. The geographic cards are stacked against its ultimate success. In the long run it is exactly as effective as buying half an umbrella.

Integration

It has always been admitted that the several kinds of conservation should be integrated with each other, and with other economic land uses. The theory is that one and the same oak will grow sawlogs, bind soil against erosion, retard floods, drop acorns to game, furnish shelter for song birds, and cast shade for picnics; that one and the same acre can and should serve forestry, watersheds, wild life, and recreation simultaneously. It required the open money-bags of 1933, however, to demonstrate what a disparity still exists between this paper ideal and the actual performance of a field-foreman turned loose with a crew and a circular of instructions on how to do some one particular kind of conservation work. There was, for example, the road crew cutting a grade along a clay bank so as permanently to roil the troutstream which another crew was improving with dams and shelters; the silvicultural crew felling the "wolf trees" and border shrubbery needed for game food; the roadside-cleanup crew burning all the down oak fuel wood available to the fireplaces being built by the recreation-ground crew; the planting crew setting pines all over the only open clover-patch available to the deer and partridges; the fire-line crew burning up all the hollow snags on a wild-life refuge, or worse yet, felling the gnarled veterans which were about the only scenic thing along a "scenic road." In short, the ecological and esthetic limitations of "scientific" technology were revealed in all their nakedness.

Such crossed wires were frequent, even in the CCC camps where crews were directed by brainy young technicians, many of them fresh from conservation schools, but each schooled only in his particular "specialty." What atrocities prevailed in the more ephemeral organizations like the CWA, he who runs may read. The instructive part of this experience is not that cub foremen should lack omniscience in integrating conservation, but that the high-ups (of which I was one) *did not anticipate* these conflicts of interest, sometimes did not see them when they occurred, and were ill-prepared to adjust them when seen. The plain lesson is that to be a practitioner of conservation on a piece of land takes more brains, and a wider range of sympathy, forethought, and experience, than to be a specialized forester, game manager, range manager, or erosion expert in a college or a conservation bureau. Integration is easy on paper, but a lot more important and more difficult in the field than any of us foresaw. None of us had ever had enough volume and variety of field labor simultaneously at work to be fully aware of either its pitfalls or its possibilities. If the *accouchement* of conservation in 1933 bore no other fruits, this sobering experience would alone be worth its pains and cost.

If trained technicians on public lands find it no small task to integrate the diverse public interests in land-use, what shall we say of the private landowner, scrambling for a hard-earned living, who has not even been told what these public interests are?

Legislation

It is a conspicuous fact that almost all our present laws and appropriations are single-track measures dealing with a single aspect of land-use. During the summer of 1933, it became an equally conspicuous fact that when applied to the soil these measures frequently clash, or at best, fail to dovetail with each other.

Take, for example, a hypothetical Wisconsin farm, and count the geeings and hawings which result from having a dozen drivers for a single horse.

First we have the AAA paying the farmer a bonus for taking land out of corn or tobacco. Is the farmer encouraged to reorganize his layout of fields so as to divert this idle acreage permanently to game, forestry, or erosion control? No—that is not the business of the AAA. On the contrary, he is free to clear new woods, or push his pastures further up the hill, to the actual detriment of forestry, game, and erosion.

Again we have the CCC, building free check-dams in the farmer's gullies, and doing a splendid job of it. But does the CCC stipulate that he must pull his cows down off the steep slopes, and so revise his farming that new gullies will not form? To a very limited extent, and only in the most flagrant cases. The single-track approach is virtually precluded from revising other land uses so as to give permanence to the benefits it confers.

Again we have the Forest Crop Law, offering a tax rebate to those who practice forestry. Does the timber owner who makes his woods produce not only timber but also game, erosion-control, fur, or wildflowers, gain any preferred status thereby? Not at all, despite the fact that he may benefit the public ten times as much as he who practices forestry only, and despite the fact that the legislature which passed the law, and the conservation commission which administers it, are equally interested in these "side-issues."

This hypothetical farm may be in a fire-protection district which receives federal aid from the Clarke-McNary Law. The district may qualify as to fire, but be a public menace as to wild life, or recreation. These things, however, cannot sway the inspector who passes on compliance with fire-control standards. He must listen only to the rigid single-track definition of conservation embalmed in his particular single-track statute.

These bewilderments, of course, extend far beyond the conservation field. The public game farm restocks the coverts which the public highway crew has just burned up or cut down. Congress is about to tax duck hunters to restore the marshes which its own agents have caused to be drained. The Agricultural Colleges preach fences for the public grazing ranges—the Interior Department prohibits them. Not all of these reversals are preventable—hindsight is better than foresight, and always will be; sincere public servants disagree on what is sound public policy, and always will. The list, however, is sufficiently impressive to raise these basic questions:

(a) Does the rigid statutory single-track definition of conservation attain even its own limited object? History so far answers: seldom.
(b) Can the private landowner be expected to integrate these uncoördinated definitions into a single system of land use? Not, I think, if government experts find it difficult to do so.
(c) When the taxpayer learns what poor teamwork exists between the various conservation dollars, will he be satisfied to roll more of them down the same old rut? I doubt it.
(d) If single-track subsidy or compulsion will not work, and if the alternative of public acquisition is not a solution, then what is the solution?

Economics

In attempting to throw light on this question, we must first examine briefly the time-honored supposition that conservation is profitable, and that the profit-incentive is sufficient to motivate its practice.

Forestry and erosion-control are often profitable *if started before deterioration sets in*—seldom if started later. Advanced erosion is always unprofitable to control if regarded from the local viewpoint, but if one adds the cost of handling the floods and silting caused by the dislocated soil, it is cheaper to cure it at its source, even though the cost may exceed the value of the land.

Game management is profitable if some major crop carries the land and if the environment need not be rebuilt—seldom if the game alone must carry the land, or if the land is ruined.

Recreation and allied esthetic uses seldom offer direct income. They can usually be considered profitable only by the general public, and after crediting intangibles.

It is apparent even from this brief survey that:
1. Direct profits are operative only in spots.
2. Advanced deterioration usually precludes profits.

3. No balanced program can be built on profit alone. Public intervention is necessary.
4. Prevention, whatever the cost, is usually cheaper than cure.
5. Incentives are more promising than penalties, because penalties are *ex post facto*.

The wholesale public expenditures for 1933 indicate that from now on, whenever a private landowner so uses his land as to injure the public interest, *the public will eventually pay the bill,* either by buying him out, or by donating the repairs, or both. Hence the prevention of damage to the soil, or to the living things upon it, has become a first principle of public finance. Abuse is no longer merely a question of depleting a capital asset, *but of actually creating a cash liability against the taxpayer.* I hope the reader will ponder this well. It is a new frame for our picture which nullifies many pre-existing grooves of thought.

The thing to be prevented is destructive private land-use of any and all kinds. The thing to be encouraged is the use of private land in such a way as to combine the public and the private interest to the greatest possible degree. If we are going to spend large sums of public money anyhow, why not use it to subsidize desirable combinations in land use, instead of to cure, by purchase, prohibition, or repair, the headache arising from bad ones?

I realize fully that such a question qualifies me for the asylum for political and economic dreamers. Yet I submit that the proposal is actually less radical politically, and possibly cheaper in economic cost, than the stampede for public ownership in which our most respectable conservatives have now joined.

Let me illustrate. Last summer I participated in the building of hundreds of erosion check-dams, each string of dams costing a sum the interest on which is greater than the taxes from the land they protect. These dams were "cures," necessary ones. But how about prevention of land uses creating more gullies needing more dams? If the farmer or stockman had, in the first place, been offered a differential tax of, let us say, 25 per cent in favor of conservative use, perhaps no dams need ever have been built. The economic saving would have been 75 per cent. Politically, is it any more radical to offer careless farmers a differential tax than to offer them free dams?

The CCC camps are planting forests on many burned-over acres at a cost as yet unannounced, but it is certainly not less than the commercial cost of $5–$10 per acre. Would the dollar or half-dollar interest on this "cure," offered as a differential tax, have prevented the lumberman who originally cut the timber from allowing the fire to run? If the present forest tax laws do not offer sufficient inducement to prevent a repetition of the tragedy, is it not logical to consider "raising the ante," or even remitting taxes altogether, on such forest properties as safeguard the public interest? Is it necessarily cheaper or better to wait and buy the charred remains as a public forest?

Our game departments are artificially restocking grazed-out or burned-over coverts year after year at $2.50 per bird, and often to no effect. How about paying the same sum to the farmer, in the form of differential taxes or shooting fees, for fencing cover spots, for feeding, and for posting the land? I know a thousand places where $2.50 worth of fence or feed will produce not one, but *ten* birds per year, ad infinitum. It would require $2,500, plus an annual bill for custodian service, to get the same results by public land purchase.

Let me at this point also plead for what may be called the "suppressed minorities" of conservation. The landowner whose boundaries happen to include an eagle's nest, or a heron rookery, or a patch of ladyslippers, or a remnant of native prairie sod, or an historical oak, or a string of Indian mounds—such a landowner is the custodian of a public interest, to an equal or sometimes greater degree than one growing a forest, or one fighting a gully. We already have such a welter of single-track statutes that new and separate prohibitions or subsidies for each of these "minority interests" would be hard to enact, and still harder to enforce or administer. Perhaps this impasse offers a clue to the whole broad problem of conservation policy. It suggests the need for some comprehensive fusion of interests, some sweeping simplification of conservation law, which sets up for each parcel of land a single criterion of land-use: "Has the public interest in *all* its resources been protected?" which motivates that criterion by a single incentive, such as the differential tax, and which delegates the function of judging compliance to some single and highly trained administrative field-inspector, subject to review by the courts. Such a man would have to be a composite tax assessor, county agent, and conservation ecologist. Such a man is hard to build, but easier, I think, than to build a law specifying in cold print the hundreds of alternative ways of handling the land resources of even a single farm.

It would perhaps be unnecessary for the law itself to define the public interest, nor for the inspector to adhere to a rigid unchanging definition through a long period of years. Such an elastic regulation of private compliance with public interest is already in successful operation in the Industrial Safety Service established by the Wisconsin Industrial Commission Act (Revised Statutes, Chap. 101).

I have administered land too long to have any illusion, or to wish to create one, that this idea of preventive subsidy is as simple as it sounds, but I doubt if it would be as complicated as the cures on which we are now embarked. Differential taxes, I realize, must reach far enough back into national finance to forestall the mere local shifting of the tax burden, and must be based on some workable criterion of good vs. bad land use. How to define it? Who to define it? Are differential taxes the best, or even a possible vehicle? I don't know. I do know that it would be hard to find a less workable criterion of that composite thing called conservation than the single track statutes we now employ. Some of them may be tolerable as a definition of the single land-use with which each deals, but as criteria of the combination of conflicting or coöperating uses which constitute the actual land problem, they seem hopeless.

I am no economist, and no jurist. It seems clear, however, even to a layman that previous to 1933 the entire search for economic mechanisms was confined within the pre-existing limits sanctioned in our political and economic law and custom. It suddenly appears that those limits are too narrow.

Is this, after all, surprising? Our legal and economic structure was evolved on a terrain (central and western Europe) inherently more resistant to abuse than any other part of the earth's surface, and at a time when our engines for subjugating the soil were still too weak to ruin it. We have transplanted that structure to a new terrain, at least half of which is set on a hair-trigger of ecologic balance. We have invented engines of unprecedented coarseness and power, and placed them freely in the hands of ignorant men. I do not regret this social experiment—it is creation's most daring attempt to mitigate the rigors of tooth-and-claw evolution—but I assert we should be surprised, not that the pre-existing structure needs widening, but that it will serve at all.

One of the symptoms of inadequacy in our now existing structure is the perennial stalemate over the public domain. How can we keep it without a huge expansion of federal machinery? How can we give it away without the certainty of misuse? There is indeed scant choice between the horns of this dilemma. But would there be a dilemma if there were such a thing as *contingent* possession, or else a differential tax exerting a constant positive pressure in favor of good use?

This paper forecasts that conservation will ultimately boil down to rewarding the private landowner who conserves the public interest. It asserts the new premise that if he fails to do so, his neighbors must ultimately pay the bill. It pleads that our jurists and economists anticipate the need for workable vehicles to carry that reward. It challenges the efficacy of single-track land laws, and the economy of buying wrecks instead of preventing them. It advances all these things, not with any illusion that they are truth, but out of a profound conviction that the public is at last ready to do something about the land problem, and that we are offering it twenty competing answers instead of one. Perhaps the cerebration induced by a blanket challenge may still enable us to grasp our opportunity.

Helping Ourselves [1934]

Putatively coauthored with farmer Reuben Paulson, this essay published in *Field and Stream* bore the subtitle "Being the adventures of a farmer and a sportsman who produced their own shooting ground." It describes a cooperative venture of a dozen farmers and urban sportsmen to improve habitat for hunting on a contiguous block of farmland south of Madison. Begun in 1931, as Leopold was trying to defend the feasibility of his recommendations in the American Game Policy, the Riley Game Cooperative continued throughout Leopold's career at the University of Wisconsin. During that period the cooperative also served as one of Leopold's research experimental areas in which he attempted to put into practice the principles expounded in *Game Management*. Indeed it continued even after his death.

The senior author of this narrative is a sportsman who had grown tired of asking suspicious farmers for permission to hunt, hike or train dogs on gameless farms. The junior author is a farmer who had grown tired of spending his Sundays ejecting miscellaneous unpermitted "rabbit hunters" from his quail coverts.

Like other outdoorsmen, both of us had listened patiently to the fair words of the prophets of conservation, predicting the early restoration of outdoor Wisconsin. We both had noticed, though, that as prophecies became thicker and thicker open seasons for hunting became shorter and shorter, and wild life scarcer and scarcer.

Three years ago, when we first met, to flush a rabbit was the biggest adventure one might hope to fall upon in a day's hike on the Paulson farm. One snowy Sunday, when we were bemoaning this scarcity of living things on the land, there came to us jointly a flickering recollection of that first theorem of social justice: The Lord helps those who help themselves. Whereupon was born the "Riley Game Cooperative."

Riley, be it known, is a flag-station and a post-office near the Paulson farm. This definition of Riley is meticulously and literally correct.

The term "game cooperative" was not quite so accurate. It was a "cooperative," all right, with one farmer and one sportsman constituting its then membership. But it was more than "game," both of us contributing to the enterprise an incurable interest in all wild things, great and small, shootable and non-shootable. However, we both had an eye cocked on the future, and decided to title only the main issue.

Paulson gathered unto himself six contiguous neighbors. Leopold gathered up five Madison sportsmen, all mutual friends and of the sort whose game pockets contain no quail feathers in pheasant season. Then we moved that the nominations be closed. The idea is that any enduring relationship between sportsmen and farmers must be based on personal confidence, and nobody can have that if the crowd is so large as to need identification tickets. We of the cooperative can name any other member across the marsh by noting the decrepitude of his particular hunting coat, or by watching the gait or ear-floppings of his particular dog.

Now it so happens that in that same winter of our discontent, when the first theorem of social justice was revealed to us, some senator or assemblyman likewise saw the burning bush. We admit that legislators seldom do this, either in Wisconsin or elsewhere, but this one did. There emerged, as out of a cloud, all duly enacted, the "Wisconsin Shooting Preserve Law," which declared that citizens who owned or controlled land and planted pheasants thereon might shoot, when duly licensed, three-quarters of the number planted, during an all-fall open season, provided there be affixed to the leg of each pheasant so shot a nonreusable metal tag, to be issued by the Conservation Commission, etc. Furthermore, the law prohibited trespass by other citizens on the premises so licensed.

The law specifies pheasants, because these can be raised artificially; and when they are counted out of the coop by the local game warden, the state knows what three-quarters is. The state gets the other quarter "on the hoof," as a private donation, to chalk up to the credit of its restocking program.

We of the cooperative are no more interested in pheasants than in other game, and still less in shooting pheasants recently let out of a coop. Be it noted, however, that the new law restricts shooting to three-quarters of the number released, not to three-quarters of the identical birds released. We saw in this a chance to build up a wild population and to do our shooting on these wild birds, releasing sufficient tame ones to satisfy the requirements of the law.

Therefore, we took out a shooting-preserve license, posted the seven member farms, and released twenty-five pen-raised pheasants as a starter. None of us shot them, or wanted to, but we all had a lot of fun that first winter maintaining feeding stations "for the succour of said beasts," and, to be honest, for the purpose of holding them on our grounds. It was a mild winter, and these "tame" pheasants soon grew too big and wild to be in need of much "succour."

It was, however, the patronage extended to our feeding stations by nonshootable game which made them fun. Paulson had planted soy-beans under his silage corn. An aftermath of these beans had matured after the corn harvest. At the very first heavy snow these soy-beans drew, out of nowhere, a pack of forty big, husky prairie chickens. No chicken had been seen on the Paulson farm for a decade.

Likewise out of nowhere came a covey of quail. They tried to establish legal residence at one of the pheasant stations, but it was soon evident, from the lawsuits recorded in the snow, that the pheasants disputed their emigration papers, and not always by peaceable means. So we promptly erected an additional station for the quail, and henceforth each species stayed in its own bailiwick.

Before the winter was over a second quail covey, doubtless starved out of some near-by farm, appeared and waxed fat at our expense. Only it shouldn't be called expense—none of us for years had so enjoyed our winter Sundays. As for rabbits, everyone within a mile of our boundaries promptly applied for membership in the corn supply of the Riley Cooperative, and when winter was over they stayed to set up housekeeping.

It was a pleasant thing that first spring, as we strolled over these formerly gameless farms, to hear quail whistling in every fence-row and pheasant cocks crowing all over the Sugar Creek marsh. We estimated that our first six months of operation had netted us a respectable pheasant population (some strayed to the "public domain," as predicted by the law) plus an unearned increment of thirty quail, plus bunnies *ad infinitum*. Our chickens left us for parts unknown after the last snow had melted, but we knew that they would be back.

It was now time to do our stuff under the preserve law. Buying grown pheasants at $2.50 each was too expensive; so we bought 150 eggs, and Mrs. Paulson hatched them under hens. When the game warden came around in August, he counted 70 half-grown birds, which had the free run of the orchard but returned to roost in the brooder-coop with their foster-mothers. This count entitled us to 53 shooting tags (three-quarters of 70), plus those unused last year. These tags are equally distributed among all our members who care to shoot, including farm members.

At this point we hear our sporting readers emit a loud snort at the prospect of shooting these half-tame "artificials." Just a minute, please. These tame hen-raised birds were all headquartered near the farmhouse, around which we blocked off an 80-acre refuge on which no shooting is allowed. Outside this refuge, ever since the corn was cut last fall, we had been training our dogs on several coveys of big wild birds, the progeny of last year's plantings. It was these wild birds that we hunted when our season opened in October. By late fall the "artificials" had gone wild and spread, by slow degrees, off the refuge, and we probably shot some of them, but at no time had we either the desire or the opportunity to shoot an immature or tame pheasant. Our refuge automatically prevents it. The sketch map on the opposite page shows how this refuge works.

At this writing, our stock of game compares with the three preceding years as follows:

	Nov. 1930 (before management)	Nov. 1931	Nov. 1932	Nov. 1933
Pheasants	none	25	100	125
Quail	15	30	90	150
Prairie Chickens (winter only)	none	40	60	30
Rabbits (estimate)	50	100	125	125
Total head	65	195	375	430

Our main loss of pheasants was due to birds flying off the preserve when shot at. In 1932 we killed twenty-five, but doubtless lost twice that number from scattering. Last year we seem to have partly overcome this by more refuges, more feeding stations, and postponing the shooting till November and December. At this season the outside range is bare and unfed, and hence less attractive to the birds.

Feeding is our main bid for permanent residence. We have had no luck with food-patches because our farmers turn their hogs loose in the fall, and they make short work of any grain left standing. However, a dozen lean-to shelters, built of poles and thatched with marsh hay and each containing a hopper or a wire crib, are on duty all winter as feeding stations. A hundred-yard circle around each feeding shelter is closed to shooting.

We settled, by a tribal moot held under Paulson's oak tree, the financial relations of town and farm members. All cost are to be shared equally. When a farm member or his wife raises pheasants, he or she is to be paid, by the town members jointly, for each pheasant counted out by the warden, half the game-farm price. When a farm member leaves grain for a feeding station, he is to be paid, by the town members jointly, half its market value. All this proceeds on the theory that about half the "keep" of a game crop is the land on which it ranges, which land the farm member is furnishing free.

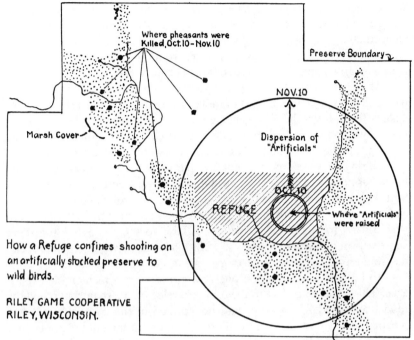

Leopold's sketch map published with "Helping Ourselves" in *Field and Stream*, August, 1934.

All shooting privileges are pooled and equally divided among those members who wish to shoot. Some of the farmers do not care to shoot, but they are none the less enthusiastic members of the Cooperative, by reason of the pleasure they derive from seeing the game, the insect-control services rendered them by the game, and the protection from irresponsible trespassers derived from the posting of the preserve. Nor should it be forgotten that the state is likewise a beneficiary. Its tangible reward is the $10 we pay the Conservation Department for a preserve license, plus at least fifty unshot pheasants, worth $2 apiece, which have spread over the country. Its intangible reward is the manifold increase we have brought about in the quail and song birds which use our feeding stations and coverts. We of the Cooperative have likewise decreased by twelve men that growing army of shooters who have no place to shoot and nothing to shoot at.

So far we have found it unnecessary to control any predators except cats. We encourage local trappers to trim down our foxes and minks, but select those individuals whom we think can be trusted to let feathers alone.

We have horned owls and red-tailed hawks, but there is no evidence that they are getting birds. Our abundant rabbits seem to serve as buffers. Skunks are plentiful, but their droppings in summer are a solid mass of June-bug wings. Hence our farmers, whose pastures suffer from the grubs, want the skunks left undisturbed until actual nest-robbing becomes evident.

As our bird stock builds up, the predator situation may change, but we will molest no predators until they molest us. If and when we take measures against them, it will be with genuine regret. It is pleasant to hear the owls hoot as we tramp carward from a day's hunting, and on winter Sundays the fox tracks in the snow add interest to our rounds of the feeding stations.

During the years we have shot over our preserve, the total of pheasants killed has never reached half of the number we were entitled to shoot. The birds killed were nearly all cocks, although our own rules allow one hen for each two cocks. Nobody has killed more than one bird in a day, although perfectly free to shoot his whole quota. All this goes to prove the astonishing conservatism of the "blood-thirsty" hunter once he begins to feel a proprietary interest in the future game crop of a particular piece of land.

To kill one's first cock on one's own grounds is a memorable experience. This cock, according to the trail unraveled by the spaniel pup and the food later found in his crop, had breakfasted in the cornfield and then repaired to a willow-bush in the marsh to repose in the mild October sun. Was this a "tame" pheasant? Not very. He hurtled out of that willow-bush, in all his bronze and violet glory, like some indignant Genghis Khan disturbed at his nap, the marsh resounding to his profane cacklings and the great tattoo of his broad, strong wings. It took both barrels before he collapsed into the marsh, and when the pup proudly emerged from the grass bearing his limp and shining burden you could see only his eyes and paws. The rest was all pheasant. One of these big, wild scarlet-jowled white-collared cocks is enough flavoring for any Sunday afield.

We want other farmer-sportsman groups to set up game production around our boundaries. This cooperative idea has all outdoors to spread in. We have not so far been accused of monopolizing shooting privileges, but we will be; and when we are accused, this will be our answer:

"There are 12,000,000 acres of farmland in southern Wisconsin, capable of carrying at least 6,000,000 game birds, of which 2,000,000 to 3,000,000 could safely be shot annually, if and when they are brought into existence. There are possibly 100,000 shooters on this area, which means that each could shoot 20 to 30 birds per year, if and when the farmers and sportsmen will get together, under shooting-preserve licenses or otherwise, and provide food and cover so that the birds can multiply. The state cannot provide food and cover. Farmers can. The Lord helps those who help themselves."

The Arboretum and the University [1934]

Leopold spoke at the dedication ceremony of the University of Wisconsin Arboretum and Wild Life Refuge on June 17, 1934. His remarks were later published in *Parks and Recreation*. They highlight another of Leopold's pioneering contributions to professional conservation, in the enterprise now called restoration ecology. Both as director of animal research at the university arboretum and as private owner of a sand country farmstead, Leopold attempted to reconstruct samples of the original Wisconsin landscape.

For twenty centuries and longer, all civilized thought has rested upon one basic premise: that it is the destiny of man to exploit and enslave the earth.

The biblical injunction to "go forth and multiply" is merely one of many dogmas which imply this attitude of philosophical imperialism.

During the past few decades, however, a new science called ecology has been unobtrusively spreading a film of doubt over this heretofore unchallenged "world view." Ecology tells us that no animal—not even man—can be regarded as independent of his environment. Plants, animals, men, and soil are a community of interdependent parts, an organism. No organism can survive the decadence of a member. Mr. Babbitt is no more a separate entity than is his left arm, or a single cell of his biceps. Neither are those aggregations of men and earth which we call Madison, or Wisconsin, or America. It may flatter our ego to be called the sons of man, but it would be nearer the truth to call ourselves the brothers of our fields and forests.

The incredible engines wherewith we now hasten our world-conquest have, of course, not heard of these ecological quibblings; neither, perhaps, have the incredible engineers. These engines are double-edged swords. They can be used for ecological coöperation. They are being used for ecological destruction on a scale almost geological in magnitude. In Wisconsin, for example, the northern half of the state has been rendered partially uninhabitable for the next two generations by man-made fire, while the southwestern quarter has been deteriorated for the next century by man-made erosion. In central Wisconsin a single fire in 1930 burned the soil off the better part of two counties.

It can be stated as a sober fact that the iron-heel attitude has already reduced by half the ability of Wisconsin to support a coöperative community of men, animals, and plants during the next century. Moreover, it has saddled us with a repair bill, the magnitude of which we are just beginning to appreciate.

If some foreign invader attempted such loot, the whole nation would resist to the last man and the last dollar. But as long as we loot ourselves, we charge the indignity to "rugged individualism," and try to forget it. But we cannot quite. There is a feeble minority called conservationists, who are indignant about something. They are just beginning to realize that their task involves the reorganization of society, rather than the passage of some fish and game laws.

What has all this to do with the Arboretum? Simply this: If civilization consists of coöperation with plants, animals, soil, and men, then a university which attempts to define that coöperation must have, for the use of its faculty and students, places which show what the land was, what it is, and what it ought to be. This Arboretum may be regarded as a place where, in the course of time, we will build up an exhibit of what was, as well as an exhibit of what ought to be. It is with this dim vision of its future destiny that we have dedicated the greater part of the Arboretum to a reconstruction of original Wisconsin, rather than to a "collection" of imported trees.

The iron-heel mentality is, of course, indifferent to what Wisconsin was. This is exactly the reason why the University cannot be. I am here to say that the invention of a harmonious relationship between men and land is a more exacting task than the invention of machines, and that its accomplishment is impossible without a visual knowledge of the land's history. Take the grass marsh here under our view: From the recession of the glacier until the days of the fur trade, it was a tamarack bog—stems and stumps are still imbedded there. In its successive layers of peat are embalmed both the pollens which record the vegetation of the bog and the surrounding countryside, and also the bones of its animals. During some drouth, man-caused fires burned off the tamarack, which gave place first to grass and brush, and then, under continual burning and grazing, to straight grass. This is the history and status of a thousand other marshes. What will happen if the decomposed surface peat is all burned off? At what stage of the retrogression from forest to meadow is the marsh of greatest use to the animal community? How is that desirable stage to be attained and maintained? What is the role of drainage? These questions are of national importance. They determine the future habitability of the earth, materially and spiritually. They are just as important as whether to join the League of Nations—it is only our iron-heel inheritance which makes the comparison ludicrous. The scientist does not know the answer—he has been too busy inventing machines. The time has come for science to busy itself with the earth itself. The first step is to reconstruct a sample of what we had to start with. That, in a nutshell, is the Arboretum.

Land Pathology [1935]

Dated April 15, 1935, this text of a lecture given to the University of Wisconsin chapter of Sigma Xi was carefully edited in Leopold's hand. As in *Sand County Almanac*, he ambitiously envisions here a unified approach to conservation which would balance private taste and initiative with public interests and incentives. Mindful of economic exigencies, he stresses ethical and esthetic values. In this essay Leopold apparently turned the phrase "land ethic" for the first time.

The properties of animal and plant populations are now to some extent known. Their interactions with environment are becoming predictable. Ecological predictions are made with such certainty as to be used daily in farm, factory, and hospital.

The properties of human populations and their interactions with land are still imperfectly understood. Predictions of behavior are made, but with much uncertainty, and hence are seldom used. Economists, conservationists, and planners are just beginning to become aware that there is a basic ecology involved.

Philosophers have long since claimed that society is an organism, but with few exceptions they have failed to understand that the organism includes the land which is its medium. The properties of human populations, which are the joint domain of sociologist, economist, and statesman, are all conditioned by land.

We may never put society and its land into a test tube, but some of their interactions are discernible by ordinary observation. This paper attempts to define and discuss those which pertain to land conservation.

Conservation is a protest against destructive land use. It seeks to preserve both the utility and beauty of the landscape. It now invokes the aid of science as a means to this end. Science has never before been asked to write a prescription for an esthetic ailment of the body politic. The effort may benefit scientists as well as laymen and land.

Conservationists are sharply divided into groups, interested respectively in soil fertility, soil erosion, forests, parks, ranges, water flows, game, fish, fur, non-game animals, landscape, wild flowers, etc.

These divergent foci of interest clearly arise from individual limitations of taste, knowledge, and experience. They also reflect the age-old conflict between utility and beauty. Some believe the two can be integrated, on the same land, to mutual advantage. Others believe their opposing claims must be fought out and settled by exclusive dedication of each parcel of land to either the one use or the other.

This paper proceeds on two assumptions. The first is that there is only one soil, one flora, one fauna, one people, and hence only one conservation problem. Each acre should produce what it is good for and no two are alike. Hence a certain acre may serve one, or several, or all of the conservation groups.

The second is that economic and esthetic land uses can and must be integrated, usually on the same acre. To segregate them wastes land, and is unsound social philosophy. The ultimate issue is whether good taste and technical skill can both exist in the same landowner. This is a challenge to agricultural education.

When we examine the history of interactions between society and land, there emerge at once a series of observational deductions. We cannot check their accuracy by controlled experiments, but they may at that be more dependable than deductions drawn by historians and statesmen who commonly know nothing of ecology in the lower organisms. These are:

(1) Before the machine age, destructive interactions between society and land tended to right themselves by automatic adjustments similar to those now seen to exist in animal communities. These include population cycles, emigration, starvation, interpredation, etc.
(2) The early phases of machine civilization occurred on land especially resistant to abuse. Northwestern Europe, for example, seems to possess extraordinary recuperative capacity, i.e. capacity, when disturbed, to establish new and stable equilibria between soil, plants, and animals.
(3) Destructive interactions probably contributed to the decay of some early societies even before the machine age. Semi-arid climates such as the eastern Mediterranean, and continental climates such as the Chinese interior, are possibly especially susceptible to upsets of equilibrium. All this, however, is conjectural, due to the possible masking effect of climatic change.
(4) America presents the first instance of a society, heavily equipped with machines, invading a terrain in large part set on a hair-trigger.

The accelerating velocity of destructive interactions is unmistakable and probably unprecedented. Recuperative mechanisms either do not exist, or have not had time to get under way. The mechanism of these interactions in such resources as soil, forests, ranges, and wild life has been traced, at least in its grosser aspects, and found to be strongly inter-connected.

(5) Not all the destruction is wrought directly by machines. The machines release natural forces, such as fire, erosion, floods, and disease, and give them an unnatural play, devoid of checks and balances. Machines also, in one way or another, nullify the checks and balances on domestic animals.

These five assertions may perhaps have weathered enough history to be called deductions. Of equal interest, however, is a further series of opinions based on very recent events. These are:

(6) Remedial practices are being worked out but are not being applied except on public land or at public expense. This presents no sufficient solution because of the universal geographic dispersion of the destructive processes. Public action cannot become universal without breaking down the tax-base which supports it.

(7) The present legal and economic structure, having been evolved on a more resistant terrain (Europe) and before the machine age, contains no suitable ready-made mechanisms for protecting the public interest in private land. It evolved at a time when the public had no interest in land except to help tame it.

(8) The unprecedented velocity of land-subjugation in America involved much hardship, which in turn created traditions which ignore esthetic land uses. The subsequent growth of cities has permitted a re-birth of esthetic culture, but in landless people who have no opportunity to apply it to the soil. The large volume and low utility of conservation legislation may be attributed largely to this maladjustment; also the dissentious character of the conservation movement.

(9) Rural education has been preoccupied with the transplantation of machinery and city culture to the rural community, latterly in the face of economic conditions so adverse as to evict the occupants of submarginal soils. The net result has been to intensify destructive forces on the abandoned land, and to further defer any rebirth of land esthetics in landowners.

With this background, we may now pose the question: What can the social and physical sciences, as now mobilized in this or other universities, do toward hastening the needed adjustment between society as now equipped, and land-use as now practiced?

We may, perhaps, first narrow the field by one exclusion. For the moment, at least, it would seem safe to conclude that all those remedies which hinge upon public purchase, or the extension of existing types of law or administration need no particular stimulation. Their momentum is already great.

We may also conjecture, from recent history, that it will require the injection of some new and potent forces to effect any real change.

In my opinion, there are two possible forces which might operate *de novo*, and which universities might possibly create by research. One is the formulation of mechanisms for protecting the public interest in private land. The other is the revival of land esthetics in rural culture.

The further refinement of remedial practices is equally important, but need not here be emphasized because it already has some momentum.

Out of these three forces may eventually emerge a land ethic more potent than the sum of the three, but the breeding of ethics is as yet beyond our powers. All science can do is to safeguard the environment in which ethical mutations might take place.

The possible ethic, and the philosophical basis for predicting its emergence, has been discussed in several recent publications.[1] Land esthetics lies outside the scope of this paper. A preliminary discussion of vehicles for public influence on private land-use has been published, but will here be restated from a different angle.[2]

A convenient way to open up the subject is to review the sequence of ideas and experiences which led to the present situation.

It was at first assumed that the profit motive would impel landowners to conserve. This expectation is so far frustrated, and we can now see at least three reasons why.

One is that in the presence of excess land, it was cheaper, or at least appeared cheaper, to exploit new land than to conserve old.

Another is that the profit motive operates only during the early stages of land deterioration. It would often pay the individual owner to reclaim slightly damaged land, but in this early stage he does not yet know it is damaged. By the time he sees the damage, it is beyond his means to cure it. It has become a community damage, and thus a charge against the public treasury.

Another is that the competition of synthetic materials, usually of min-

1. Oscar De Beaux, "Biological Ethics," *Italian Mail & Tribune* (Florence, 1932); Aldo Leopold, "The Conservation Ethic," *Journal of Forestry* 31:6 (Oct 1933), 634–643.

2. Aldo Leopold, "Conservation Economics," *Journal of Forestry* 32:5 (May 1934), 537–544.

eral origin, has destroyed confidence in the future of such products as lumber.

When private conservation for profit failed to materialize, legislative compulsion was advanced as an alternative. By this time, however, science had shown good land-use to require much positive skill as well as negative abstention. Compulsion was never tried on any scale.

The recent Lumber Code was a self-imposed compulsion of great promise, but is now thrown out by the Courts. It collided in some of its implications with certain older doctrines of great and massive stability.

Confronted by this succession of obstacles, conservation has now turned to government ownership and subsidy as the way out. The fallacy inherent in this policy has already been pointed out: There is nothing to prevent *all* our vulnerable land from eventually running through the same sequence of private deterioration followed by public repairs.

The system contains the seeds of its own eventual breakdown. It lacks some way to prevent the beginnings of the landslide—some mechanism for checking deterioration while costs are still low. This critical point lies *ipso facto* on private holdings; the government holds only the wrecks. It would cost the government less to prevent these wrecks than repair them. But how prevent them? A vehicle for rewarding good private practices, and penalizing bad ones, is a possible answer, and also the only visible way to prevent the public repairs policy from dying, like the dinosaur, of its own bigness.

Incidentally, such a vehicle could also be used to encourage the conservation of landscape beauty. There never has been even any initial assumption as to how else this could be done. A few parcels of outstanding scenery are immured as parks, but under the onslaughts of mass transportation their possible function as "outdoor universities" is being impaired by the very human need which impelled their creation. Parks are over-crowded hospitals trying to cope with an epidemic of esthetic rickets; the remedy lies not in hospitals, but in daily dietaries. The vast bulk of land beauty and land life, dispersed as it is over a thousand hills, continues to waste away under the same forces as are undermining land utility. The private owner who today undertakes to conserve beauty on his land, does so in defiance of all man-made economic forces from taxes down—or up. There is much beauty left—animate and inanimate—but its existence, and hence its continuity, is almost wholly a matter of accident.

I plead, in short, for positive and substantial public encouragement, economic and moral, for the landowner who conserves the public values—economic or esthetic—of which he is the custodian. The search for practicable vehicles to carry that encouragement is a research problem, and I think a soluble one. A solution apparently calls for a synthesis of biological, legal, and economic skills, or, if you will, a social application of the physical sciences of the sort now sought by this university's "Science Inquiry."

I might say, defensively, that such a vehicle would not necessarily imply regimentation of private land-use. The private owner would still decide what to use his land for; the public would decide merely whether the net result is good or bad for its stake in his holdings.

Those charged with the search for such a vehicle must first seek to intellectually encompass the whole situation. It may mean something far more profound than I have foreseen. Any remedy may imply corollary commitments and changes.

One of these I can see plainly. Every American has tatooed on his left breast the basic premise that manifestations of economic energy are inherently beneficent. Yet there is one which to me seems malignant, not inherently, but because a good thing has outrun its limits of goodness. We learn, in ecology at least, that all truths hold only within limits. Here is a good thing—the improvement in economic tools. It has exceeded the speed, or degree, within which it was good. Equipped with this excess of tools, society has developed an unstable adjustment to its environment, from which both must eventually suffer damage or even ruin. Regarding society and land collectively as an organism, that organism has suddenly developed pathological symptoms, i.e. self-accelerating rather than self-compensating departures from normal functioning. The tools cannot be dropped, hence the brains which created them, and which are now mostly dedicated to creating still more, must be at least in part diverted to controlling those already in hand. Granted that science can invent more and more tools, which might be capable of squeezing a living even out of a ruined countryside, yet who wants to be a cell in that kind of a body politic? I for one do not.

Coon Valley: An Adventure in Cooperative Conservation [1935]

Leopold's concern about soil erosion, developed while he was in the Southwest, found focus also in Wisconsin. With colleagues in the university's College of Agriculture, he helped persuade H. H. Bennett, chief of the newly established U.S. Soil Erosion Service (now the Soil Conservation Service), to establish the nation's first demonstration area for erosion control on the Coon Valley watershed in southwestern Wisconsin in 1933. The project was a cooperative effort among federal technicians, university specialists, and local farmers. It integrated erosion control with other land uses and land values. As such, it was one of the few New Deal efforts that Leopold could celebrate, as he does in this article published in *American Forests*.

There are two ways to apply conservation to land. One is to superimpose some particular practice upon the pre-existing system of land-use, without regard to how it fits or what it does to or for other interests involved.

The other is to reorganize and gear up the farming, forestry, game cropping, erosion control, scenery, or whatever values may be involved so that they collectively comprise a harmonious balanced system of land-use.

Each of our conservation factions has heretofore been so glad to get any action at all on its own special interest that it has been anything but solicitous about what happened to the others. This kind of progress is probably better than none, but it savors too much of the planless exploitation it is intended to supersede.

Lack of mutual cooperation among conservation groups is reflected in laws and appropriations. Whoever gets there first writes the legislative ticket to his own particular destination. We have somehow forgotten that all this unorganized avalanche of laws and dollars must be put in order before it can permanently benefit the land, and that this onerous job, which is evidently too difficult for legislators and propagandists, is being wished upon the farmer and upon the administrator of public properties. The farmer is still trying to make out what it is that the many-voiced public wants him to do. The administrator, who is seldom trained in more than one of the dozen special fields of skill comprising conservation, is growing gray trying to shoulder his new and incredibly varied burdens. The stage, in short, is all set for somebody to show that each of the various public interests in land is better off when all cooperate than when all compete with each other. This principle of integration of land uses has been already carried out to some extent on public properties like the National Forests. But only a fraction of the land, and the poorest fraction at that, is or can ever become public property. The crux of the land problem is to show that integrated use is possible on private farms, and that such integration is mutually advantageous to both the owner and the public.

Such was the intellectual scenery when in 1933 there appeared upon the stage of public affairs a new federal bureau, the United States Soil Erosion Service. Erosion-control is one of those new professions whose personnel has been recruited by the fortuitous interplay of events. Previous to 1933 its work had been to define and propagate an idea, rather than to execute a task. Public responsibility had never laid its crushing weight on their collective shoulders. Hence the sudden creation of a bureau, with large sums of easy money at its disposal, presented the probability that some one group would prescribe its particular control technique as the panacea for all the ills of the soil. There was, for example, a group that would save land by building concrete check-dams in gullies, another by terracing fields, another by planting alfalfa or clover, another by planting slopes in alternating strips following the contour, another by curbing cows and sheep, another by planting trees.

It is to the lasting credit of the new bureau that it immediately decided to use not one, but all, of these remedial methods. It also perceived from the outset that sound soil conservation implied not merely erosion control, but also the integration of all land crops. Hence, after selecting certain demonstration areas on which to concentrate its work, it offered to each farmer on each area the cooperation of the government in installing on his farm a reorganized system of land-use, in which not only soil conservation and agriculture, but also forestry, game, fish, fur, flood-control, scenery, songbirds, or any other pertinent interest were to be duly integrated. It will probably take another decade before the public appreciates either the novelty of such an attitude by a bureau, or the courage needed to undertake so complex and difficult a task.

The first demonstration area to get under way was the Coon Valley watershed, near LaCrosse, in west-central Wisconsin. This paper attempts a thumbnail sketch of what is being done on the Coon Valley Erosion Project. Coon Valley is one of the innumerable little units of the Mississippi Valley which collectively fill the national dinner pail. Its particular contribution is butterfat, tobacco, and scenery.

When the cows which make the butter were first turned out upon the hills which comprise the scenery, everything was all right because there were more hills than cows, and because the soil still retained the humus which the wilderness vegetation through the centuries had built up. The trout streams ran clear, deep, narrow, and full. They seldom overflowed. This is proven by the fact that the first settlers stacked their hay on the creekbanks, a procedure now quite unthinkable. The deep loam of even the steepest fields and pastures showed never a gully, being able to take on any rain as it came, and turn it either upward into crops, or downward into perennial springs. It was a land to please everyone, be he an empire-builder or a poet.

But pastoral poems had no place in the competitive industrialization of pre-war America, least of all in Coon Valley with its thrifty and ambitious Norse farmers. More cows, more silos to feed them, then machines to milk them, and then more pasture to graze them—this is the epic cycle which tells in one sentence the history of the modern Wisconsin dairy farm. More pasture was obtainable only on the steep upper slopes, which were timber to begin with, and should have remained so. But pasture they now are, and gone is the humus of the old prairie which until recently enabled the upland ridges to take on the rains as they came.

Result: Every rain pours off the ridges as from a roof. The ravines of the grazed slopes are the gutters. In their pastured condition they cannot resist the abrasion of the silt-laden torrents. Great gashing gullies are torn out of the hillside. Each gully dumps its load of hillside rocks upon the fields of the creek bottom, and its muddy waters into the already swollen streams. Coon Valley, in short, is one of the thousand farm communities which, through the abuse of its originally rich soil, has not only filled the national dinner pail, but has created the Mississippi flood problem, the navigation problem, the overproduction problem, and the problem of its own future continuity.

The Coon Valley Erosion Project is an attempt to combat these national evils at their source. The "nine-foot channel" and endless building of dykes, levees, dams and harbors on the lower river, are attempts to put a halter on the same bull after he has gone wild.

The Soil Erosion Service says to each individual farmer in Coon Valley: "The government wants to prove that your farm can be brought back. We will furnish you free labor, wire, seed, lime, and planting stock, if you will help us reorganize your cropping system. You are to give the new system a 5-year trial." A total of 315 farmers, or nearly half of all the farms in the watershed, have already formally accepted the offer. Hence we now see foregathered at Coon Valley a staff of technicians to figure out what should be done; a C.C.C. camp to perform labor; a nursery, a seed warehouse, a lime quarry, and other needed equipments; a series of contracts with farmers, which, collectively, comprise a "regional plan" for the stabilization of the watershed and of the agricultural community which it supports.

The plan, in a nutshell, proposes to remove all cows and crops from steep slopes, and to use these slopes for timber and wildlife only. More intensive cultivation of the flat lands is to make up for the retirement of the eroding hillsides. Gently sloping fields are to be terraced or strip-cropped. These changes, plus contour farming, good crop rotations, and the repair of eroding gullies and stream banks, constitute the technique of soil restoration.

The steep slopes now to be used for timber and game have heretofore been largely in pasture. The first visible evidence of the new order on a Coon Valley farm is a C.C.C. crew stringing a new fence along the contour which marks the beginning of forty per cent gradients. This new fence commonly cuts off the upper half of the pasture. Part of this upper half still bears timber, the rest is open sod. The timbered part has been grazed clear of undergrowth, but with protection this will come back to brush and young timber and make range for ruffed grouse. The open part is being planted, largely to conifers—white pine, Norway pine, and Norway spruce for north slopes, Scotch pine for south slopes. The dry south slopes present a special problem. In pre-settlement days they carried hazel, sumac, and bluestem rather than timber, the grass furnishing the medium for quick hot fires. Will these hot dry soils, even under protection, allow the planted Scotch pine to thrive? I doubt it. Only the north slopes and coves will develop commercial timber, but all the fenced land can at least be counted upon to produce game and soil cover and cordwood.

Creek banks and gullies, as well as steep slopes, are being fenced and planted. Despite their much smaller aggregate area, these bank plantings will probably add more to the game carrying capacity of the average farm than will the larger solid blocks of plantings on slopes. This prediction is based on their superior dispersion, their higher proportion of deciduous species, and their richer soils.

The bank plantings have showed up a curious hiatus in our silvicultural knowledge. We have learned so much about the growth of the noble conifers that we employ higher mathematics to express the profundity of our information, but at Coon Valley there have arisen, unanswered, such sobering elementary questions as this: What species of willow grow from cuttings? When and how are cuttings made, stored, and planted? Under what conditions will sprouting willow logs take root? What shrubs combine thorns, shade tolerance, grazing resistance, capacity to grow from cuttings, and the production of fruits edible by wild life? What are the comparative soil-binding properties of various shrub and tree roots? What shrubs and trees allow an understory of grass to grow, thus affording both shallow and deep rootage? How do native shrubs or grasses compare with cultivated grasses for rootbinding terrace outlets? What silvicultural treatment favors an ironwood understory to furnish buds for grouse? Can white birch for budding be planted on south slopes? Under what conditions do oak sprouts retain leaves for winter game cover?

Forestry and fencing are not the alpha and omega of Coon Valley technique. In odd spots of good land near each of the new game coverts, the observer will see a newly enclosed spot of a half-acre each. Each of these little enclosures is thickly planted to sorghum, kaffir, millet, proso, sunflower. These are the food patches to forestall winter starvation in wild life. The seed and fence were furnished by the government, the cultivation and care by the farmer. There were 337 such patches grown in 1934—the largest food-patch system in the United States, save only that found on the Georgia Quail Preserves. There is already friendly rivalry among many farmers as to who has the best food patch, or the most birds using it. This feeding system is, I think, accountable for the fact that the population of quail in 1934–35 was double that of 1933–34, and the pheasant population was quadrupled. Such a feeding system, extended over all the farms of Wisconsin, would, I think, double the crop of farm game in a single year.

This whole effort to rebuild and stabilize a countryside is not without its disappointments and mistakes. A December blizzard flattened out most of the food-patches and forced recourse to hopper feeders. The willow cuttings planted on stream banks proved to be the wrong species and refused to grow. Some farmers, by wrong plowing, mutilated the new terraces just built in their fields. The 1934 drouth killed a large part of the plantings of forest and game cover.

What matter, though, these temporary growing pains when one can cast his eyes upon the hills and see hard-boiled farmers who have spent their lives destroying land now carrying water by hand to their new plantations? American lumbermen may have become so steeped in economic determinism as actually to lack the personal desire to grow trees, but not Coon Valley farmers! Their solicitude for the little evergreens is sometimes almost touching. It is interesting to note, however, that no such pride or tenderness is evoked by their new plantings of native hardwoods. What explains this difference in attitude? Does it arise from a latent sentiment for the conifers of the Scandinavian homeland? Or does it merely reflect that universal urge to capture and domesticate the exotic which found its first American expression in the romance of Pocohontas, and its last in the Americanization of the ringnecked pheasant?

Most large undertakings display, even on casual inspection, certain policies or practices which are diagnostic of the mental attitude behind the whole venture. From these one can often draw deeper inferences than from whole volumes of statistics. A diagnostic policy of the Coon Valley staff is its steadfast refusal to straighten streams. To those who know the speech of hills and rivers, straightening a stream is like shipping vagrants—a very successful method of passing trouble from one place to the next. It solves nothing in any collective sense.

Not all the sights of Coon Valley are to be seen by day. No less distinctive is the nightly "bull session" of the technical staff. One may hear a forester expounding to an engineer the basic theory of how organic matter in the soil decreases the per cent of run-off; an economist holds forth on tax rebates as a means to get farmers to install their own erosion control. Underneath the facetious conversation one detects a vein of thought—an attitude toward the common enterprise—which is strangely reminiscent of the early days of the Forest Service. Then, too, a staff of technicians, all under thirty, was faced by a common task so large and so long as to stir the imagination of all but dullards. I suspect that the Soil Erosion Service, perhaps unwittingly, has recreated a spiritual entity which many older conservationists have thought long since dead.

Review of Elton, *Exploring the Animal World* [1935]

During his career as a university professor, Leopold published reviews of dozens of books in the fields of natural history and conservation, many of them gems of the genre. In this brief review of a work by the famed British zoologist Charles Elton, published in Bird-Lore, *the magazine of the National Audubon Society, Leopold pays tribute to Elton's extraordinary skill at communicating complex ecological principles to nonprofessionals. Leopold and Elton, who had met at a scientific conference several years earlier, had their own mutual admiration society grounded in appreciation of each other's ability to conceptualize new advances in ecology and express them in language any thinking individual could grasp.*

Exploring the Animal World. By Charles Elton, Director of the Bureau of Animal Populations, University of Oxford. George Allen and Unwin, Ltd., Museum St., London, 1933.

Sciences, like lodges, luncheon clubs, and legions, tend to become mutual admiration societies. One quiet form of mutual flattery prevalent in scientific groups is the tacit assumption by each that its particular subject matter is too deep, difficult, and devious to be understood by laymen. Unfortunately this appears to be often actually the case.

Until I read Elton's recent book, I for one believed it to be impossible to present the science of ecology to laymen. Now it is clear enough that it was impossible only because the right man had never tried it. After all, the intelligent layman can understand any science that can hang its subject matter on his mental hatrack. Elton has discovered that human politics, sociology, and economics are admirable hatracks for animal ecology, and he uses them to excellent effect. He has produced an authentic picture of ecological mechanisms light enough to be read for pure entertainment and withal substantial enough to earn a place in solid scientific libraries.

Exploring the Animal World is, to my thinking, a sort of key that may unlock, for lay inspection, the true contents of many a seemingly dull volume, and the true merit of many a seemingly lost cause.

For example, we need look no farther than the moot question of predator-control. We have bombarded the layman for years with statistics of food habits of predators, and we have assured him that natural balances are too complex to be lightly tampered with. We have told him that he may injure rather than benefit himself by shooting the Hawk that displays an interest in his game or chickens, but we have never given him, in lay language, a convincing picture of the elaborate natural mechanisms by virtue of which these things are true. Elton has come nearer doing this than any other recent writer.

His pages are completely devoid of propaganda or of preoccupation with any particular animal or issue. This somehow enhances their value as propaganda for that attitude toward Nature which he himself epitomizes in one sentence of characteristic simplicity. "We ought," he says, "to feel continually pleased and excited that so many kinds of animals are able to live with us on the same island."

Wilderness [1935]

Of the several papers forthcoming from Leopold's three-month trip to Germany in 1935 to study German methods of forestry and wildlife management, "Deer and *Dauerwald* in Germany" and "*Naturschutz* in Germany" are the best known. Of them all, however, this undated, handwritten draft of a speech is the most evocative and intimate. It is less about wilderness, or even wildness, than about its absence in Germany. This realization haunted Leopold and strengthened his determination to avoid the same fate in America. His German audience would have recognized his reference in the last sentence to the *Erlkönig*, a figure from German folklore immortalized in a poem by Goethe.

To an American conservationist, one of the most insistent impressions received from travel in Germany is the lack of wildness in the German landscape.

Forests are there—interminable miles of them, spires of spruce on the skyline, glowering thickets in ravines, and many a quick glimpse "where the yellow pines are marching straight and stalwart up the hillside where they gather on the crest." Game is there—the skulking roebuck or even a scurrying *Rudel* of red-deer is to be seen any evening, even from a train-window. Streams and lakes are there, cleaner of cans and old tires than our own, and no worse beset with hotels and "bide-a-wee" cottages. But yet, to the critical eye, there is something lacking that should not be lacking in a country which actually practices, in such abundant measure, all of the things we in America preach in the name of "conservation." What is it?

Let me admit to begin with the obvious difference in population density, and hence in population pressure, on the economic mechanisms of landuse. I knew of that difference before coming over, and think I have made allowance for it. Let it further be clear that I did not hope to find in Germany anything resembling the great "wilderness areas" which we dream and talk about, and sometimes briefly set aside, in our National Forests and Parks. Such monuments to wilderness are an esthetic luxury which Germany with its timber deficit and the evident land-hunger of its teeming millions, cannot afford. I speak rather of a certain quality which should be but is not found in the ordinary landscape of producing forests and inhabited farms, a quality which still in some measure persists in some of the equivalent landscapes of America, and which we I think tacitly assume will be enhanced by rather than lost in the hoped-for practice of conservation. I speak specifically to the question of whether and under what limitations that assumption is correct.

It may be well to first inquire whether the Germans themselves, who know and love their rocks and rills with an intensity long patent to all the world, admit any such esthetic deficit in their countryside. "Yes" and "no" are of course worthless as criteria of such a question. I offer in evidence, first, the existence of a very vigorous esthetic discontent, in the form of a "Naturschutz" (nature-protection) movement, the equivalent of which preceded the emergence of the wilderness idea in America. This impulse to save wild remnants is always, I think, the forerunner of the more important and complex task of mixing a degree of wildness with utility. I also submit that the Germans are still reading Cooper's "Leatherstocking" and Parkman's "Oregon Trail," and still flock to the wild-west movies. And when I asked a forester with a philosophical bent why people did not flock to his forest to camp out, as in America, he shrugged his shoulders and remarked that perhaps the tree-rows stood too close together for convenient tenting! All of which, of course, does not answer the question. Or does it?

And this calls to mind what is perhaps the first element in the German deficit: their former passion for unnecessary outdoor geometry. There is a lag in the affairs of men—the ideas which were seemingly buried with the cold hard minds of the early-industrial era rise up out of the earth today for us to live with. Most German forests, for example, though laid out over a hundred years ago, would do credit to any cubist. The trees are not only in rows and all of a kind, but often the various age-blocks are parallelograms, which only an early discovery of the ill-effects of wind saved from being rectangles. The age-blocks may be in ascending series—1, 2, 3—like the proverbial stepladder family. The boundary between wood and field tends to be sharp, straight, and absolute, unbroken by those charming little indecisions in the form of draw, coulee, and stump-lot, which, especially in our "shiftless" farming regions, bind wood and field into an harmonious whole. The Germans are now making a determined effort to get away from cubistic forestry—experience has revealed that in about the third successive crop of conifers in "pure" stands the microscopic flora of the soil becomes upset and the trees quit growing, but it will be another generation before the new policy emerges in landscape form.

Not so easily, though, will come any respite from what the geometrical mind has done to the German rivers. If there were only room for them, it would be a splendid idea to collect all the highway engineers in the world, and also their intellectual kith and kin the Corps of Army Engineers, and settle them for life upon the perfect curves and tangents of some "improved" German river. I am aware, of course, that there are weighty commercial reasons for the canalization of the larger rivers, but I also saw many a creek and rivulet laid out straight as a dead snake, and with masonry banks to boot. I am depressed by such indignities, and I have black misgivings over the swarm of new bureaus now out to improve the American countryside. It is, I think, an historical fact that no American bureau equipped with money, men, and machines ever refused *on principle* to straighten a river, save only one—the Soil Conservation Service.

Another more subtle (and to the average traveller, imperceptible) element in the deficit of wildness is the near-extirpation of birds and animals of prey. I think it was Stewart Edward White who said that the existence of one grizzly conferred a flavor to a whole county. From the German hills that flavor has vanished—a victim to the misguided zeal of the game-keeper and the herdsman. Even the ordinary hawks are nearly gone—in four months travel I counted only ____. And the great owl or "Uhu"—without whose vocal austerity the winter night becomes a mere blackness—persists only in the farthest marches of East Prussia. Before our American sportsmen and game keepers and stockmen have finished their self-appointed task of extirpating our American predators, I hope that we may begin to realize a truth already written bold and clear on the German landscape: that success in most over-artificialized land-uses is bought at the expense of the public interest. The game-keeper buys an unnatural abundance of pheasants at the expense of the public's hawks and owls. The fish-culturist buys an unnatural abundance of fish at the expense of the public's herons, mergansers, and terns. The forester buys an unnatural increment of wood at the expense of the soil, and in that wood maintains an unnatural abundance of deer at the expense of all palatable shrubs and herbs.

This effect of too many deer on the ground flora of the forest deserves special mention because it is an illusive burglary of esthetic wealth, the more dangerous because unintentional and unseen. Forest undergrowth consists of many species, some palatable to deer, others not. When too dense a deer population is built up, and there are no natural predators to trim it down, the palatable plants are grazed out, whereupon the deer must be artificially fed by the game-keeper, whereupon next year's pressure on the palatable species is still further increased, etc. ad infinitum. The end result is the extirpation of the palatable plants—that is to say an unnatural simplicity and monotony in the vegetation of the forest floor, which is still further aggravated by the too-dense shade cast by the artificially crowded trees, and by the soil-sickness already mentioned as arising from conifers. One is put in mind of

Shakespeare's warning that "virtue, grown into a pleurisy, dies of its own too-much." Be that as it may, the forest landscape is deprived of a certain exuberance which arises from a rich variety of plants fighting with each other for a place in the sun. It is almost as if the geological clock had been set back to those dim ages when there were only pines and ferns. I never realized before that the melodies of nature are music only when played against the undertones of evolutionary history. In the German forest—that forest which inspired the *Erlkönig*—one now hears only a dismal fugue out of the timeless reaches of the carboniferous.

Threatened Species [1936]

This essay, published in *American Forests*, documents Leopold's one-hundred-eighty-degree reversal on the question of predators. It also demonstrates that Leopold was keenly aware of another, less heralded, shift of perspective paralleling the reevaluation of predators: the shift from "game" management to "wildlife" management. Here Leopold pleads strongly and eloquently for a broad public commitment not only to the preservation of predators but also to the preservation of other rare and endangered wild species—flora as well as fauna. Here too he outlines the structural means for a coordinated assault on the problem, describing everything from forming an independent joint committee which would oversee the inventory and management of threatened species to redefining the mission of the National Parks and eliciting the cooperation of far-flung chapters of private conservation organizations and their individual members.

The volume of effort expended on wildlife conservation shows a large and sudden increase. This effort originates from diverse courses, and flows through diverse channels toward diverse ends. There is a widespread realization that it lacks coordination and focus.

Government is attempting to secure coordination and focus through reorganization of departments, laws, and appropriations. Citizen groups are attempting the same thing through reorganization of associations and private funds.

But the easiest and most obvious means to coordination has been overlooked: explicit definition of the immediate needs of particular species in particular places. For example: Scores of millions are being spent for land purchase, C.C.C. labor, fences, roads, trails, planting, predator-control, erosion control, poisoning, investigations, water developments, silviculture, irrigation, nurseries, wilderness areas, power dams, and refuges, within the natural range of the grizzly bear.

Few would question the assertion that to perpetuate the grizzly as a part of our national fauna is a prime duty of the conservation movement. Few would question the assertion that any one of these undertakings, at any time and place, may vitally affect the restoration of the grizzly, and make it either easy or impossible of accomplishment. Yet no one has made a list of the specific needs of the grizzly, in each and every spot where he survives, and in each and every spot where he might be reintroduced, so that conservation projects in or near that spot may be judged in the light of whether they *help or hinder* the perpetuation of the noblest of American mammals.

On the contrary, our plans, departments, bureaus, associations, and movements are all focused on abstract categories such as recreation, forestry, parks, nature education, wildlife research, more game, fire control, marsh restoration. Nobody cares anything for these except as means toward ends. What ends? There are of course many ends which cannot and many others which need not be precisely defined at this time. But it admits of no doubt that the immediate needs of threatened members of our fauna and flora must be defined now or not at all.

Until they are defined and made public, we cannot blame public agencies, or even private ones, for misdirected effort, crossed wires, or lost opportunities. It must not be forgotten that the abstract categories we have set up as conservation objectives may serve as alibis for blunders, as well as ends for worthy work. I cite in evidence the C.C.C. crew which chopped down one of the few remaining eagle's nests in northern Wisconsin, in the name of "timber stand improvement." To be sure, the tree was dead, and according to the rules, constituted a fire risk.

Most species of shootable non-migratory game have at least a fighting chance of being saved through the process of purposeful manipulation of laws and environment called management. However great the blunders, delays, and confusion in getting management of game species under way, it remains true that powerful motives of local self-interest are at work in their behalf. European countries, through the operation of these motives, have saved their resident game. It is an ecological probability that we will evolve ways to do so.

The same cannot be said, however, of those species of wilderness game which do not adapt themselves to economic land-use, or of migratory birds which are owned in common, or of non-game forms classed as predators, or of rare plant associations which must compete with economic plants and livestock, or in general of all wild native forms which fly at large or have only an esthetic and scientific value to man. These, then, are the special and immediate concern of this inventory. Like game, these forms depend for their perpetuation on protection and a favorable environment. They need "management"—the perpetuation of good habitat—just as game does, but the ordinary motives for providing it are lacking. They are the threatened element in outdoor America—the crux of conservation policy. The new organizations which have now assumed the name "wildlife" instead of "game," and which aspire to implement the wildlife movement, are I think obligated to focus a substantial part of their effort on these threatened forms.

This is a proposal, not only for an inventory of threatened forms in each of their respective places of survival, but an inventory of the information, techniques, and devices applicable to each species in each place, and of local human agencies capable of applying them. Much information exists, but it is scattered in many minds and documents. Many agencies are or would be willing to use it, if it were laid under their noses. If for a given problem no information exists, or no agency exists, that in itself is useful inventory.

For example, certain ornithologists have discovered a remnant of the Ivory-billed Woodpecker—a bird inextricably interwoven with our pioneer tradition—the very spirit of that "dark and bloody ground" which has become the locus of the national culture. It is known that the Ivory-bill requires as its habitat large stretches of virgin hardwood. The present remnant lives in such a forest, owned and held by an industry as reserve stumpage. Cutting may begin, and the Ivory-bill may be done for at any moment. The Park Service has or can get funds to buy virgin forests, but it does not know of the Ivory-bill or its predicament. It is absorbed in the intricate problem of accommodating the public which is mobbing its parks. When it buys a new park, it is likely to do so in some "scenic" spot, with the general objective of making room for more visitors, rather than with the specific objective of perpetuating some definite thing to visit. Its wildlife program is befogged with the abstract concept of inviolate sanctuary. Is it not time to establish particular parks or their equivalent for particular "natural wonders" like the Ivory-bill?

You may say, of course, that one rare bird is no park project—that the Biological Survey should buy a refuge, or the Forest Service a National Forest, to take care of the situation. Whereupon the question bounces back: the Survey has only duck money; the Forest Service would have to cut the timber. But is there anything to prevent the three possible agencies concerned from getting together and agreeing whose job this is, and while they are at it, a thousand other jobs of like character? And how much each would cost? And just what needs to be done in each case? And can anyone doubt that the public, through Congress, would support such a program? Well—this is what I mean by an inventory and plan.

Some sample lists of the items which need to be covered are wilderness and other game species, such as grizzly bear, desert and bighorn sheep, caribou, Minnesota remnants of spruce partridge, masked bobwhite, Sonora deer, peccary, sagehen; predator and allied species, such as the wolf, fisher, otter, wolverine and Condor; migratory birds, including the trumpeter swan, curlews, sandhill crane, Brewster's warbler; plant associations, such as prairie floras, bog floras, Alpine and swamp floras.

In addition to these forms, which are rare everywhere, there is the equally important problem of preserving the attenuated edges of species common at their respective centres. The turkey in Colorado, or the ruffed grouse in Missouri, or the antelope in Nebraska, are rare species within the meaning of this document. That there are grizzlies in Alaska is no excuse for letting the species disappear from New Mexico.

It is important that the inventory represent not merely a protest of those privileged to think, but an agreement of those empowered to act. This means that the inventory should be made by a joint committee of the conservation bureaus, plus representatives of the Wildlife Conference as representing the states and the associations. The plan for each species should be a joint commitment of what is to be done and who is to do it. The bureaus, with their avalanche of appropriations, ought to be able to loan the necessary expert personnel for such a committee, without extra cost. To sift out any possible imputation of bureaucratic, financial, or clique interest, the inter-bureau committee should feed its findings to the public through a suitable group in the National Research Council, and subject to the Council's approval. The necessary incidental funds for a secretary, for expense of gathering testimony and maps, and for publications might well come from the Wildlife Institute, or from one of the scientific foundations.

There is one cog lacking in the hoped-for machine: a means to get some kind of responsible care of remnants of wildlife remote from any bureau or its field officers. Funds can hardly be found to set up special paid personnel for each such detached remnant. It is of course proved long ago that closed seasons and refuge posters without personnel are of no avail. Here is where associations with their far-flung chapters, state officers or departments, or even private individuals can come to the rescue. One of the tragedies of contemporary conservation is the isolated individual or group who complains of having no job. The lack is not of jobs, but of eyes to see them.

The inventory should be the conservationist's eye. Every remnant should be definitely entrusted to a custodian—ranger, warden, game manager, chapter, ornithologist, farmer, stockman, lumberjack. Every conservation meeting—national, state, or local—should occupy itself with hearing their annual reports. Every field inspector should contact their custodians—he might often learn as well as teach. I am satisfied that thousands of enthusiastic conservationists would be proud of such a public trust, and many would execute it with fidelity and intelligence.

I can see in this set-up more conservation than could be bought with millions of new dollars, more coordination of bureaus than Congress can get by new organization charts, more genuine contacts between factions than will ever occur in the war of the inkpots, more research than would accrue from many gifts, and more public education than would accrue from an army of orators and organizers. It is, in effect, a vehicle for putting Jay Darling's concept of "ancestral ranges" into action on a quicker and wider scale than could be done by appropriations alone.

Means and Ends in Wild Life Management [1936]

One of the most heavily edited pencil drafts found in Leopold's desk file of unpublished manuscripts after his death is this highly significant paper apparently prepared for delivery at the Chamberlain Science Club of Beloit College on May 5, 1936. Leopold tried four different titles and did not address all the topics in his original outline—which may help to explain the unsatisfactory ending. Nevertheless, the paper is included here as an illustration of Leopold's effort to grapple with the problem of means and ends in wildlife management, the profession that he pioneered during the period of his career when he underwent the most profound transmutation of his intellectual outlook. As recently as 1933 in "The Conservation Ethic" and in *Game Management*, he had been optimistic about the possibilities of a controlled environment, but he was now telling students that wildlife managers had discovered that they were unable to replace natural equilibria with artificial ones by means of scientific "controls" and would not want to even if they could.

When a member of a scientific group tries to uproot himself and describe his own undertakings with the objective pen of a spectator the task is liable to put quite a strain both on his modesty and on his sense of humor. Sometimes he exceeds the elastic limit of these somewhat friable materials.

Hence, out of caution, I will claim no importance for the new profession of wild-life management save only this: that its services are vital to those atavistic few for whom a world without wild things would be no world at all—even though every citizen in it were given a Ford airplane, a Wright house, and a Townsend pension. I do not know how numerous these atavists are or how long it will take for them to die out. The evolutionary probability would indicate a pretty long time.

Peculiarities of Wild Life Research

As a research group we are repeating the history of agriculture, in that we employ biology to build up a cultural technique for which trial-and-error proved too slow. We are peculiar, though, in that our tools are scientific whereas our output is weighed in esthetic satisfaction, rather than in economic pounds or dollars. An echelon of wild geese is economic only in the sense that an actress is. Both have physical substance, and people will pay to see or hear them. But the measure of their worth is wholly qualitative and inheres in "personality." Strangely enough, though, our senses do not differentiate individual personality in animals. Our perceptions of character are of the species. In edible game there is an incidental meat value, but its triviality is conveyed by the reflection that a single inedible Carolina Parakeet today may be worth more than a million edible pheasants.

In Audubon's day, before there were any pheasants, but with parakeets in every burdock patch, the opposite may have been true. It would appear, then, that the value of wild things is in part a scarcity value, like that of gold. It is also in part an artistic value, like that of a painting. The final arbiter of both is that elusive entity known as "good taste." There is, though, this residual difference: a painting might conceivably be re-created, but an extinct species never.

In any event, there is no standard scale wherein the wildlife manager may weigh his output. Consequently there is the widest conceivable range of opinion as to what is worth trying and as to which of two conflicting ends is the more important. There seem to be few fields of research where the means are so largely of the brain, but the ends so largely of the heart. In this sense the wild life manager is perforce a dual personality. Whether he achieves any degree of consistency as between his tools and his objectives, we must leave for others to judge.

As in agriculture, the husbandry of animals deals mostly with the husbandry of plants. In the construction of a plant environment, though, we have a shorter tether, both in the choice of plants and the methods of growing them. Cultivated food and cover are used for wild animals, but only within certain esthetic and economic limits. Pheasants, quail, and Hungarian partridges could conceivably be grown in a wholly cultivated plant habitat, but it would cost too much, and, moreover, it would be, by tacit agreement, "unsuitable." On the other hand, grizzly bears and mountain sheep tolerate no cultivation at all, unless it be those nearly invisible cultures known as extensive forestry and range management. The strong esthetic element in our evaluation of wildlife is reflected in the higher unit worth accorded, by common consent, to these intolerant animals or "wilderness game."

Scientific wild life management, while far younger than scientific agriculture, has perhaps forged ahead of it in one point of its philosophy: the recognition of invisible interdependencies in the biotic community. When some women's club protests against the "control" of game-killing hawks, or the poisoning of stock-killing carnivores, or of crop-eating rodents, they are raising—whether they know it or not—a new and fundamental issue in human land-use. Agriculture has assumed that by the indefinite pyramiding of new "controls," an artificial plant-animal community can be substituted for the natural one. There are omens that this assumption may be false. Pests and troubles in need of control seem to be piling up even faster than new science and new dollars for control work. Ecologists like Weaver are discovering that on the plains the physical structure and moisture regimen of the soil deteriorate under even the best agriculture. Granulation and moisture equilibrium can be restored, he thinks, by restoring the native vegetation. Perhaps a periodic reversion to prairie is to be the price of farming the inland empire.

Wild life management, however, has already admitted its inability to replace natural equilibria with artificial ones, and its unwillingness to do so even if it could. Forestry, range-management, and landscape architecture—all "semi-wild" cultures—are rapidly falling in line. Like wild-life management, their dependency on natural equilibria tends to rest on esthetic as well as economic arguments. Even erosion-control, the newest and most important of the pyramid of "fixatives" created by research, is tending to lean on natural techniques wherever they suffice. How far agriculture itself is to be invaded by this naturalistic philosophy of land-use is one of the unanswered issues of the future.

Our inability to synthesize wild-life environments carries with it an inability to isolate variables in research. The food of a hen, being handmade, can be studied *ad lib* under controlled laboratory conditions, but the food of a quail cannot. The annual dietary of a quail may cover a thousand species, of which hundreds may be available at a single time and place. An individual quail can be kept alive on a diet as simple as a hen's, but a quail population, in order to display the full complexity of that pattern of behavior which is the subject-matter of management, must apparently have available the whole gamut of its foods, coverts, and enemies. There are now grounds for suspecting that a given item of food may fluctuate seasonally in value, without any visible change in physical condition. Thus quail may riddle a sorghum patch in November when the choice of alternatives is large, but reject the same sorghum in February when the alternative foods are limited.

The difficulty of isolating variables leads the game manager to place first dependence on observation in-the-wild. From these observed behaviors he tries to select items which are susceptible of experimental verification, but the unnatural simplicity of his controlled tests forces him to be suspicious of even the most carefully verified results. Few "conclusions" in wild-life research stay put for a long period.

The Cycle: A Problem Defying the Experimental Method

Hares and grouse in northern latitudes display alternating periods of abundance and scarcity. Until a decade ago, these fluctuations were assumed to be local and sporadic, and were ascribed to weather or other ordinary disturbances.

When improving communications showed them to be synchronous over large areas, the word "cycle" entered the popular vocabulary, and speculation began as to "causes." Invasions of arctic raptors, parasites, bacterial diseases, and "sunspots" were successively postulated as cycle-mechanisms. While all are evidently associated with the cycle, none appears to afford a satisfactory explanation, for in the course of examining them it became evident that a large variety of birds and mammals show cycles of different lengths, and that migratory movements as well as population levels are affected. While the original search was for causes of mortality, the present search is for causes of fluctuation in reproduction and in mobility. At least for the moment, lethal agents tend to be regarded as effects rather than causes.

Laboratory procedures were of course invoked to study disease, but to devise a laboratory procedure which will isolate the factors of fluctuation in reproduction rate is a more difficult matter. What species shall be used? Grouse are difficult to breed at all in captivity. The quails and pheasants which breed easily are, so far as known, not cyclic. The best chance for controlled work seems to lie in hares or small rodents, but even there the difficulties are great. The cycle mechanism is probably tied in with food in some way, hence natural foods must be used. What are they, and how shall they be fed to simulate nature? Moreover the intolerances associated with crowding, which are probably not involved in cycles, may nevertheless mask the factors which are. If we use large enclosures to get away from crowding, how is predation by aerial predators to be controlled or kept down to normal? These unanswered questions represent the present impasse of cycle research. They arise from the peculiarity already noted: that the wild animal population does not display its normal behavior except in the presence of its normal environment.

I suppose one way out is to split up the question into smaller parts, some which may be tested on domesticated species. A virgin field awaits the investigator who is able to do this splitting. It seems to call for more and better hypotheses in a field already surfeited with them.

Conservationist in Mexico [1937]

Leopold's personal transmutation of values in the mid-1930s was triggered in large part by the fortuitous juxtaposition of his travels to Germany in 1935 and to the Chihuahua Sierra of Mexico in 1936. Although several years would pass before he would appreciate the full impact of that experience on his evolving concept of land health, he realized immediately that the Sierra Madre represented an intact, stable landscape unlike anything he had seen in the United States. This essay, published in *American Forests*, describes that undisturbed ecosystem south of the border. Here Leopold's changed attitude toward predators and fire is evident, and his more mature appreciation of the value of wilderness to science is anticipated. Also, his earlier analyses of land abuse in the Southwest are here united with his later stress on the importance of individual ethics as the cornerstone of an effective conservation policy.

The predatory Apache of our Southwest was early rounded up and confined in reservations, whereas across the line in Mexico he was, until his recent near-extinction, allowed to run at large. Therefore our southwestern mountains are now badly gutted by erosion, whereas the Sierra Madre range across the line still retains the virgin stability of its soils and all the natural beauty that goes with that enviable condition.

This seemingly disconnected reasoning will appear absurd only to those who still believe that the world is composed of a number of things, the inter-relationships of which are obvious or nearly so.

As a matter of fact, the statement is substantially accurate. This article aims to explain why and to philosophize on the irony of it. For it is ironical that Chihuahua, with a history and a terrain so strikingly similar to southern New Mexico and Arizona should present so lovely a picture of ecological health, whereas our own states, plastered as they are with National Forests, National Parks and all the other trappings of conservation, are so badly damaged that only tourists and others ecologically color-blind, can look upon them without a feeling of sadness and regret.

Let me hasten to add that this enviable contrast holds good only for the mountains. The low country on both sides of the line has been equally abused and spoiled. The Sierras escaped because of the mutual fear and hatred between Apaches and Mexicans. So great was the fear of Indians that the Sierras were never settled, hence never grazed, hence never eroded. This holds true up to Pancho Villa's revolution of 1916. During the revolution bandits performed the same ecological function as Indians. Since then, depression and unstable land policies have served to keep the mountains green.

It is this chain of historical accidents which enables the American conservationist to go to Chihuahua today and feast his eyes on what his own mountains were like before the Juggernaut. To my mind these live oak-dotted hills fat with side oats grama, these pine-clad mesas spangled with flowers, these lazy trout streams burbling along under great sycamores and cottonwoods, come near to being the cream of creation. But on our side of the line the grama is mostly gone, the mesas are spangled with snakeweed, the trout streams are now cobble-bars.

Somehow the watercourse is to dry country what the face is to human beauty. Mutilate it and the whole is gone. The rest of the organism may survive and even do useful work. The economist, the engineer, or the forester may feel there has been no great loss and adduce statistics of production to prove it. But there are those who know, nevertheless, that a great wrong has been committed—perhaps the greatest of all wrongs, and the sadder because both unintentional and irretrievable.

The Chihuahua Sierras burn over every few years. There are no ill effects, except that the pines are a bit farther apart than ours, reproduction is scarcer, there is less juniper, and there is much less brush, including mountain mahogany—the cream of the browse feed. But the watersheds are intact, whereas our own watersheds, sedulously protected from fire, but mercilessly grazed before the forests were created, and much too hard since, are a wreck. If there be those who do not yet know they are a wreck, let them read Will C. Barnes' history of the San Simon valley of Arizona in the October issue of *American Forests*.

The Chihuahua Sierras have been grazed only near the Mormon colonies. The Mormons were not afraid of Apaches and they sprinkled many a mountain valley with their brick ranch houses. Near the colony I visited—Colonia Pacheco—overgrazing and erosion have not progressed as far as they had in the White Mountains of Arizona in 1910. But the colonies are microscopic when compared with the bulk of the mountain area, which from my observation is for the most part ungrazed.

Very recently the Mexican "Resettlement Administration" has scattered landless voters over many a non-irrigable mountain valley, to dry-farm if the

Lord sent rain and to get along somehow in any event. The only improvement over our own Act of June 11, 1906, is that the scattering is done only where there is enough land for a community and that the settlers have no guns.

These forest homesteaders are "deadening" the pines, scratching corn into the thin soil and day-herding their goats on the nearest hillside, a type of agriculture intermediate between an Appalachian hill-farm, a Philippine caigin, and a New Mexico "Small Holding Claim." I recognize the land pressure which forces the adoption of such a policy, but I also recognize the inevitable ruin which will follow. One can tell when nearing one of these settlements by the thinning sod, the thickening weeds, the browsed-off willows, and the oaks skinned for tanbark. Just so were our own dry canyons sent to their death.

But these resettlements are also as yet microscopic when compared with the bulk of the mountain area. They occur only near roads, and roads are as yet poor and far between. Engineers would call the mountains roadless.

In Arizona and New Mexico there are in general two kinds of deer range, the overstocked and nearly empty. Most of the herds are very thin, but every few years some new spot flares up with a sudden overpopulation of deer. The Kaibab was the first of these, but there has been a new one every year or two for a decade. Often, before the heavy wheels of legislative adjustment can turn, the range is severely injured. Most laymen have no comprehension of what a serious thing it is to overtax a browse range, especially in an arid climate. Recovery is a matter of decades, rather than of years. Some ranges wash away before they can recover.

Deer irruptions are by no means confined to the Southwest. They are breaking out from Georgia to Wisconsin, and from California to Pennsylvania. Why? Have deer always fluctuated from scarcity to overabundance? History would hardly so indicate.

In Chihuahua one can glean, by comparison, a hint of what may be the matter with our deer. Whitetail deer are abundant in the Sierras, but not excessive. So are wild turkeys. In nine days of hard hunting, two of us saw 187 deer, fifty of them bucks of two or more prongs. Deer irruptions are unknown. Mountain lions and wolves are still common. I doubt whether the lion-deer ratio is much different from that of Coronado's time. There are no coyotes in the mountains, whereas with us there is universal complaint from Alaska to New Mexico that the coyote has invaded the high country to wreak havoc on both game and livestock.

I submit for conservationists to ponder the question of whether the wolves have not kept the coyotes out? And whether the presence of a normal complement of predators is not, at least in part, accountable for the absence of irruption? If so, would not our rougher mountains be better off and might we not have more normalcy in our deer herds, if we let the wolves and lions come back in reasonable numbers?

At the very least, the Sierras present to us an example of an abundant game population thriving in the midst of its natural enemies. Let those who habitually ascribe all game scarcity to predators or who prescribe predator control as the first and inevitable step in all game management, take that to heart.

On the dry tops of the highest mesas, in the bottoms of the roughest and wildest canyons, anywhere in fact where a short watershed is intercepted by a ledge, dyke, or other favorable spot for impounding soil, the traveler in the Sierras finds loose-masonry dams constructed by the hand of man. There are hundreds of them.

How old are they? Who built them? What for? The first two questions find a ready answer. Not infrequently a 200-year-old pine is found growing behind the dam, its root-collar flush with the surface of the impounded soil. Obviously the dam is older than the tree. Unless Coronado and his captains had an unsuspected weakness for laying rock, and also more time and manpower than their journals indicate, these dams were built by prehistoric Indians.

In one case I saw the rocks of the dam clutched tightly in the roots of a great tree. Nobody stuck them there to fool tourists. Moreover there are dams in spots no white man has ever looked upon.

What were the dams for? This question is not so easy to answer. Some local residents say "erosion control." It might be conceivable that the Indians built dams to protect their more valuable soils—say in irrigated valleys—against erosion. But many of the dams I am describing are found around the edges of high mesas a thousand feet above the nearest permanent water. If such a spot ever showed erosion, the natural thing would be to seek a new spot, rather than to laboriously check a gully with rocks.

One is forced back to the theory that these dams were built to create little fields or food patches. The purpose was to impound soil where it would be irrigated by the runoff from slight rainfalls. The choice of locations strongly substantiates this belief. Short watersheds composed mostly of bare rock were especially favored, provided there was a ledge or dyke or narrow place offering secure footing for the dam. In such spots the lightest rain produced runoff and irrigated the field, whereas the heaviest rain could not gather headway enough to tear out the dam.

What crops were raised in these little fields? This, to me, is a perplexing question. Their small size and the wide dispersion seems to preclude constant patrol against game, while the absence of metal tools seems to preclude game-proof fencing. Surely there were deer, turkey, and bears enough in those days to wreck any crop of plants palatable to them. The clue must lie in plants palatable to Indians but not to animals. Corn, it appears, is not molested by game until the ears form, but after that I fail to see how it could get by. Squash and melons would have the same weakness. Beans would seemingly be vulnerable at all times. Potatoes, peppers, and tobacco might possibly qualify as game-proof. I wonder if the archeologists have considered game-damage in reconstructing their picture of prehistoric Indian agriculture?

Everybody in Mexico has heard of the new motor road to Mexico City and is hoping for one like it to his village. The tourist-promotion policy of the present government is well known. It appears then that funds alone will limit the rate at which the Sierra Madre is opened up. The policy of settling the landless in the mountain valleys will, if it persists, add further velocity to the road-building process and it will scatter livestock, as well as hunters and tourists, over the mountain country. The end result will be bad, unless Mexico does a better job than we have done in the regulation of grazing.

I sometimes wonder whether semi-arid mountains can be grazed at all without ultimate deterioration. I know of no arid region which has ever survived grazing through long periods of time, although I have seen individual ranches which seemed to hold out for shorter periods. The trouble is that where water is unevenly distributed and feed varies in quality, grazing usually means overgrazing.

With the extension of roads, recreation so-called will of course repeat the now familiar process of losing in quality as it gains in quantity of human service. Mexican citizens protest that they are going strong on National Parks and Forests. They are particularly proud of the International Park at Big Bend. They do not realize that these devices, laudable and necessary as they are, have not exempted us from the inexorable process of losing quality to gain quantity.

Mexico's experience with American hunters is an illuminating example of the limitations inherent in conservation formulae. It is no secret that until recently many visiting American hunters made pigs of themselves. Neither is it any secret that they were often aided and abetted in so doing by commercial guides. Mexico in self-defense has adopted the formula of clapping on a high license fee, and of limiting non-resident hunting to members of bonded "clubs." The theory is to call the bond for any misbehavior.

But how does the formula actually work? The bonded hunter is careful enough to stay within the law, but after such outlays he is, I think, equally careful to take all the law allows. In other words, he helps himself pretty generously and the drain on the game is probably not much less than it was in the lawless days.

I point no moral except that we seem ultimately always thrown back on individual ethics as the basis of conservation policy. It is hard to make a man, by pressure of law or money, do a thing which does not spring naturally from his own personal sense of right and wrong.

Our own Southwest was pretty badly misused before the idea of conservation was born. As a result, our own conservation program for the region has been in a sense a post-mortem cure. There are, however, two magnificent semi-arid regions in which settlement came later than the conservation idea. One is South Africa and the other is the Mexican mountains. Hence both are of world-wide interest as laboratories in which conservation can be given a full and fair test. Can they arrest and control the wasteful and predatory nature of what we call "development?" The self-defeating nature of mass-use of outdoor resources? Or are these evils inherent in industrial civilization? The next few decades will probably bring us the answer.

Perhaps a clear answer to these complex questions of policy is too much to hope for, but in any event the Sierra Madre offers us the chance to describe, and define, in actual ecological measurements, the lineaments and physiology of an unspoiled mountain landscape. What is the mechanism of a natural forest? A natural watershed? A natural deer herd? A natural turkey range? On our side of the line we have few or no natural samples left to measure. I can see here the opportunity for a great international research enterprise which will explain our own history and enlighten the joint task of profiting by its mistakes.

Chukaremia [1938]

As Leopold came more and more to appreciate the inner workings of the natural system, he had less and less patience with the imported species that were the pride of so many state game departments. Here, in an editorial contributed to *Outdoor America*, Leopold deftly diagnoses the newest sportsmen's disease, Chukaremia.

Sportsmen, as well as rabbits, have their cycles of disease. Tularemia discolors the rabbit's liver, but Chukaremia distorts the sportsman's point of view.

A rabbit recovering from tularemia becomes immune to reinfection, but few sportsmen ever become immune to the idea that foreign game birds are the answer to the "more game" problem.

Tularemia may kill the rabbit, but Chukaremia never kills the sportsman. It never even makes him sick. The only damage which can, up to this date, be charged to Chukaremia is that it has depleted the game funds of 48 states for half a century, and has served as a perfect alibi for postponing the practice of game management.

Mild local outbreaks of Chukaremia date back to revolutionary times, when Richard Bache, son-in-law of Benjamin Franklin, first planted Hungarian partridges in New Jersey. The first continental epidemic hit in 1905, when everybody began planting Hungarians as a substitute for quail. By 1911, a hundred thousand had been imported at three dollars each. Perhaps five percent of the territory planted now contains Hungarians.

Perhaps half of the territory originally planted with ringneck pheasants now is populated with these birds.

These, however, were our two successful ventures, and we have occasion to remember them. We have forgotten the plantings of Tinamou, Curassow, Chacalaca, ocellated turkey, guinea fowl, ptarmigan, willow grouse, black grouse, capercailzie, hazel grouse, elegant quail, redleg partridge, francolin, bamboo partridge, painted quail, Egyptian quail, Chinese quail, and a score or so of assorted pheasants, doves, and pigeons which have died out, but which have served to put off for fifty years the day when we shall face the question of doing something real for the game species already in our coverts.

And now the Chukar.

There may be regions where the planting of Chukars is necessary and wise because restoration of native birds in shootable quantity is impossible. But has anybody really studied the question of where those regions are? How many of the game departments now rushing pell-mell into full-scale Chukar production have made advance tests to determine whether they will survive? (One that I know of: Missouri.)

How many of them have conducted researches and demonstrations to find out just what can be expected of birds already present if put under management? (Perhaps a dozen.)

How many have told their sportsmen that there is no man living who can *predict* the behavior of an importation? Instead of warning sportsmen as to the probable failure of most Chukar plants, we see departments issuing such optimistic statements as this: "If the experiment is successful, sportsmen may see a new game bird in the field and one that is a delight for the hunter to work with his dog." The public will not notice the "if," nor remember the fact that exactly similar press releases date back to the 80s.

In most states a dollar invested in game management will produce more long-time results and less public applause than a dollar invested in Chukars. Must we conclude then, that the average administrator is more interested in applause than in results? I think not. I think he is entirely unconscious of the distinction. By keeping himself uninformed about wildlife history and about wildlife ecology, the average administrator is able to entertain the same genuine enthusiasm for nostrums as exists in the public mind.

A profession is a body of men who voluntarily measure their work by a higher standard than their clients demand. To be professionally acceptable, a policy must be sound as well as salable. Wildlife administration, in this respect, is not yet a profession.

Letter to a Wildflower Digger [1938]

If he became angry enough, Leopold could fire off an acerbic letter. This one, in defense of a yellow ladyslipper stolen from the University Arboretum, he sent to the unknown thief through pages of the local newspaper, the *Wisconsin State Journal*. In it he pays incidental homage to John Muir.

This letter is addressed, through the columns of the *State Journal*, to that unknown person who last week dug up the only remaining yellow ladyslipper in the Wingra woods.

While your name is unknown, your action sufficiently portrays the low estate of either your character or your education. On the chance that the latter rather than the former is at fault, I address to you this letter. I address it also to all whose gardens at this season suddenly blossom forth with new wildflowers lifted from other people's woods.

When John Muir came to the Madison region two generations ago, the woods and marshes were studded with millions of ladyslippers of a score of species. Today, what with drainage, fire, cow, plow, and wildflower diggers—like yourself—a dozen of these species are extinct, and the remainder are so rare that the average citizen has never seen one.

Now John Muir got something pleasant and valuable from his wildflowers. He became a great man, and it seems likely that his wildflowers had something to do with it. It is reasonable to suppose that the present generation might get something pleasant and valuable from them, too—if there were any. But no one, even yourself, is going to get anything valuable from this ladyslipper languishing in your backyard.

The University of Wisconsin has got the notion, perhaps a foolish one, that the privilege of seeing a ladyslipper woods has got something to do with education. For this reason it is acquiring an arboretum. It wants to take its botany students out there and show them what Wisconsin looked like in its youth—in John Muir's youth. It hopes that this will make them dissatisfied with what Wisconsin looks like now. But now, thanks to you, the Wingra woods is one step nearer looking like all the rest of the state. Perhaps, after all, our students would learn a lot if we took them out there and said:

"Here is where we used to have a ladyslipper."

Then, if you will consent to the invasion of your privacy, we would like to take them to your backyard and show them where you have planted it, and how it is thriving in its new home.

In respect of thriving, here are some things you may not know:

Only one man has ever succeeded in germinating the seeds of this species in artificial surroundings. It takes a high-powered chemist to reproduce the conditions necessary for its germination. Wild woods sometimes allow of reproduction, but backyards never. After the seedling has been born, it takes four years to reach the age of flowering. Do you think your ladyslipper will reproduce its kind in your backyard?

One of our ambitions for the arboretum is to apply the newly discovered chemistry for germinating the species, i.e., to start a "ladyslipper nursery" out of which the Wingra woods, and all other Wisconsin woods not yet graved to death, may be abundantly restocked. To this end we have hired the only living man who knows how to do it, and he is ready to make the attempt. But now you have taken his source of seed. We can find other plants, to be sure, but it will not be long, what with the thousands of other wildflower diggers like yourself, before the goose with golden eggs is dead. We had better hurry.

I invite your attention to the fact that this ladyslipper is not the only public property which you might lift for the embellishment of your home. There are numerous paintings in the Memorial Union which you could cut out of their frames while nobody is looking. They are, I admit, less beautiful than your flower, but their loss could be more easily replaced. In the historical museum are any number of things as irreplaceable as your flower—why not add some of them to your collection?

I anticipate your reply and tell you why not: because you, and also your friends and neighbors, would recognize your act as vandalism. You do not recognize your theft of the ladyslipper as vandalism. I will leave it to you to decide whether it is.

Yours truly,

—Aldo Leopold
Research Director
University of Wisconsin Arboretum

Engineering and Conservation [1938]

The text of a lecture that Leopold delivered to the University of Wisconsin College of Engineering is printed here from a revised typescript, dated April 11, 1938, edited in Leopold's hand. It wryly points out the untoward ecological consequences of civil engineering, treating engineering not only as a profession with goals that often overlap and conflict with those of conservation but also as an emblem of the public state of mind and the dominant idea of the industrial age. Withal, Leopold achieves a remarkably diplomatic, positive tone.

The public mind is a mirror into which every vocation reflects its image. That image may flatter its subject, or the contrary, depending upon accumulated public impressions of the group and how its members live, think, and work.

A decade ago the public image of labor was a rather pleasing one. Since the advent of CIO it has become much harder to look at.

In the writings of Alexander Hamilton and Thomas Jefferson we find the word "industrialist" used as a term of high honor. Today one uses the term guardedly.

The banker's picture has of late suffered an unflattering distortion, culminating in the newspaper epithet "bankster" in the early 1930s.

Not long ago the railroads had cloven hooves; now what with rate reductions, streamliners, and 35-cent dinners they have acquired merit and may soon sprout wings.

It is clear that, in general, the underdog tends to be uppermost in public favor. Conversely, when a profession becomes important or powerful, it has need to look to its laurels.

The engineer, from Kitchener to Herbert Hoover, enjoyed a public image of ever-increasing comeliness. The reasons are too well known to need comment. At the present moment, however, the word "engineer" in the minds of some conservationists is associated with an attitude toward natural resources which they dislike. It evokes in them a mental image of marshes needlessly drained, of rivers expensively channelized to revive an expiring navigation, of floods aggravated by stream straightening and by constricting levees, of irrigation reservoirs silted before the maturity of their bonds, and of a veritable mycelium of roads at least a part of which are built regardless of cost or need.

This tendency to challenge the engineer is admittedly confined to that small group preoccupied with the biological aspects of public policy. As a member of this group I here attempt to shed some light on their reactions. That these reactions are just and fair I cannot certify, but the avowal that they exist may be a useful first step toward clearing the issue.

We may perhaps strike at the root of the matter by this generalization: the engineer believes, and has taught the public to believe, that a constructed mechanism is inherently preferable to a natural one. The conservationist believes the contrary.

All generalizations are inaccurate, including this one. A few cases may help clarify the intended meaning.

Consider the Columbia River dams. As between abundant power and abundant salmon, priority automatically went to power. The dams were started before the probable destruction of the salmon resource was seriously debated. It made no difference that the need for power was questionable, the fate of the salmon was nearly certain. By an axiom long in the making, the man-made resource must be superior to the natural one. I do not know whether the engineers built the axiom or the axiom built the engineers. The result is the same.

The Mississippi dams involve a more subtle issue. That the great river is sick all will agree. Treatment can be applied either to the channel where the symptoms are most conspicuous, or to the deranged watershed which gives rise to the symptoms. The engineers started to bandage the channel with steel and concrete before giving ear to the question of what ails the organism as a whole. The case of course involves many other issues which I do not here discuss. I point out merely the seeming assumption that skillful structures can solve our water problems, and (by implication) exempt us from the penalties of bungling land use.

The history of irrigation reservoirs in the West presents a similar question. In many instances the silting life of a storage basin was assumed during the promotion stage to be perpetual. During the construction stage it would be scaled down to a century, and during the pay-up stage it would finally appear as a generation. Isolated errors in predicting the life of reservoirs would be natural enough, but their repetition through forty years of experience forces the observer to conclude that the profession as a whole is not yet conscious of that organic disintegration which has afflicted nearly all semi-arid watersheds since their occupation by livestock. (There are brilliant individual exceptions to this rule. Olmstead's report on the Gila River is one such.)

When some inventor comes out with a new alloy the engineers lose no time making a path to his door. But discoveries outside the engineering field may have an equal bearing on the responsibilities of the engineering profession. Take, for example, Lowdermilk's formulation, in terms of physical chemistry, of the basic mechanism by which plants influence runoff. This reorients the old controversy about the influence of forests and presents a challenging opportunity for joint research by soil chemists, engineers, and botanists. But who is doing such research? I here criticize all three parties for inaction.

Again, take Weaver's discovery that the composition of the plant community determines the ability of soils to retain their granulation, and hence their stability. If finally verified, this new principle may necessitate the revision of our entire system of thought on flood control and erosion control. I do not hear it discussed among engineers (nor, for that matter, among economists, business men, or statesmen).

The cases I have cited all involve big and complex issues of national importance. Consider now, for contrast, a small and local one. In the sand counties of central Wisconsin are many defunct drainage districts. In 1933 the government began to buy out the surviving farmers and convert the area into a wildlife reservation. Travel in the area had always followed "sandtracks." There were hundreds of miles of these tracks; unimproved but passable routes winding picturesquely through the jack pines and scrub oaks.

I believe it is an engineering fact that in sand a semi-sodded track is the best possible road short of a surfaced turnpike. But the engineers could not resist the temptation of soft yardage, abundant CCCs, and government gas. Today the area is geometrically gridironed with graded sandpiles, expensively inferior to the old tracks. It looks as if some new glacier had acquired the knack of laying down eskers with a transit. The drainage of this region was, by hindsight, a mistake, but now in our effort to give it back to the birds, we must give it one last mutilating gouge with power tools.

This same propensity for carving soft landscapes perhaps accounts for the recent drainage of nearly the whole Atlantic tidal marsh from Maine to Alabama. This was done with relief labor, in the name of mosquito control, over the protests of wildlife interests. These marshes are the wintering ground of many species of migratory waterfowl and the breeding ground of others. The effectiveness of such drainage as a mosquito control measure is at least debatable. Biological methods of mosquito control are known but were not tried. The project was not led by engineers and is chargeable to engineering only in the sense that it shows what the mechanical idea of landscaping can do when combined with too much haste, too much government money, a resort-owner's chamber of commerce, and the prevalent unconsciousness of biological equilibria. I suspect that the real impulse behind the whole venture is the local realtor's solicitude for silk stockings on his beaches.

I mention last what to me seems the least discussed but most regrettable instance of short-sighted engineering—the wholesale straightening of small rivers and creeks. This is done to hasten the runoff of local flood waters, and of course aggravates the piling up of flood peaks in major streams. It is, on its face, a process of pushing trouble downstream, of seeking benefit for the locality at the expense of the community. In justice the stream-straightener should indemnify the public for damage; in practice I fear the public may at times subsidize him with relief labor.

I know of at least one engineering group which has foresworn stream-straightening—the Soil Conservation Service. I salute them.

The interplay of engineering and ecological evils is an insidious thing. I know a locality in western Dane County where erosion is gradually destroying the upland cornfields. The farmers must have corn; their only recourse is the marshy creek bottoms. These, however, are subject to flashy floods. To raise corn on the bottoms the floods will have to be prodded downstream by straightening, which in turn will aggravate the flashy runoff and augment erosion. Thus the cycle of misuse.

Incidentally these marshy bottoms contain the only wildlife cover and are now good pasture. The cover will disappear with straightening, and the pasture will have to move back to the eroded uplands.

These cases collectively imply, but I will now specifically admit, certain qualifications which, in justice, I must attach to my criticism of the engineer.

First of all, let me admit that in some cases the biological professions seem just as remiss as the engineering group.

Secondly, let me admit that the engineer is to me a symbol for a state of the public mind, as well as a professional man who has made mistakes. The cited instances of error are chargeable to voters and politicians as well as engineers. The Columbia dams, the Mississippi dams, the irrigation reservoirs, the needless roads and the mosquito drainage were backed by strong local booster and even pork-barrel interests. Every professional man must, within limits, execute the jobs people are willing to pay for. But every profession in the long run writes its own ticket. It does so through the emergence of leaders who can afford to be skeptical out loud and in public—professors, for example. What I here decry is not so much the prevalence of public error in the use of engineering tools as the scarcity of engineering criticism of such misuse. Perhaps that criticism exists *in camera*, but it does not reach the interested layman.

I admit, too, that the engineer is not the only focus for biological discontent. The chemist scattering new comforts with one hand and new pollutions with the other, evokes in us the same disquiet. Both professions exemplify priority for the synthetic over the natural, a certain atrophy of esthetic discrimination, a yearning for prosperity and comfort at any cost. I do not claim that we, the disaffected, disdain the prosperity and the comforts. Our only contribution is the idea that the cost is large, unnecessarily large.

With these qualifying admissions I now summarize my criticism: The engineer has respect for mechanical wisdom because he created it. He has disrespect for ecological wisdom, not because he is contemptuous of it, but because he is unaware of it. We have, in short, two professions whose responsibilities for land use overlap much, but whose respective zones of awareness overlap only a little. What can we say about their future relationship? About the direction of possible adjustments?

All history shows this: that civilization is not the progressive elaboration of a single idea, but the successive dominance of a series of ideas. Greece, Rome, the Renaissance, the industrial age, each had a new and largely distinct zone of awareness. The people of each lived not in a better, nor a worse, but in a new and different intellectual field. Progress, if there be any, is the slender hoard of fragments retained from the whole intellectual succession.

Engineering is clearly the dominant idea of the industrial age. What I have here called ecology is perhaps one of the contenders for a new order. In any case our problem boils down to increasing the overlap of awareness between the two.

This may prove less difficult than appears on the surface, for the ecologist is in many ways an engineer. The biotic mechanism is too complex to enable him to predict its reactions; therefore he advocates what an engineer would in like case: go slow, cut and try.

He feels an engineer's admiration for this complexity which defies science, and an engineer's aversion for discarding any of its parts. The real difference lies in the ecologist's conviction that to govern the animate world it must be led rather than coerced. To me this is engineering wisdom; the reason the engineer does not display it is unawareness of the animate world.

The tools which the engineer has given the public are so crude and powerful that they invite coercive use. It is not likely that the public will lay them down. The only alternative is the pooling of engineering and ecological skills for wiser use of those tools. Is this pooling under way? Perhaps. We now see engineers and ecologists jointly attacking the soil erosion problem, but only after the resource reached an advanced stage of deterioration. Need we always await the willy-nilly pressure of wrecked resources before professional cooperation begins?

We end, I think, at what might be called the standard paradox of the twentieth century: our tools are better than we are, and grow better faster than we do. They suffice to crack the atom, to command the tides. But they do not suffice for the oldest task in human history: to live on a piece of land without spoiling it.

The Farmer as a Conservationist [1939]

In this masterpiece, originally a talk delivered at the University's Farm and Home Week in February 1939, Leopold distinguishes between conservation understood negatively as restraint and that understood positively as skill. Narrowly economic and utilitarian desiderata are contrasted with wider, less quantifiable human values. And the familiar refrain of conservation by government versus ecologically informed and esthetically and ethically motivated conservation by landowners is beautifully illustrated in a brief idyll of enlightened husbandry. After distribution as an extension circular, this essay was revised and published in *American Forests*.

Conservation means harmony between men and land.

When land does well for its owner, and the owner does well by his land; when both end up better by reason of their partnership, we have conservation. When one or the other grows poorer, we do not.

Few acres in North America have escaped impoverishment through human use. If someone were to map the continent for gains and losses in soil fertility, waterflow, flora, and fauna, it would be difficult to find spots where less than three of these four basic resources have retrograded; easy to find spots where all four are poorer than when we took them over from the Indians.

As for the owners, it would be a fair assertion to say that land depletion has broken as many as it has enriched.

It is customary to fudge the record by regarding the depletion of flora and fauna as inevitable, and hence leaving them out of the account. The fertile productive farm is regarded as a success, even though it has lost most of its native plants and animals. Conservation protests such a biased accounting. It was necessary, to be sure, to eliminate a few species, and to change radically the distribution of many. But it remains a fact that the average American township has lost a score of plants and animals through indifference for every one it has lost through necessity.

What is the nature of the process by which men destroy land? What kind of events made it possible for that much-quoted old-timer to say: "You can't tell me about farming; I've worn out three farms already and this is my fourth"?

Most thinkers have pictured a process of gradual exhaustion. Land, they say, is like a bank account: if you draw more than the interest, the principal dwindles. When Van Hise said "Conservation is wise use," he meant, I think, restrained use.

Certainly conservation means restraint, but there is something else that needs to be said. It seems to me that many land resources, when they are used, get out of order and disappear or deteriorate before anyone has a chance to exhaust them.

Look, for example, at the eroding farms of the cornbelt. When our grandfathers first broke this land, did it melt away with every rain that happened to fall on a thawed frost-pan? Or in a furrow not exactly on contour? It did not; the newly broken soil was tough, resistant, elastic to strain. Soil treatments which were safe in 1840 would be suicidal in 1940. Fertility in 1840 did not go down river faster than up into crops. Something has got out of order. We might almost say that the soil bank is tottering, and this is more important than whether we have overdrawn or underdrawn our interest.

Look at the northern forests: did we build barns out of all the pineries which once covered the lake states? No. As soon as we had opened some big slashings we made a path for fires to invade the woods. Fires cut off growth and reproduction. They outran the lumberman and they mopped up behind him, destroying not only the timber but also the soil and the seed. If we could have kept the soil and the seed, we should be harvesting a new crop of pines now, regardless of whether the virgin crop was cut too fast or too slow. The real damage was not so much the overcutting, it was the run on the soil-timber bank.

A still clearer example is found in farm woodlots. By pasturing their woodlots, and thus preventing all new growth, cornbelt farmers are gradually eliminating woods from the farm landscape. The wildflowers and wildlife are of course lost long before the woodlot itself disappears. Overdrawing the interest from the woodlot bank is perhaps serious, but it is a bagatelle compared with destroying the capacity of the woodlot to yield interest. Here again we see awkward use, rather than over-use, disordering the resource.

In wildlife the losses from the disordering of natural mechanisms have, I suspect, far exceeded the losses from exhaustion. Consider the thing we call "the cycle," which deprives the northern states of all kinds of grouse and rabbits about seven years out of every ten. Were grouse and rabbits always and everywhere cyclic? I used to think so, but I now doubt it. I suspect that cycles are a disorder of animal populations, in some way spread by awkward land-use. We don't know how, because we do not yet know what a cycle is. In the far north cycles are probably natural and inherent, for we find them in the untouched wilderness, but down here I suspect they are not inherent. I suspect they are spreading, both in geographic sweep and in number of species affected.

Consider the growing dependence of fishing waters on artificial restocking. A big part of this loss of toughness inheres in the disordering of waters by erosion and pollution. Hundreds of southerly trout streams which once produced natural brook trout are stepping down the ladder of productivity to artificial brown trout, and finally to carp. As the fish resource dwindles, the flood and erosion losses grow. Both are expressions of a single deterioration. Both are not so much the exhaustion of a resource as the sickening of a resource.

Consider deer. Here we have no exhaustion; perhaps there are too many deer. But every woodsman knows that deer in many places are exterminating the plants on which they depend for winter food. Some of these, such as white cedar, are important forest trees. Deer did not always destroy their range. Something is out of kilter. Perhaps it was a mistake to clean out the wolves; perhaps natural enemies acted as a kind of thermostat to close the "draft" on the deer supply. I know of deer herds in Mexico which never get out of kilter with their range; there are wolves and cougars there, and always plenty of deer but never too many. There is substantial balance between those deer and their range, just as there was substantial balance between the buffalo and the prairie.

Conservation, then, is keeping the resource in working order, as well as preventing over-use. Resources may get out of order before they are exhausted, sometimes while they are still abundant. Conservation, therefore, is a positive exercise of skill and insight, not merely a negative exercise of abstinence or caution.

What is meant by skill and insight?

This is the age of engineers. For proof of this I look not so much to Boulder Dams or China Clippers as to the farmer boy tending his tractor or building his own radio. In a surprising number of men there burns a curiosity about machines and a loving care in their construction, maintenance, and use. This bent for mechanisms, even though clothed in greasy overalls, is often the pure fire of intellect. It is the earmark of our times.

Everyone knows this, but what few realize is that an equal bent for the mechanisms of nature is a possible earmark of some future generation.

No one dreamed, a hundred years ago, that metal, air, petroleum, and electricity could coordinate as an engine. Few realize today that soil, water, plants, and animals are an engine, subject, like any other, to derangement. Our present skill in the care of mechanical engines did not arise from fear lest they fail to do their work. Rather was it born of curiosity and pride of understanding. Prudence never kindled a fire in the human mind; I have no hope for conservation born of fear. The 4-H boy who becomes curious about why red pines need more acid than white is closer to conservation than he who writes a prize essay on the dangers of timber famine.

This necessity for skill, for a lively and vital curiosity about the workings of the biological engine, can teach us something about the probable success of farm conservation policies. We seem to be trying two policies, education and subsidy. The compulsory teaching of conservation in schools, the 4-H conservation projects, and school forests are examples of education. The woodlot tax law, state game and tree nurseries, the crop control program, and the soil conservation program are examples of subsidy.

I offer this opinion: these public aids to better private land use will accomplish their purpose only as the farmer matches them with this thing which I have called skill. Only he who has planted a pine grove with his own hands, or built a terrace, or tried to raise a better crop of birds can appreciate how easy it is to fail; how futile it is passively to follow a recipe without understanding the mechanisms behind it. Subsidies and propaganda may evoke the farmer's acquiescence, but only enthusiasm and affection will evoke his skill. It takes something more than a little "bait" to succeed in conservation. Can our schools, by teaching, create this something? I hope so, but I doubt it, unless the child brings also something he gets at home. That is to say, the vicarious teaching of conservation is just one more kind of intellectual orphanage; a stop-gap at best.

Thus we have traversed a circle. We want this new thing, we have asked the schools and the government to help us catch it, but we have tracked it back to its den under the farmer's doorstep.

I feel sure that there is truth in these conclusions about the human qualities requisite to better land use. I am less sure about many puzzling questions of conservation economics.

Can a farmer afford to devote land to woods, marsh, pond, windbreaks? These are semi-economic land uses—that is, they have utility but they also yield non-economic benefits.

Can a farmer afford to devote land to fencerows for the birds, to snag-trees for the coons and flying squirrels? Here the utility shrinks to what the chemist calls "a trace."

Can a farmer afford to devote land to fencerows for a patch of ladyslippers, a remnant of prairie, or just scenery? Here the utility shrinks to zero.

Yet conservation is any or all of these things.

Many labored arguments are in print proving that conservation pays economic dividends. I can add nothing to these arguments. It seems to me, though, that something has gone unsaid. It seems to me that the pattern of the rural landscape, like the configuration of our own bodies, has in it (or should have in it) a certain wholeness. No one censures a man who loses his leg in an accident, or who was born with only four fingers, but we should look askance at a man who amputated a natural part on the grounds that some other is more profitable. The comparison is exaggerated; we had to amputate many marshes, ponds and woods to make the land habitable, but to remove any natural feature from representation in the rural landscape seems to me a defacement which the calm verdict of history will not approve, either as good conservation, good taste, or good farming.

Consider a single natural feature: the farm pond. Our godfather the Ice-king, who was in on the christening of Wisconsin, dug hundreds of them for us. We have drained ninety and nine. If you don't believe it, look on the original surveyor's plot of your township; in 1840 he probably mapped water in dozens of spots where in 1940 you may be praying for rain. I have an undrained pond on my farm. You should see the farm families flock to it of a Sunday, everybody from old grandfather to the new pup, each bent on the particular aquatic sport, from water lilies to bluegills, suited to his (or her) age and waistline. Many of these farm families once had ponds of their own. If some drainage promoter had not sold them tiles, or a share in a steam shovel, or some other dream of sudden affluence, many of them would still have their own water lilies, their own bluegills, their own swimming hole, their own redwings to hover over a buttonbush and proclaim the spring.

If this were Germany, or Denmark, with many people and little land, it might be idle to dream about land-use luxuries for every farm family that needs them. But we have excess plowland; our conviction of this is so unanimous that we spend a billion out of the public chest to retire the surplus from cultivation. In the face of such an excess, can any reasonable man claim that economics prevents us from getting a life, as well as a livelihood, from our acres?

Sometimes I think that ideas, like men, can become dictators. We Americans have so far escaped regimentation by our rulers, but have we escaped regimentation by our own ideas? I doubt if there exists today a more complete regimentation of the human mind than that accomplished by our self-imposed doctrine of ruthless utilitarianism. The saving grace of democracy is that we fastened this yoke on our own necks, and we can cast it off when we want to, without severing the neck. Conservation is perhaps one of the many squirmings which foreshadow this act of self-liberation.

The principle of wholeness in the farm landscape involves, I think, something more than indulgence in land-use luxuries. Try to send your mind up in an airplane; try to see the *trend* of our tinkerings with fields and forests, waters and soils. We have gone in for governmental conservation on a huge scale. Government is slowly but surely pushing the cutovers back into forest; the peat and sand districts back into marsh and scrub. This, I think, is as it should be. But the cow in the woodlot, ably assisted by the ax, the depression, the June beetle, and the drouth, is just as surely making southern Wisconsin a treeless agricultural steppe. There was a time when the cessation of prairie fires added trees to southern Wisconsin faster than the settlers subtracted them. That time is now past. In another generation many southern counties will look, as far as trees are concerned, like the Ukraine, or the Canadian wheatlands. A similar tendency to create *monotypes*, to block up huge regions to a single land-use, is visible in many other states. It is the result of delegating conservation to government. Government cannot own and operate small parcels of land, and it cannot own and operate good land at all.

Stated in acres or in board feet, the crowding of all the timber into one place may be a forestry program, but is it conservation? How shall we use forests to protect vulnerable hillsides and riverbanks from erosion when the bulk of the timber is up north on the sands where there is no erosion? To shelter wildlife when all the food is in one county and all the cover in another? To break the wind when the forest country has no wind, the farm country nothing but wind? For recreation when it takes a week, rather than an hour, to get under a pine tree? Doesn't conservation imply a certain interspersion of land-uses, a certain pepper-and-salt pattern in the warp and woof of the land-use fabric? If so, can government alone do the weaving? I think not.

It is the individual farmer who must weave the greater part of the rug on which America stands. Shall he weave into it only the sober yarns which warm the feet, or also some of the colors which warm the eye and the heart? Granted that there may be a question which returns him the most profit as an individual, can there be *any* question which is best for his community? This raises the question: is the individual farmer capable of dedicating private land to uses which profit the community, even though they may not so clearly profit him? We may be over-hasty in assuming that he is not.

I am thinking, for example, of the windbreaks, the evergreen snow-fences, hundreds of which are peeping up this winter out of the drifted snows of the sandy counties. Part of these plantings are subsidized by highway funds, but in many others the only subsidy is the nursery stock. Here then is a dedication of private land to a community purpose, a private labor for a public gain. These windbreaks do little good until many land-owners install them; much good after they dot the whole countryside. But this "much good" is an undivided surplus, payable not in dollars, but rather in fertility, peace, comfort, in the sense of something alive and growing. It pleases me that farmers should do this new thing. It foreshadows conservation. It may be remarked, in passing, that this planting of windbreaks is a direct reversal of the attitude which uprooted the hedges, and thus the wildlife, from the entire cornbelt. Both moves were fathered by the agricultural colleges. Have the colleges changed their mind? Or is an Osage windbreak governed by a different kind of economics than a red pine windbreak?

There is still another kind of community planting where the thing to be planted is not trees but thoughts. To describe it, I want to plant some thoughts about a bush. It is called bog-birch.

I select it because it is such a mousy, unobtrusive, inconspicuous, uninteresting little bush. You may have it in your marsh but have never noticed it. It bears no flower that you would recognize as such, no fruit which bird or beast could eat. It doesn't grow into a tree which you could use. It does no harm, no good, it doesn't even turn color in fall. Altogether it is the perfect nonentity in bushes; the complete biological bore.

But is it? Once I was following the tracks of some starving deer. The tracks led from one bog-birch to another; the browsed tips showed that the deer were living on it, to the exclusion of scores of other kinds of bushes. Once in a blizzard I saw a flock of sharptail grouse, unable to find their usual grain or weed seeds, eating bog-birch buds. They were fat.

Last summer the botanists of the University Arboretum came to me in alarm. The brush, they said, was shading out the white ladyslippers in the Arboretum marsh. Would I ask the CCC crews to clear it? When I examined the ground, I found the offending brush was bog-birch. I cut the sample shown on the left of the drawing. Notice that up to two years ago rabbits had mowed it down each year. In 1936 and 1937 the rabbits had spared it, hence it grew up and shaded the ladyslippers. Why? Because of the cycle; there were no rabbits in 1936 and 1937. This past winter of 1938 the rabbits mowed off the bog-birch, as shown on the right of the drawing.

It appears, then, that our little nonentity, the bog-birch, is important after all. It spells life or death to deer, grouse, rabbits, ladyslippers. If, as some think, cycles are caused by sunspots, the bog-birch might even be regarded a sort of envoy for the solar system, dealing out appeasement to the rabbit, in the course of which a suppressed orchid finds its place in the sun.

The bog-birch is one of hundreds of creatures which the farmer looks at, or steps on, every day. There are 350 birds, ninety mammals, 150 fishes, seventy reptiles and amphibians, and a vastly greater number of plants and insects native to Wisconsin. Each state has a similar diversity of wild things.

Disregarding all those species too small or too obscure to be visible to the layman, there are still perhaps 500 whose lives we might know, but don't. I have translated one little scene out of the life-drama of one species. Each of the 500 has its own drama. The stage is the farm. The farmer walks

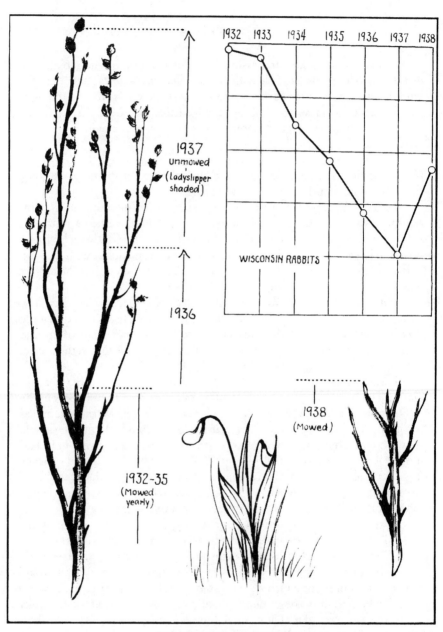

THE STORY OF A CYCLE

A mousy, unobtrusive, inconspicuous little bush, the bog-birch, plays an important role in the ups and downs of plant and animal life. Here is illustrated how it spells life or death to deer, grouse, rabbits, and ladyslippers in Wisconsin.

In 1932 to 1935 rabbits were abundant and ate down the bog-birches each winter, giving the ladyslippers the sun. During 1936 and 1937 the cycle decimated the rabbits and the bog-birches grew high and shaded out the ladyslippers. In 1938 the rabbits recovered, mowed down the birches and the ladyslippers regained their place in the sun.

Leopold's drawing as it appeared with his essay in *American Forests*, June, 1939.

among the players in all his daily tasks, but he seldom sees any drama, because he does not understand their language. Neither do I, save for a few lines here and there. Would it add anything to farm life if the farmer learned more of that language?

One of the self-imposed yokes we are casting off is the false idea that farm life is dull. What is the meaning of John Steuart Curry, Grant Wood, Thomas Benton? They are showing us drama in the red barn, the stark silo, the team heaving over the hill, the country store, black against the sunset. All I am saying is that there is also drama in every bush, if you can see it. When enough men know this, we need fear no indifference to the welfare of bushes, or birds, or soil, or trees. We shall then have no need of the word conservation, for we shall have the thing itself.

The landscape of any farm is the owner's portrait of himself.

Conservation implies self-expression in that landscape, rather than blind compliance with economic dogma. What kinds of self-expression will one day be possible in the landscape of a cornbelt farm? What will conservation look like when transplanted from the convention hall to the fields and woods?

Begin with the creek: it will be unstraightened. The future farmer would no more mutilate his creek than his own face. If he has inherited a straightened creek, it will be "explained" to visitors, like a pock-mark or a wooden leg.

The creek banks are wooded and ungrazed. In the woods, young straight timber-bearing trees predominate, but there is also a sprinkling of hollow-limbed veterans left for the owls and squirrels, and of down logs left for the coons and fur-bearers. On the edge of the woods are a few wide-spreading hickories and walnuts for nutting. Many things are expected of this creek and its woods: cordwood, posts, and sawlogs; flood-control, fishing and swimming; nuts and wildflowers; fur and feather. Should it fail to yield an owl-hoot or a mess of quail on demand, or a bunch of sweet william or a coon-hunt in season, the matter will be cause for injured pride and family scrutiny, like a check marked "no funds."

Visitors when taken to the woods often ask, "Don't the owls eat your chickens?" Our farmer knows this is coming. For answer, he walks over to a leafy white oak and picks up one of the pellets dropped by the roosting owls. He shows the visitor how to tear apart the matted felt of mouse and rabbit fur, how to find inside the whitened skulls and teeth of the bird's prey. "See any chickens?" he asks. Then he explains that his owls are valuable to him, not only for killing mice, but for excluding other owls which *might* eat chickens. His owls get a few quail and many rabbits, but these, he thinks, can be spared.

The fields and pastures of this farm, like its sons and daughters, are a mixture of wild and tame attributes, all built on a foundation of good health. The health of the fields is their fertility. On the parlor wall, where the embroidered "God Bless Our Home" used to hang in exploitation days, hangs a chart of the farm's soil analyses. The farmer is proud that all his soil graphs point upward, that he has no check dams or terraces, and needs none. He speaks sympathetically of his neighbor who has the misfortune of harboring a gully, and who was forced to call in the CCC. The neighbor's check dams are a regrettable badge of awkward conduct, like a crutch.

Separating the fields are fencerows which represent a happy balance between gain in wildlife and loss in plowland. The fencerows are not cleaned yearly, neither are they allowed to grow indefinitely. In addition to bird song and scenery, quail and pheasants, they yield prairie flowers, wild grapes, raspberries, plums, hazelnuts, and here and there a hickory beyond the reach of the woodlot squirrels. It is a point of pride to use electric fences only for temporary enclosures.

Around the farmstead are historic oaks which are cherished with both pride and skill. That the June beetles did get one is remembered as a slip in pasture management not to be repeated. The farmer has opinions about the age of his oaks, and their relation to local history. It is a matter of neighborhood debate whose oaks are most clearly relics of oak-opening days, whether the healed scar on the base of one tree is the result of a prairie fire or a pioneer's trash pile.

Martin house and feeding station, wildflower bed and old orchard go with the farmstead as a matter of course. The old orchard yields some apples but mostly birds. The bird list for the farm is 161 species. One neighbor claims 165, but there is reason to suspect he is fudging. He drained his pond; how could he possibly have 165?

His pond is our farmer's special badge of distinction. Stock is allowed to water at one end only; the rest of the shore is fenced off for the ducks, rails, redwings, gallinules, and muskrats. Last spring, by judicious baiting and decoys, two hundred ducks were induced to rest there a full month. In August, yellow-legs use the bare mud of the water-gap. In September the pond yields an armful of waterlilies. In the winter there is skating for the youngsters, and a neat dozen of rat-pelts for the boys' pin-money. The farmer remembers a contractor who once tried to talk drainage. Pondless farms, he says, were the fashion in those days; even the Agricultural College fell for the idea of making land by wasting water. But in the drouths of the thirties, when the wells went dry, everybody learned that water, like roads and schools, is community property. You can't hurry water down the creek without hurting the creek, the neighbors, and yourself.

The roadside fronting the farm is regarded as a refuge for the prairie flora: the educational museum where the soils and plants of pre-settlement days are preserved. When the professors from the college want a sample of virgin prairie soil, they know they can get it here. To keep this roadside in prairie, it is cleaned annually, always by burning, never by mowing or cutting. The farmer tells a funny story of a highway engineer who once started to grade the cutbanks all the way back to the fence. It developed that the poor engineer, despite his college education, had never learned the difference between a silphium and a sunflower. He knew his sines and cosines, but he had never heard of the plant succession. He couldn't understand that to tear out all the prairie sod would convert the whole roadside into an eyesore of quack and thistle.

In the clover field fronting the road is a huge glacial erratic of pink granite. Every year, when the geology teacher brings her class out to look at it, our farmer tells how once, on a vacation trip, he matched a chip of the boulder to its parent ledge, two hundred miles to the north. This starts him on a little oration on glaciers; how the ice gave him not only the rock, but also the pond, and the gravel pit where the kingfisher and the bank swallows nest. He tells how a powder salesman once asked for permission to blow up the old rock "as a demonstration in modern methods." He does not have to explain his little joke to the children.

He is a reminiscent fellow, this farmer. Get him wound up and you will hear many a curious tidbit of rural history. He will tell you of the mad decade when they taught economics in the local kindergarten, but the college president couldn't tell a bluebird from a blue cohosh. Everybody worried about getting his share; nobody worried about doing his bit. One farm washed down the river, to be dredged out of the Mississippi at another farmer's expense. Tame crops were over-produced, but nobody had room for wild crops. "It's a wonder this farm came out of it without a concrete creek and a Chinese elm on the lawn." This is his whimsical way of describing the early fumblings for "conservation."

A Biotic View of Land [1939]

This is one of Leopold's landmark papers, delivered as a plenary address to a joint meeting of the Society of American Foresters and the Ecological Society of America on June 21, 1939, in Milwaukee, Wisconsin. It was subsequently published in the *Journal of Forestry*. From the most recent contemporaneous ecological theory it abstracts an emerging new portrait of nature—the biotic or ecosystemic concept. Like "The Conservation Ethic," the essay represents a milepost on Leopold's intellectual pilgrimage, and substantial portions of it were incorporated in *Sand County Almanac's* "The Land Ethic."

In pioneering times wild plants and animals were tolerated, ignored, or fought, the attitude depending on the utility of the species.

Conservation introduced the idea that the more useful wild species could be managed as crops, but the less useful ones were ignored and the predaceous ones fought, just as in pioneering days. Conservation lowered the threshold of toleration for wildlife, but utility was still the criterion of policy, and utility attached to species rather than to any collective total of wild things. Species were known to compete with each other and to cooperate with each other, but the cooperations and competitions were regarded as separate and distinct; utility as susceptible of quantitative evaluation by research. For proof of this we need look no further than the bony framework of any campus or capitol: department of economic entomology, division of economic mammalogy, chief of food habits research, professor of economic ornithology. These agencies were set up to tell us whether the red-tailed hawk, the gray gopher, the lady beetle, and the meadowlark are useful, harmless, or injurious to man.

Ecology is a new fusion point for all the natural sciences. It has been built up partly by ecologists, but partly also by the collective efforts of the men charged with the economic evaluation of species. The emergence of ecology has placed the economic biologist in a peculiar dilemma: with one hand he points out the accumulated findings of his search for utility, or lack of utility, in this or that species; with the other he lifts the veil from a biota so complex, so conditioned by interwoven cooperations and competitions, that no man can say where utility begins or ends. No species can be "rated" without the tongue in the cheek; the old categories of "useful" and "harmful" have validity only as conditioned by time, place, and circumstance. The only sure conclusion is that the biota as a whole is useful, and biota includes not only plants and animals, but soils and waters as well.

In short, economic biology assumed that the biotic function and economic utility of a species was partly known and the rest could shortly be found out. That assumption no longer holds good; the process of finding out added new questions faster than new answers. The function of species is largely inscrutable, and may remain so.

When the human mind deals with any concept too large to be easily visualized, it substitutes some familiar object which seems to have similar properties. The "balance of nature" is a mental image for land and life which grew up before and during the transition to ecological thought. It is commonly employed in describing the biota to laymen, but ecologists among each other accept it only with reservations, and its acceptance by laymen seems to depend more on convenience than on conviction. Thus "nature lovers" accept it, but sportsmen and farmers are skeptical ("the balance was upset long ago; the only way to restore it is to give the country back to the Indians"). There is more than a suspicion that the dispute over predation determines these attitudes, rather than vice versa.

To the lay mind, balance of nature probably conveys an actual image of the familiar weighing scale. There may even be danger that the layman imputes to the biota properties which exist only on the grocer's counter.

To the ecological mind, balance of nature has merits and also defects. Its merits are that it conceives of a collective total, that it imputes some utility to all species, and that it implies oscillations when balance is disturbed. Its defects are that there is only one point at which balance occurs, and that balance is normally static.

If we must use a mental image for land instead of thinking about it directly, why not employ the image commonly used in ecology, namely the biotic pyramid? With certain additions hereinafter developed it presents a truer picture of the biota. With a truer picture of the biota, the scientist might take his tongue out of his cheek, the layman might be less insistent on utility as a prerequisite for conservation, more hospitable to the "useless" cohabitants of the earth, more tolerant of values over and above profit, food, sport, or tourist-bait. Moreover, we might get better advice from economists and philosophers if we gave them a truer picture of the biotic mechanism.

I will first sketch the pyramid as a symbol of land, and later develop some of its implications in terms of land use.

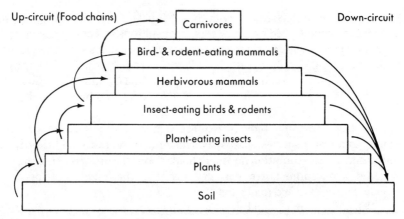

A rendering of Leopold's drawing of the biotic pyramid which appeared in the *Journal of Forestry*. The drawing depicts the plant and animal community as an energy circuit.

Plants absorb energy from the sun. This energy flows through a circuit called the biota. It may be represented by the layers of a pyramid. The bottom layer is the soil. A plant layer rests on the soil, an insect layer on the plants, and so on up through various groups of fish, reptiles, birds, and mammals. At the top are predators.

The species of a layer are alike not in where they came from, nor in what they look like, but rather in what they eat. Each successive layer depends on those below for food and often for other services, and each in turn furnishes food and services to those above. Each successive layer decreases in abundance; for every predator there are hundreds of his prey, thousands of their prey, millions of insects, uncountable plants.

The lines of dependency for food and other services are called food chains. Each species, including ourselves, is a link in many food chains. Thus the bobwhite quail eats a thousand kinds of plants and animals, i.e., he is a link in a thousand chains. The pyramid is a tangle of chains so complex as to seem disorderly, but when carefully examined the tangle is seen to be a highly organized structure. Its functioning depends on the cooperation and competition of all its diverse links.

In the beginning, the pyramid of life was low and squat; the food chains short and simple. Evolution has added layer after layer, link after link. Man is one of thousands of accretions to the height and complexity of the pyramid. Science has given us many doubts, but it has given us at least one certainty; the trend of evolution is to elaborate the biota.

Land, then, is not merely soil; it is a fountain of energy flowing through a circuit of soils, plants, and animals. Food chains are the living channels which conduct energy upward; death and decay return it to the soil. The circuit is not closed; some energy is dissipated in decay, some is added by absorption, some is stored in soils, peats, and forests, but it is a sustained circuit, like a slowly augmented revolving fund of life.

The upward flow of energy depends on the complex structure of the plant and animal community, much as the upward flow of sap in a tree depends on its complex cellular organization. Without this complexity normal circulation would not occur. Structure means the characteristic numbers, as well as the characteristic kinds and functions of the species.

This interdependence between the complex structure of land and its smooth functioning as an energy circuit is one of its basic attributes.

When a change occurs in one part of the circuit, many other parts must adjust themselves to it. Change does not necessarily obstruct the flow of energy; evolution is a long series of self-induced changes, the net result of which has been probably to accelerate the flow; certainly to lengthen the circuit.

Evolutionary changes, however, are usually slow and local. Man's invention of tools has enabled him to make changes of unprecedented violence, rapidity, and scope.

One change is in the composition of floras and faunas. The larger predators are lopped off the cap of the pyramid; food chains, for the first time in history, are made shorter rather than longer. Domesticated species are substituted for wild ones, and wild ones moved to new habitats. In this world-wide pooling of faunas and floras, some species get out of bounds as pests and diseases, others are extinguished. Such effects are seldom intended or foreseen; they represent unpredicted and often untraceable readjustments in the structure. Agricultural science is largely a race between the emergence of new pests and the emergence of new techniques for their control.

Another change affects the flow of energy through plants and animals, and its return to the soil. Fertility is the ability of soil to receive, store, and return energy. Agriculture, by overdrafts on the soil, or by too radical a substitution of domestic for native species in the superstructure, may clog the channels of flow or deplete storage. Soils depleted of their stores wash away faster than they form. This is erosion.

Waters, like soils, are part of the energy circuit. Industry, by polluting waters, excludes the plants and animals necessary to keep energy in circulation.

Transportation brings about another basic change: the plants or animals grown in one region are consumed and return to the soil in another. Thus the formerly localized and self-contained circuits are pooled on a world-wide scale.

The process of altering the pyramid for human occupation releases stored energy, and this often gives rise, during the pioneering period, to a deceptive exuberance of plant and animal life, both wild and tame. These releases of biotic capital tend to becloud or delay the penalties of violence.

This thumbnail sketch of land as an energy circuit conveys three ideas more or less lacking from the balance of nature concept:

(1) That land is not merely soil.
(2) That the native plants and animals kept the energy circuit open; others may or may not.
(3) That man-made changes are of a different order than evolutionary changes, and have effects more comprehensive than is intended or foreseen.

These ideas, collectively, raise two basic issues: Can the land adjust itself to the new order? Can violence be reduced?

Biotas seem to differ in their capacity to sustain violence. Western Europe, for example, carries a far different pyramid than Caesar found there. Some large animals are lost; many new plants and animals are introduced, some of which escape as pests; the remaining natives are greatly changed in distribution and abundance. Yet the soil is still fertile, the waters flow normally, the new structure seems to function and to persist. There is no visible stoppage of the circuit.

Western Europe, then, has a resistant biota. Its processes are tough, elastic, resistant to strain. No matter how violent the alterations, the pyramid, so far, has developed some new *modus vivendi* which preserves its habitability for man and for most of the other natives.

The semiarid parts of both Asia and America display a different reaction. In many spots there is no longer any soil fit to support a complex pyramid, or to absorb the energy returning from such as remains. A cumulative process of wastage has set in. This wastage in the biotic organism is similar to disease in an animal, except that it does not culminate in absolute death. The organism recovers, but at a low level of complexity and human habitability. We attempt to offset the wastage by reclamation, but where the regimen of soils and waters is disturbed it is only too evident that the prospective longevity of reclamation projects is short.

The combined evidence of history and ecology seems to support one general deduction: the less violent the man-made changes, the greater the probability of successful readjustment in the pyramid. Violence, in turn, would seem to vary with human population density; a dense population requires a more violent conversion of land. In this respect, America has a better chance for nonviolent human dominance than Europe.

It is worth noting that this deduction runs counter to pioneering philosophy, which assumes that because a small increase in density enriched human life, that an indefinite increase will enrich it indefinitely. Ecology knows of no density relationship which holds within wide limits, and sociology seems to be finding evidence that this one is subject to a law of diminishing returns.

Whatever may be the equation for men and land, it is improbable that we as yet know all its terms. The recent discoveries in mineral and vitamin nutrition reveal unsuspected dependencies in the up-circuit; incredibly minute quantities of certain substances determine the value of soils to plants, of plants to animals. What of the down-circuit? What of the vanishing species, the preservation of which we now regard as an aesthetic luxury? They helped build the soil; in what unsuspected ways may they be essential to its maintenance? Professor Weaver proposes that we use prairie flowers to reflocculate the wasting soils of the dust bowl; who knows for what purpose cranes and condors, otters and grizzlies may some day be used?

Can the violence be reduced? I think that it can be, and that most of the present dissensions among conservationists may be regarded as the first gropings toward a nonviolent land use.

For example, the fight over predator control is no mere conflict of interest between field-glass hunters and gun-hunters. It is a fight between those who see utility and beauty in the biota as a whole, and those who see utility and beauty only in pheasants or trout. It grows clearer year by year that violent reductions in raptorial and carnivorous species as a means of raising game and fish are necessary only where highly artificial (i.e., violent) methods of management are used. Wild-raised game does not require hawkless coverts, and the biotically educated sportsman gets no pleasure from them.

Forestry is a turmoil of naturalistic movements.

Thus the Germans, who taught the world to plant trees like cabbages, have scrapped their own teachings and gone back to mixed woods of native species, selectively cut and naturally reproduced (*Dauerwald*). The "cabbage brand" of silviculture, at first seemingly profitable, was found by experience to carry unforeseen biotic penalties: insect epidemics, soil sickness, declining yields, foodless deer, impoverished flora, distorted bird population. In their new Dauerwald the hard-headed Germans are now propagating owls, woodpeckers, titmice, goshawks, and other useless wildlife.

In America, the protests against radical "timber stand improvement" by the C.C.C. and against the purging of beech, white cedar, and tamarack from silvicultural plans are on all fours with Dauerwald as a return to nonviolent forestry. So is the growing skepticism about the ultimate utility of exotic plantations. So is the growing alarm about the epidemic of new Kaibabs, the growing realization that only wolves and lions can insure the forest against destruction by deer and insure the deer against self-destruction.

We have a whole group of discontents about the sacrifice of rare species:

condors and grizzlies, prairie flora and bog flora. These, on their face, are protests against biotic violence. Some have gone beyond the protest stage: witness the Audubon researches for methods of restoring the ivory-billed woodpecker and the desert bighorn; the researches at Vassar and Wisconsin for methods of managing wildflowers.

The wilderness movement, the Ecological Society's campaign for natural areas, the German *Naturschutz*, and the international committees for wildlife protection all seek to preserve samples of original biota as standards against which to measure the effects of violence.

Agriculture, the most important land use, shows the least evidence of discontent with pioneering concepts. Conservation, among agricultural thinkers, still means conservation of the soil, rather than of the biota including the soil. The farmer must by the nature of his operations modify the biota more radically than the forester or the wildlife manager; he must change the ratios in the pyramid and exclude the larger predators and herbivores. This much difference is unavoidable. Nevertheless it remains true that the exclusions are always more radical than necessary; that the substitution of tame for wild plants and the annual renewal of the plant succession creates a rich habitat for wildlife which has never been consciously utilized except for game management and forestry. Modern "clean farming," despite its name, sends a large portion of its energy into wild plants; a glance at the aftermath of any stubble will prove this. But the animal pyramid is so simplified that this energy is not carried upward; it either spills back directly into the soil, or at best passes through insects, rodents, and small birds. The recent evidence that rodents increase on abused soils (animal weed theory) shows, I think, a simple dearth of higher animal layers, an unnatural downward deflection of the energy circuit at the rodent layer. Biotic farming (if I may coin such a term) would consciously carry this energy to higher levels before returning it to the soil. To this end it would employ all native wild species not actually incompatible with tame ones. These species would include not merely game, but rather the largest possible diversity of flora and fauna.

Biotic farming, in short, would include wild plants and animals with tame ones as expressions of fertility. To accomplish such a revolution in the landscape, there must of course be a corresponding revolution in the landholder. The farmer who now seeks merely to preserve the soil must take account of the superstructure as well; a good farm must be one where the wild fauna and flora has lost acreage without losing its existence.

It is easy, of course, to wish for better kinds of conservation, but what good does it do when on private lands we have very little of any kind? This is the basic puzzle for which I have no solution.

It seems possible, though, that prevailing failure of economic self-interest as a motive for better private land use has some connection with the failure of the social and natural sciences to agree with each other, and with the landholder, on a common concept of land. This may not be it, but ecology, as the fusion point of sciences and all the land uses, seems to me the place to look.

New Year's Inventory Checks Missing Game [1940]

As early as 1929 Leopold began writing short how-to articles on game management and other natural history topics for farmers and other landowners. By 1938 he was publishing such pieces in the *Wisconsin Agriculturist and Farmer*—so regularly that he began adjusting the topics to the natural cycle of the seasons. Of the dozens of vignettes he produced during the next few years, several found their way into *Sand County Almanac*, and it is conceivable that his notion of arranging the book in the form of an almanac originated with this series. It is not difficult to see in this piece the germ of "January Thaw."

Some Sunday in January when the tracking is good, I like to stroll over my acres and make mental note of the birds and mammals whose sign ought to be there, but isn't. One appreciates what is left only after realizing how much has already disappeared.

Every large Wisconsin woodlot, for example, ought to show the mincing lady-like tracks of ruffed grouse, but few do. There are a dozen counties now grouseless. Why? Because we failed to reserve part of the woods from grazing.

Every woodlot, during winter thaws, ought to show the hurried wanderings of coons emerging hungry from their den trees. Few do, because few woods have any den trees left. The hollow basswoods and white oaks which formerly harbored coons have been chopped out, often by improvident coon hunters. The same lack of hollow trees has eliminated the flying squirrel, the screech owl, and the barred owl from many a woods.

Out in the corn stubble by the marsh we should find the peculiar tap-dancing tracks of the prairie chicken; instead we find only the trotting-horse stride of a racing pheasant. Why no chickens? Because years ago we plowed up their booming grounds, mowed, burned or pastured their nesting cover, and then overshot the young in fall. Today we have a dozen chicken-less counties, and if fires are not checked in the peat lands, we shall end with a chickenless state.

In the tussock swamp by the tamaracks we can look for a track few people know; the kangaroo-like springs of the jumping mouse. But if the tussocks have been drained, or too hard pastured, the jumping mouse will have disappeared, to be replaced by the prosaic meadow mouse.

In the tamaracks, if you have any, you should find the regurgitated pellets of the long-eared owl. Note well the mouse skulls; three skulls per pellet, one pellet per day, 100 days in the winter, 300 mice per owl per year. Can you afford to let some rabbit-hunter with the trigger-itch shoot him just for fun? Is it worth while to keep a few tamaracks just to have owls around?

By the river bank, if you have one, there should be, at least at rare intervals, the toboggan-slide of an otter playing in the snow. Most Wisconsin rivers are now otterless; monotonous ribbons of mud and water. A single otter will travel 20 miles of river, and to the mind of the initiated convert that long stretch of mud and water into a personality. In England, otters are common, even in densely settled districts. Why not in Wisconsin?

The State of the Profession [1940]

Delivered to the Wildlife Society on March 18, 1940, Leopold's presidential address was subsequently published in the *Journal of Wildlife Management*. Leopold ranged widely in his talk, moving from the nature of science and its connection with human values and the arts to the need for more basic research and opportunities in "private practice" for the growing cadre of wildlife professionals.

One of the ironies frequent in history is a group of men attempting one thing and accomplishing another. We are attempting to manage wildlife, but it is by no means certain that we shall succeed, or that this will be our most important contribution to the design for living.

For example, we may, without knowing it, be helping to write a new definition of what science is for.

We are not scientists. We disqualify ourselves at the outset by professing loyalty to and affection for a thing: wildlife. A scientist in the old sense may have no loyalties except to abstractions, no affections except for his own kind.

Moreover, some of us entertain heresies and doubts. We doubt whether science can claim the credit for bigger and better tools, comforts, and securities without also claiming the credit for bigger and better erosions, denudations, and pollutions. We doubt whether the good life flows automatically from the good invention.

The definitions of science written by, let us say, the National Academy, deal almost exclusively with the creation and exercise of power. But what about the creation and exercise of wonder, of respect for workmanship in nature? I see hints of such dissent, even in the writings of the scientifically elect—Fraser Darling, for example. Of course, we have always had such writers (David, Isaiah, John Muir) but they were not scientifically elect; they were only poets. Is Fraser Darling only a poet?

The peculiar pertinence of this to our profession is that we deal with science, but we have no prospect of inventing new tools or powers. Our job is to harmonize the increasing kit of scientific tools and the increasing recklessness in using them with the shrinking biotas to which they are applied. In the nature of things we are mediators and moderators, and unless we can help rewrite the objectives of science our job is predestined to failure.

I daresay few wildlife managers have any intent or desire to contribute to art and literature, yet the ecological dramas which we must discover if we are to manage wildlife are inferior only to the human drama as subject matter for the fine arts. Is it not a little pathetic that poets and musicians must paw over shopworn mythologies and folklores as media for art, and ignore the dramas of ecology and evolution?

There are straws which indicate that this senseless barrier between science and art may one day blow away, and that wildlife ecology, if not wildlife management, may help do the blowing. We have, at long last, an ecological novel—Peattie's "Prairie Grove." Darling is not the only ecologist whose scientific writings have literary quality. In our profession, and on its fringes, are a growing number of painters and photographers who are also researchers. These intergrades in human taxonomy are perhaps more important than those which so perplex the mammalogists and ornithologists. Their skulls are not yet available to the museums, but even a layman can see that their brains are distinctive.

In its external trappings of printed knowledge, our profession has attained in four years a maturity which might well have taken a decade. I refer, of course, to *The Journal of Wildlife Management* and the *Wildlife Review*. Some journal papers are still a bit thin, but the average is high and getting higher. No other conservation profession has the equal of the *Wildlife Review*.

One of the weak points in our profession is the low proportion of private employment. Even among the publicly employed the proportion dealing directly with practice on private lands is small.

I think there are more opportunities in private practice than we foresee.

A few industries are already set to go ahead. The guano industry of Peru has employed an American ornithologist to manage its birds, and thus to put guano on "sustained yield." The number of guano birds on these very small Peruvian islands is nearly as great as the number of ducks in North America. This comparison may help convey the heroic proportions of this venture, and incidentally, the low estate of our ducks.

The Hudson's Bay Company is, I think, about set for fur management in the Canadian Arctic. They must be, for the wild fur is gone. Canadians, I think, will do well to encourage the venture, else they will have more acres on relief than we have. Incidentally, a boom in beavers will do much for the ducks.

Why do large private holdings retain wildlife managers in the Southeast, but not elsewhere? If our profession can give valuable service on strings of quail preserves, why not to other game species in other regions where large holdings exist? At least one reason harks back, I think, to a lack of respect for private property. In most regions the public puts all landowners, large and small, under moral suasion not to post. The public does not realize that this is moral suasion not to manage.

Some day the hunter will learn that hunting and fishing are not the only wildlife sports; that the new sports of ecological study and observation are as free to all now as hunting was to Daniel Boone. These new sports depend on the retention of a rich flora and fauna. Management of private holdings to rebuild the fauna and flora is one of the opportunities offered our profession today. There is a growing number of private sanctuaries, private arboreta, and private research stations, all of which are gropings toward non-lethal forms of outdoor recreation. But few such gropings are skillful. Wildlife managers, acting in a consulting capacity, could help owners find what they are looking for.

Some fear that we are getting too much research and not enough management into our journals and (by implication) into our programs. I do not share in this view; in fact, I think the shoe is on the other foot. We know how to manage only a few easy species like deer and pheasants. In other species we know a few fragmentary treatments which are *probably* beneficial, but this is not enough. Until we know more it is proper that a high proportion of our professional effort should go into research.

Too much research, however, is superficial and aimed at quick returns. The high proportion of sporting funds in both the ten unit system and the new Pittman-Robertson structure tends to perpetuate this distortion. So does the low proportion of research groups which have as yet demonstrated capacity to execute more fundamental investigations. So does the series of inter-bureau treaties which confine federal research to a single bureau. Is it not just as illogical to confine wildlife research to a single bureau as it is to confine conservation to a single department?

If anyone doubts that we are trying to eat our research dessert before starting the soup-kettle, let him appraise the national situation in terms of the following questions: In how many species do we know the sex and age composition of a population, and its rate of turnover? In how many species do we have criteria of age? In how many species have we followed the behavior of a sample population for ten years?

The research program is out of balance in other respects. One is the paucity of research of an ecological nature in such groups as rodents where the problem is to manage downward rather than upward. Stockmen and farmers quite naturally want direct action in the form of control, while absentee conservationists protest at any action at all. It is probable that both are wrong, and that the eventual answer in rodents, as in game, lies in indirect environmental manipulation. But where is the research aimed to develop and implement this concept? Concepts do not help to manage land.

The research program is out of balance in that certain kinds of wildlife are omitted altogether; for example, wildflowers and other non-economic vegetation. It would be interesting to see one of the Ten Units get into a huddle with the botany department and propose to the Director of Conservation that something be found out about the management of lady-slippers in farm woodlots. Would the checkbook snap open or shut?

It is encouraging to note that one erstwhile orphan, innocent of economic utility, is no longer high-hatted by his useful conservation cousins. I refer to rare species.

Lastly, the research program pays too little attention to the history of wildlife, and our system of publications makes no provision for historical monographs. We do not yet appreciate how much historical evidence can be dug up, or how important it can be in the appraisal of contemporary ecology. I have in mind such historical work as that of A. W. Schorger in Wisconsin, most of which is not yet in print. I would like to see the Society set up a basket to receive funds for the publication of historical wildlife monographs. I believe that both the monographs and the funds would eventually be forthcoming.

One problem which now faces the profession is how to organize extension. I use that word in its agricultural sense, *i.e.,* sending out trained men to help landowners to help themselves.

Most extension efforts in wildlife have been aimed at helping sportsmen to help themselves. It represents real progress to see states like Texas and Missouri sending out young technicians to deal directly with landowners. The coordination of this new enterprise with the parallel enterprises in agriculture and forestry remains to be worked out.

The problem of teaching conservation to laymen is distinguishable from extension only in name. State after state is legislating conservation courses into the curricula of public schools. But where are the local teaching materials, and who is to teach the teachers how to use them?

In those states which have wildlife research units the production of teaching materials is presumably under way, but what of the teaching of teachers? Why do so many universities spend most of their wildlife funds and use their ablest men in training professional managers when the greater need is for wildlife courses for the general student body and for prospective teachers?

These are problems of educational policy far wider than our own profession, but the speed and skill with which they are solved will depend in large degree on the statesmanship of wildlife managers.

In this little list of unanswered problems and dilemmas there lies concealed, but I hope not undiscovered, a story of almost romantic expansion in professional responsibilities.

Our profession began with the job of producing something to shoot. However important this may seem to us, it is not very important to the emancipated moderns who no longer feel soil between their toes.

We find that we cannot produce much to shoot until the landowner changes his ways of using land, and he in turn cannot change his ways until his teachers, bankers, customers, editors, governors, and trespassers change their ideas about what land is for. To change ideas about what land is for is to change ideas about what anything is for.

Thus we started to move a straw, and end up with the job of moving a mountain.

Ecology and Politics [1941]

To those who knew him personally, Aldo Leopold was above all a teacher. Wildlife Ecology 118, a general course he began offering in 1939 for liberal arts majors as well as wildlifers, was a memorable experience for those who were fortunate enough to enroll in it. Though he usually spoke from notes, he wrote out the introductory lecture for the spring 1941 term in its entirety. Disquieted by the darkening course of world history, he argues perhaps too freely by analogy from animal to human populations. Nevertheless, the lecture stands as a remarkable effort to come to terms with the ethical implications of ecology against the background of world war. Russell Lord, editor of *The Land*, was prepared to publish a slightly revised version as of April 1941, but Leopold continued to work on it and in the end never did publish it. Like many professors, he was perhaps more willing to float new and incompletely explored ideas like these before students than to commit them to posterity.

Ecology tries to understand the interactions between living things and their environment. Every living thing represents an equation of give and take. Man or mouse, oak or orchid, we take a livelihood from our land and our fellows, and give in return an endless succession of acts and thoughts, each of which changes us, our fellows, our land, and its capacity to yield us a further living. Ultimately we give ourselves.

That this collective account between the earth and its creatures ultimately balances is implicit in the fact that both continue to live.

It does not follow, however, that each species continues to live. Paleontology is a book of obsequies for defunct species.

Man, for reasons sufficient to himself, would rather see than be one of the defunct. Fear of human extinction has been the drum beaten by every prophet, from St. John to the Los Angeles cults. But we moderns, seeing science defeat one after another of St. John's four horsemen, have, until very recently, accepted the notion that our continuity is predestined and automatic.

There remains a doubt whether war, famine, and pestilence are the only horsemen to be feared. A new one, unnamed in holy writ, is now much in the headlines: a condition of unstable equilibrium between soils and waters, and their dependent plants and animals. Ecology is the attempt to understand what makes resources stable or unstable.

Another new threat, perhaps even more serious, is the genetical deterioration of the human species. Ecology is here only a bystander, except in this sense; it offers abundant testimony that only healthy species achieve continuity.

The emergence of these new apparitions does not mean that the original and authentic four are unsaddled. War, famine, and perhaps pestilence again thunder across the continents. War is a disruption of the give and take equation. What, if anything, can ecology say about it?

Not much, except by analogy with animals. Whether such analogies are valid is anybody's guess. I shall try to sketch the human enterprise, in its relation to war, as it now appears to me.

Every environment carries not only characteristic kinds of animals, but characteristic *numbers* of each. Thus the characteristic number of Indians in virgin America was small. More Indians would either have starved or killed each other off; fewer Indians would have risked annihilation of the race in some blizzard, drouth, or epidemic. Every animal in every land has its characteristic number. That number is the carrying capacity of that land for that species.

When we arrived on the scene we raised the carrying capacity of the land for man by means of tools. Tools enable us to extract more livelihood from fewer acres, i.e. they change the "take" side of our biotic equation. But in so doing, they also change our "give" side, and also the equations for every fellow creature. Technology has wrought marvels in increasing our take, but it has largely ignored our give, and it has almost entirely ignored the adjustments forced on other animals and plants. Its fundamental assumption, so far, is that take can be increased indefinitely, and with it human populations.

It is probably true that food can, by drawing on aerial and geological stores of fertilizer, be increased indefinitely. But it is far from true, as Darwin once postulated, that animal populations are limited mainly by food. One of the most emphatic lessons of ecology is that animal populations are usually self-limiting; that the mechanisms for limitation are diverse, even for a single species; and that they often shift inexplicably from one kind to another; that the usual sequence is for some limitation to act before the end of the current food supply is in sight.

For example: the characteristic number of bobwhites on 3200 acres near Prairie du Sac is 300. Six times in the last eleven years bobwhites have increased beyond 300, only to be cut down, by one force or another, to 300 or less. The forces which did the cutting were predation, storms, eviction to submarginal habitats, and starvation. These mechanisms of limitation worked in various combinations, never twice the same, but *always with the same result.* When they cut the population below 300, it rebounded, usually in one year, but always within two or three.

A still clearer case is presented in muskrat marshes, which sometimes become overpopulated before predators discover that fact. Such rats, despite abundant food, begin to fight each other. Wounds become infected with fungous disease, which promptly reduces the population. This closely parallels the human case.

Such self-limiting mechanisms are an integral part of nature, and are probably as immutable as the color, form, and habits of the individual creature. That is to say, the tendency for too many muskrats to fight, become infected, and die is probably a constant attribute of muskrat *populations*, just as the tendency to have red teeth is a constant attribute of the individual muskrat. We have not learned to think of populations as having fixed attributes; they become visible only in large stretches of time and space. But these attributes exist, and it is the task of ecology to see and understand them.

Return now to man: having suspended the laws of carrying capacity by inventing tools, he next suspended the laws of predation by inventing ethics. These two manipulations of the natural order are highly interdependent. Tools cannot be made or used without peace; peace cannot be sustained without tools, for men who are hungry, either for food or other necessities, automatically fight.

Ethics are an adaptation without parallel in animal history. The success of ethical restraints depends entirely on mutuality of acceptance. We know, to our cost, that this mutuality of acceptance periodically breaks down, and that such breakdowns are followed by a reversion to the ancestral predatory order. Each such reversion becomes more destructive than the last, for organized predation, backed by tools, is far more fearsome than the unorganized individualistic combat prevailing in animals.

Come back, now, to the fundamental assumptions of technology. The technologists' cure for war is more technology, to the end that we may increase take, thus raise standards of living, and thus promote ethics. Incidentally, should a neighbor lapse from his acceptance of ethics, the accretions of technology will help defend us against his predations.

There are few savages today who are not aware that this technological recipe for civilization is, at least for the moment, a failure. Nations fight over *who shall take charge* of increasing the take and *to whom* the better life shall accrue. Even in peace-time the energies of mankind are directed not toward *creating* the better life, but toward *dividing* the materials supposedly necessary for it. From president to parlor-pink, from economist to stevedore, all are preoccupied with dividing the means rather than building the end.

As for ethics, each seems to write his code to fit his material needs, rather than vice versa. Each political or economic group has such powerful tools that each lives in terror lest his neighbor use them.

The task of appraising the right and wrong of this tragic dilemma is beyond my powers. It may be fitting, though, for an ecologist to appraise the probable soundness of the assumptions on which the whole modern structure is built. If science cannot lead us to wisdom as well as power, it is surely no science at all.

We may begin by admitting that the technological formula, in its early stages, actually succeeded in raising carrying capacity, standards of living, and ethics. But this is no evidence that it will continue to do so indefinitely through greater and greater elaborations of the same idea. All ecology is replete with laws which begin to operate at a threshold, and cease operating at a ceiling. No one law holds good through the entire gamut of time and circumstance. Religions teach the existence of absolute laws, but science cannot find them.

If there is a doubt about the upper limits of the technological formula, what, if anything, can be done to define the limit or amend the formula?

Over this dilemma the ecologist speculates in this wise: tools have actually raised carrying capacity, and ethics have at times suspended predation, but perhaps this is possible only within certain limits of population density. Perhaps the present world-revolution is the sign that we have exceeded that limit, or that we have approached it too rapidly. If so, instead of calling a moratorium on science, as some have proposed, why not call a moratorium on human increase? Why not seek for quality in place of ciphers in human populations? Why not bend science more toward new understandings, less toward new machines?

But these, I fear, may be mere words. Self-limitation of population, like ethics, depends upon unanimity for its success. Hitler and Mussolini are advocating competitive multiplication, obviously with a view to bigger and better predations. That is to say, their remedy for the overpopulation of Europe is more overpopulation in Germany and Italy. Any child should comprehend the fallacy of such doctrine, which is utterly illogical without the corollary assumption that all cultural values repose in these expanding and predatory groups.

On the other hand, the collapse of France raises the question whether voluntary self-limitation of numbers is not automatically followed by decay of moral fibre. If so, one is forced to the conclusion that technological civilization is inherently self-terminating, and can exist only temporarily in new and underpopulated habitats.

One arrives at this same disquieting thought by scrutinizing institutions and governments from the ecological angle. It is commonly assumed that men *select* their form of government. It may be argued with equal logic that the form is in the long run *dictated* by habitat, by population density, and by pressure of predatory neighbors. Democracy, like ethics, may depend on mutuality of acceptance, i.e. on absence of attack. Certainly it is militarily inefficient, and when liberals fear that total defense of democracy may destroy democracy, they are voicing this same apprehension.

These puzzlements raise the question: do other animals select their "form of government" to fit their adaptations, or does circumstance dictate the form? Both seem to play a part. Certainly adaptations seem to determine the broad outlines of the social structure for each species, but circumstance may cause sudden and violent changes in it. Thus many ungulates assume "defense formations" when attacked, but in some species this takes place only when the herd is of a certain size. Thus antelope fight off wolves by forming a defensive ring, but only when a dozen or so are present. Smaller groups do not join in mutual defense, but simply flee.

Let me return again to the fundamental assumptions of technological culture. We assume, I think naively, that increasing "take" (i.e. more extraction, conversion, and consumption of resources) always raises standards of living. Sometimes it merely raises population levels. Perhaps this is a bear chasing his own tail. When the British ameliorated the hard lot of South African natives (by medical service, better farming, etc.), the response was more natives rather than higher standards, and more strain on an already overcrowded range. Feeding starving deer is a close analogy. Deer starve because their range is unbalanced or overtaxed, and the price of ameliorating their lot is more deer, more need of feeding, more damage to the range, and eventual malnutrition and deterioration of the herd. Perhaps only animals capable of qualitative self-improvement and quantitative self-limitation can be safely ameliorated.

Again, we overwork the assumption that better living makes higher ethics. There is much evidence against, as well as for, this universal thesis of technological culture. Perhaps ethics are too complex to follow automatically in the wake of newer Fords and shinier bathtubs.

Sewall Wright's theory of plant and animal variants presents a curious human analogy. He postulates that survival of a species depends not on the small exigencies which beset it frequently, but on the catastrophes which occur at long intervals. No population ever survives a catastrophe, but rather only those individuals whose deviations from "normal" happen to enable them to. Since successive catastrophes are seldom twice alike, it follows that survival depends on the constant presence of individuals deviating from "normal" in many respects. Thus a population containing "individualists" in respect of cold, heat, drouth, starvation, predation, and disease may survive, whereas a similar population homogeneous in these respects may perish.

I see in this an evolutionary mandate for individualism. Perhaps the deviations from physical and mental pattern which are tolerated in our social organization, but frowned upon or persecuted in more regimented societies, are an evolutionary "safety device" which may one day determine our continuity.

While the ecological view of politics offers no sure and certain path to a better future, and indeed casts doubt on some guide-signs heretofore accepted as reliable, it nevertheless offers a few assurances on what is the right direction.

There can be no doubt that better human stocks, both as to inheritance and environment, are more likely to find a *modus vivendi* than poorer ones. It is a truism that our education has lagged behind our tools, and you can't build better people with poorer materials.

There can be no doubt that a society rooted in the soil is more stable than one rooted in pavements. Stability seems to vary inversely to the mental distance from fields and woods. The disruptive movements which now threaten the continuity of human culture are born not on the land where the take originates, but in the factories and offices where it is processed and distributed, and in the capitols where the rules of division are written. If courses like this one can decrease our mental distance from fields and woods, they are worth taking, and worth giving.

Wilderness as a Land Laboratory [1941]

Leopold's concerns for integrated land management—the artful husbandry of productive land and the restoration of damaged land—and for wilderness preservation are directly related in this article published in the *Living Wilderness*. Wilderness is necessary to science, Leopold avers, as an index of normal structure and function to which used and abused land may be compared. As such it needs to represent a full range of biotic provinces. Leopold worked portions of this piece into "Wilderness" in *Sand County Almanac*.

The recreational value of wilderness has been often and ably presented, but its scientific value is as yet but dimly understood. This is an attempt to set forth the need of wilderness as a base-datum for problems of land-health.

The most important characteristic of an organism is that capacity for internal self-renewal known as health.

There are two organisms in which the unconscious automatic processes of self-renewal have been supplemented by conscious interference and control. One of these is man himself (medicine and public health). The other is land (agriculture and conservation).

The effort to control the health of land has not been very successful. It is now generally understood that when soil loses fertility, or washes away faster than it forms, and when water systems exhibit abnormal floods and shortages, the land is sick.

Other evidences are generally known as facts, but not as symptoms of land-sickness. The disappearance of plant and animal species without visible causes despite efforts to protect them, and the irruption of others as pests, despite efforts to control them, must, in the absence of simpler explanations, be regarded as symptoms of derangement in the land-organism. Both are occurring too frequently to be dismissed as normal evolutionary changes.

The status of thought on these ailments of the land is reflected in the fact that our treatments for them are still prevailingly local.

Thus when a soil loses fertility we pour on fertilizer, or at best alter its tame flora and fauna, without considering the fact that its wild flora and fauna, which built the soil to begin with, may likewise be important to its maintenance. It was recently discovered, for example, that good tobacco crops depend, for some unknown reason, on the pre-conditioning of the soil by wild ragweed. It does not occur to us that such unexpected chains of dependency may have wide prevalence in nature.

When prairie dogs, ground squirrels, or mice increase to pest levels we poison them, but we do not look beyond the animal to find the cause of the irruption. We assume that animal troubles must have animal causes. The latest scientific evidence points to derangements of the *plant* community as the real seat of rodent irruptions, but few or no explorations of this clue are being made.

Many forest plantations are producing one-log or two-log trees on soil which originally grew three-log and four-log trees. Why? Advanced foresters know that the cause probably lies not in the tree, but in the micro-flora of the soil, and that it may take more years to restore the soil flora than it took to destroy it.

Many conservation treatments are obviously superficial. Flood control dams have no relation to the cause of floods. Check dams and terraces do not touch the cause of erosion. Refuges and propagating plants to maintain animals do not explain why the animal fails to maintain itself.

In general, the trend of the evidence indicates that in land, just as in the human body, the symptom may lie in one organ and the cause in another. The practices we now call conservation are, to a large extent, local alleviations of biotic pain. They are necessary, but they must not be confused with cures. The art of land-doctoring is being practiced with vigor, but the science of land-health is a job for the future.

A science of land health needs, first of all, a base-datum of normality, a picture of how healthy land maintains itself as an organism.

We have two available norms. One is found where land physiology remains largely normal despite centuries of human occupation. I know of only one such place: northeastern Europe. It is not likely that we shall fail to study it.

The other and most perfect norm is wilderness. Paleontology offers abundant evidence that wilderness maintained itself for immensely long periods; that its component species were rarely lost, neither did they get out of hand; that weather and water built soil as fast or faster than it was carried away. Wilderness, then, assumes unexpected importance as a land-laboratory.

One cannot study the physiology of Montana in the Amazon; each biotic province needs its own wilderness for comparative studies of used and unused land. It is of course too late to salvage more than a lop-sided system of wilderness remnants, and most of these remnants are far too small to retain their normality. The latest report* from Yellowstone Park, for example, states that cougars and wolves are gone. Grizzlies and mountain sheep are probably going. The irruption of elk following the loss of carnivores has damaged the plant community in a manner comparable to sheep grazing. "Hoofed locusts" are not necessarily tame.

I know of only one wilderness south of the Canadian boundary which retains its full flora and fauna (save only the wild Indian) and which has only one intruded species (the wild horse). It lies on the summit of the Sierra Madre in Chihuahua. Its preservation and study, as a norm for the sick lands on both sides of the border, would be a good neighborly act well worthy of international consideration.

All wilderness areas, no matter how small or imperfect, have a large value to land-science. The important thing is to realize that recreation is not their only or even their principal utility. In fact, the boundary between recreation and science, like the boundaries between park and forest, animal and plant, tame and wild, exists only in the imperfections of the human mind.

*Adolph Murie, *Ecology of the Coyote in the Yellowstone*, Fauna Series no. 4 of the National Parks of the United States.

The Last Stand [1942]

Though the essays Leopold selected for the almanac are all characterized by a timeless quality, he wrote others as finely honed that were clearly topical and even propagandist, written to inspire immediate action. One such essay is "The Last Stand," published in *Outdoor America* at a time when the last stand of old-growth northern hardwoods in the Porcupine Mountains of Michigan's Upper Peninsula was threatened with wartime cutting. An editorial sidebar explained pending legislation and urged readers to write their congressmen. As a result of continued prodding by Leopold and others, the Michigan legislature in 1943 appropriated one million dollars to purchase the area for a state park.

Sometime in 1943 or 1944 an axe will bite into the snowy sapwood of a giant maple. On the other side of the same tree a crosscut saw will talk softly, spewing sweet sawdust into the snow with each repetitious syllable. Then the giant will lean, groan, and crash to earth: the last merchantable tree of the last merchantable forty of the last virgin hardwood forest of any size in the Lake States.

With this tree will fall the end of an epoch.

There will be an end of cheap, abundant, high-quality sugar maple and yellow birch for floors and furniture. We shall make shift with inferior stuff, or with synthetic substitutes.

There will be an end of cathedral aisles to echo the hermit thrush, or to awe the intruder. There will be an end of hardwood wilderness large enough for a few days' skiing or hiking without crossing a road. The forest primeval, in this region, will henceforward be a figure of speech.

There will be an end of the pious hope that America has learned from her mistakes in private forest exploitation. Each error, it appears, must continue to its bitter end; conservation must wait until there is little or nothing to conserve.

Finally, there will be an end of the best schoolroom for foresters to learn

The Last Stand

By ALDO LEOPOLD

SOMETIME in 1943 or 1944 an axe will bite into the snowy sapwood of a giant maple. On the other side of the same tree a crosscut saw will talk softly, spewing sweet sawdust into the snow with each repetitious syllable. Then the giant will lean, groan, and crash to earth: the last merchantable tree of the last merchantable forty of the last virgin hardwood forest of any size in the Lake States.

WITH this tree will fall the end of an epoch.

There will be an end of cheap, abundant, high-quality sugar maple and yellow birch for floors and furniture. We shall make shift with inferior stuff, or with synthetic substitutes. There will be an end of cathedral aisles to echo the hermit thrush, or to awe the intruder. There will be an end of hardwood wilderness large enough for a few days' skiing or hiking without crossing a road. The forest primeval, in this region, will henceforward be a figure of speech.

There will be an end of the pious hope that America has learned from her mistakes in private forest exploitation. Each error, it appears, must continue to its bitter end; conservation must wait until there is little or nothing to conserve.

Finally, there will be an end of the best schoolroom for foresters to learn what remains to be learned about hardwood forestry: the mature hardwood forest. We know little, and we understand only part of what we know.

This last stand of the northern hardwoods is in the Porcupine Mountain region of the Upper Peninsula of Michigan. Fifty years ago northern hardwoods covered seven million acres in the Lake States. Five years ago the main remnant in the Porcupine region still comprised 170,000 acres. By 1941 this had shrunk to 140,000 acres. Last winter's cuttings were extra

• The bills referred to are H. R. 3793 (Hook Bill), now before the House Committee on Agriculture, Hon. Hampton P. Fulmer, chairman; and companion bill S. 1131, now before the Senate Committee on Agriculture and Forestry, Hon. Ellison D. Smith, chairman. They do not specifically provide for the avowed objective, but include incidentally the Porcupine Mountain area in a much vaster area for sustained yield timber management. Furthermore, we seriously question the advisability at the present time, when America must concentrate its resources both financial and material on vigorous prosecution of the war, of pressing for such a large amount of money ($30,000,000) when a small fraction of that amount would suffice for the specific objective.

Ten per cent of the money authorized by the bill would likely be ample for purchasing the Porcupine Mountains and the timbered areas immediately adjacent to the south and west—a solid block of approximately 100,000 acres. Half of that money, or five per cent, might well suffice for the immediately pressing acreage. However, we have told Congress that we are not interested in mechanics. We have stated the objective and urge that Congress, in its wisdom, mold the legislation as needed to attain this objective—and that next year will be too late. You can help save the Porcupine area by making your wishes known to Congressman Fulmer and Senator Smith.

large due to war demand. At the present rate of cutting, only stands too rocky or poor to repay the operator have much chance to outlive the next two years. After that fires are likely to polish up the slashings, leaving a nice pile of brushy rocks as a monument to our generation.

There are, of course, odd bits of uncut hardwoods left elsewhere. The largest bit (10,000 acres) is owned by a private club, and is kept to look at. It is ironical that this club may in the end outscore the combined efforts of the Congress of the United States, the U. S. Forest Service, the sovereign state of Michigan, and the mighty lumber industry as a conserver of virgin forest.

The sugar maple is as American as the rail fence or the Kentucky rifle. Generations have been rocked in maple cradles, clothed from maple spinning wheels, and fed with maple-sweetened cakes served on maple tables before maple fires. Yet the demise of the maple forest brings us less regret than the demise of an old tire. Like the shrew who burrows in maple woods, we take our environment for granted while it lasts. Unlike the shrew, we make shift with substitutes. The poorest is the European "Norway maple", a colorless fast-growing tree persistently used by misguided suburbanites to kill lawns. Wisconsin has used Norway maples to shade its capitol. No governor and no citizen has protested this affront to the peace and dignity of the state.

Maple boards, like maple shade, take time to grow. We have lots of prospective maple lumber in second-growth stands. It is doubtful whether these regrowths will ever achieve the quality or volume of the original stands, first because we shall lack the patience to wait for them to mature; secondly because the maple forest is one of the most highly organized communities on earth; hence the slashing likely injures its future capacity to produce.

Lake of the Clouds
(Reprinted from the May-June, 1942 issue of Outdoor America)

Title page of "The Last Stand" in *Outdoor America*, May–June 1942.

what remains to be learned about hardwood forestry: the mature hardwood forest. We know little, and we understand only part of what we know.

This last stand of the northern hardwoods is in the Porcupine Mountain region of the Upper Peninsula of Michigan. Fifty years ago northern hardwoods covered seven million acres in the Lake States. Five years ago the main remnant in the Porcupine region still comprised 170,000 acres. By 1941 this had shrunk to 140,000 acres. Last winter's cuttings were extra large due to war demand. At the present rate of cutting, only stands too rocky or poor to repay the operator have much chance to outlive the next two years. After that fires are likely to polish up the slashings, leaving a nice pile of brushy rocks as a monument to our generation.

There are, of course, odd bits of uncut hardwoods left elsewhere. The largest bit (10,000 acres) is owned by a private club, and is kept to look at. It is ironical that this club may in the end outscore the combined efforts of the Congress of the United States, the U.S. Forest Service, the sovereign state of Michigan, and the mighty lumber industry as a conserver of virgin forest.

The sugar maple is as American as the rail fence or the Kentucky rifle. Generations have been rocked in maple cradles, clothed from maple spinning wheels, and fed with maple-sweetened cakes served on maple tables before maple fires. Yet the demise of the maple forest brings us less regret than the demise of an old tire. Like the shrew who burrows in maple woods, we take our environment for granted while it lasts. Unlike the shrew, we make shift with substitutes. The poorest is the European "Norway maple," a colorless fast-growing tree persistently used by misguided suburbanites to kill lawns. Wisconsin has used Norway maples to shade its capitol. No governor and no citizen has protested this affront to the peace and dignity of the state.

Maple boards, like maple shade, take time to grow. We have lots of prospective maple lumber in second-growth stands. It is doubtful whether these regrowths will ever achieve the quality or volume of the original stands, first because we shall lack the patience to wait for them to mature; secondly because the maple forest is one of the most highly organized communities on earth; hence the slashing likely injures its future capacity to produce.

Few laymen realize that the penalties of violence to a forest may far outlast its visible evidence. I know a hardwood forest called the Spessart, covering a mountain on the north flank of the Alps. Half of it has sustained cuttings since 1605, but was never slashed. The other half was slashed during the 1600's, but has been under intensive forestry during the last 150 years. Despite this rigid protection, the old slashing now produces only mediocre pine, while the unslashed portion grows the finest cabinet oak in the world; one of those oaks fetches a higher price than a whole acre of the old slashings. On the old slashings the litter accumulates without rotting, stumps and limbs disappear slowly, natural reproduction is slow. On the unslashed portion litter disappears as it falls, stumps and limbs rot at once, natural reproduction is automatic. Foresters attribute the inferior performance of the old slashing to its depleted microflora, meaning that underground community of bacteria, molds, fungi, insects, and burrowing mammals which constitute half the environment of a tree.

The existence of the term microflora implies, to the layman, that science knows all the citizens of the underground community, and is able to push them around at will. As a matter of fact, science knows little more than that the community exists, and that it is important. In a few simple communities like alfalfa, science knows how to add certain bacteria to make the plants grow. In a complex forest, science knows only that it is best to let well enough alone.

But industry doesn't know this. I fear that the present mistreatment of the northern hardwoods may be pondered more seriously in 2042 than in 1942. Industries wince with pain when fixers and planners lay violent hands on their highly organized economic community, yet these same industries fix their forests to death with never a flicker of recognition that the same principle is involved. In neither case do we understand all the intricacies of internal adjustment. Communities are like clocks, they tick best while possessed of all their cogs and wheels.

While the northern hardwood forest, like the Spessart, is injured by violence, it is known to stand up under gentle intelligent use to an extraordinary degree. You can cut a third of the volume of a 200-year-old stand and come back every 20 years and take as much again. The reason inheres in the extreme shade-tolerance of the sugar maple and its associated species. Under each mature veteran stand a dozen striplings, full-height and ready to lay on wood the year after the felled veteran bequeaths to them his place in the sun. This method of quick turnover utilization is called selective logging. Its technology has been fully explored by the research branch of the Forest Service. It differs from slash logging in that the mature trees are cut periodically instead of simultaneously, and the striplings are left to grow instead of to burn in the next fire.

How has industry, with its ear ever cocked for new technology, received this innovation? The answer is written on the face of the hills. Industry, with the notable exception of a half-dozen companies, is slashing as usual. The reason given is that most mills are so nearly cut out anyhow that they cannot await the deferred returns of selective logging; they prefer to die quickly in their accustomed shower of sawdust, rather than to live forever on a reduced annual budget of boards.

One is apt to make the error of assuming that a corporation possesses the attributes of a prudent person. It may not. It is a new species of animal, created by mutation, with a morphology of its own and a behavior pattern which will unfold with time. One can only say that its behavior pattern as an owner of forests is so far not very prudent.

Years ago, when the green robe of the Porcupines still spread over much of upper Michigan, bills were introduced in Congress to buy the area as a National Forest. At that time, the proceeds from selective logging would have paid for the land, and left the growing forest to boot. Nothing was done.

Today, when the green robe of the Porcupines has shrunk to the dimensions of a barely respectable necktie, bills are still before Congress. I suppose Congress hesitates to buy, fearing catcalls from patriotic constituents who assume that all internal problems can wait. Most of them doubtless can and should, but not this one. The war will surely outlast this remnant of forest.

I doubt whether public acquisition, as a means of assuring the national timber supply, is a satisfactory substitute for forestry practice by private owners. The job is too big. When government takes over a small area for decent use, it aims to educate by example, but I fear it also generates a false assurance that things are on the mend. In any event the Porcupine necktie is now too small to be of any consequence as a source of timber. But the Porcupine necktie is more than timber; it is a symbol. It portrays a chapter in national history which we should not be allowed to forget. When we abolish the last sample of the Great Uncut, we are, in a sense, burning books. I am convinced that most Americans of the new generation have no idea what a decent forest looks like. The only way to tell them is to show them. To preserve a remnant of decent forest for public education is surely a proper function of government, regardless of one's views on the moot question of large-scale timber production. Moreover, the Porcupines offer the only steep topography available to the public in the snow-belt of the Lake States; they have a future as a ski area, provided they are not further denuded. The necktie is worth keeping for this purpose alone.

I would like to see the Porcupine region acquired and preserved as an act of national contrition, as the visible reminder of an unsolved problem, as a token of things hoped for. To this end it had best be kept roadless, axeless, hotel-less, and open only to ski or foot travel. The mere existence of such a token-forest might hasten the day when the green robe again spreads over the Lake States, and when the cutting and using of mature timber becomes an act of normal land-cropping, rather than an act of land-pillage.

Land-Use and Democracy [1942]

Published in *Audubon Magazine* at a time when war was the national preoccupation, this essay artfully weaves together Leopold's by then fully formed understanding of the principal elements of a sound public conservation policy. In addition to his emphasis on the esthetic tastes and ethical obligations of the private land owner, summed up by the word *husbandry*, this essay stresses the moral responsibility for conservation that we all bear as consumers. This broadly targeted piece clearly spells out what government can as well as cannot do for the conservation cause and comes to a focus on the role of the citizen.

Conservation education appeared, before December 7, to be making considerable headway.

Now, against a background of war, it looks like a milk-and-water affair. War has defined the issue: we must prove that democracy can use its land decently. At our present rate of progress we *might* arrive at decent land-use a century or two hence. That is too little and too late.

Conservation education, in facing up to its task, reminds me of my dog when he faces another dog too big for him. Instead of dealing with the dog, he deals with a tree bearing his trademark. Thus he assuages his ego without exposing himself to danger.

Just so we deal with bureaus, policies, laws, and programs, which are the *symbols* of our problem, instead of with resources, products, and land-users, which *are* the problem. Thus we assuage our ego without exposing ourselves to contact with reality.

The symbols of conservation have become complex and confusing, but its essential reality has the devastating simplicity of the needle's eye. If we don't like the way landowner X is using the natural resources of which he is owner, why do we buy his products? Why do we invest in his securities? Why do we accord him the same social standing as landowner Y, who makes an honest attempt to use his land as if he were its trustee? Why do we tell our government to reform Mr. X, instead of doing it ourselves? The answer must be either that we do not know the limits of what government can do, or that we don't care deeply enough to risk personal action or danger.

When the Audubon Society killed the millinery feather trade in 1913, what was its real weapon, the prohibitory law or the refusal of intelligent women to buy wild bird plumage? The answer is plain. The law was merely the symbol of a conviction in the mind of a minority. That conviction was so strong and unequivocal that it was willing to risk direct action, danger of ridicule, and even danger of mistakes to achieve the common good.

I have no illusion that all of the products of land-abuse are as easy to identify, or as easy to do without, as a wild bird-skin on a hat. I do assert that many products of land-abuse can be identified as such, and can be discriminated against, given the conviction that it is worth the trouble. Conversely, the products of good land-use can often be singled out and favored.

Back now to education: who is to be educated? By tacit consent it is the *coming generation*; we have only to teach them why and how to act. Here is the dog again, addressing the symbol, walking around the problem instead of facing up to it.

"Children are like grown-ups: they understand what others *do* better than what others are *saying*. Unless the grown-up world shows itself willing to practice conservation, that practice will be hard for the younger generation to adopt." With these two sentences, Paul Sears demolishes the "let posterity do it" school of education.

There is lacking only a simple formula by which we, and posterity, may act to make America a permanent institution instead of a trial balloon. The formula is: learn how to tell good land-use from bad. Use your own land accordingly, and refuse aid and comfort to those who do not.

Isn't this more to the point than merely voting, petitioning, and writing checks for bigger and better bureaus, in order that our responsibilities may be laid in bigger and better laps?

Such an approach may be implemented with cases that present an intellectual gradient suitable for all ages and all degrees of land-use education. No one person, young or old, need feel any obligation to act beyond his own personal range of vision.

For example, a case visible to all who ride and read: does a good American shave with soaps that plaster rock and rill with signs, hiding bad manners behind a barrage of puns? Can the legislature abate this nuisance while the voter rewards its impudence with his custom?

Again, a little less obvious: does a good American accept gifts of stolen goods, or credit scores made by cheating? To wit: ducks bought from the pusher, or refrigerated beyond the legal date? Venison hung up by the guide? Wildflowers pilfered without consent? Can society prevent by law what it condones by social usage?

A little harder: Does one buy Christmas trees that should have been left to grow? How does one tell trees representing legitimate thinnings from trees representing exploitation and robbery? Both are for sale; neither is labeled. Could they be?

Up a step: Dairy X buys milk from steep eroding pastures, which spill floods on the neighbors, and ruin streams. It also buys milk from careful farmers, and mixes the two, so that conservation milk is indistinguishable from exploitation milk. What should the conscientious buyer do? What can the careful farmer do? Could farmers form pools to regulate their own pasture practices, as they now regulate butterfat and bacteria?

Still harder: Lumberman X claims to practice forestry. His boards are necessarily knottier than those offered by lumberman Y, who is still skinning the illimitable (?) woods where rolls the Oregon, and hears no sound save his own dashings. Which board do you buy? Should you buy the honest board, even at a higher price?

Simple, but really tough: Should one accept pay for doing the decent thing to land, as most landowners do when the AAA pays them for lowering their ratio of corn and cotton to legumes and grass? If this is defensible for a year, is it for a decade?

Again: Newspaper X buys its paper from a sulphite mill which turns its wastes into the river that moves in majesty. All other mills do the same, and all other newspapers. Each editorial on conservation sends its additional spurt of offal into the public waters. Your cousin draws his paycheck from the mill, and your brother draws his from the newspaper. "There ought to be a law"; in fact there is a law, but it is not enforced; it can't be. An extra penny for each newspaper would pay the cost of reclaiming the offal, and thus break the whole vicious circle. Whose penny is going to break it? How? When? Has this anything to do with the struggle between democracy and fascism?

Finally: Nearly all American wheat is the product of exploitation. Behind your breakfast toast is the burning strawstack, feeding the air with nitrogen belonging in the soil. Behind your birthday cake is the eroding Palouse, the over-wheated prairies, feeding the rivers with silt for army engineers to push around with dredge and shovel, at your expense; for irrigation engineers to fill their dams with, at the expense of the future. Behind each loaf of (inedible) baker's bread is the "ever normal" granary, the roar of the combine, the swish of the gang-plow, ravaging the land they were built to feed, because it is cheaper to raise wheat by exploitation than by honest farming. It wouldn't be cheaper if exploitation wheat lacked a market. You are the market, but transportation has robbed you of all power to discriminate. If you want conservation wheat, you will have to raise it yourself.

These are samples of the easy, the possible, the difficult, and the insoluble realities of conservation, presented as problems for the citizen. Education, so far, presents them only as problems for his agents.

When the ecologist sees any given force at work in the animal community, he can safely predict that it will operate only up to the length of its tether, after which some other force will take over.

Conservation is our attempt to put human ecology on a permanent footing. Milk-and-water education has convinced people that such an attempt should be made, and they have told their government to act for them. Some other force must now persuade them to act for themselves.

Money-minded people think they are acting when they pay taxes. This hallucination, during the "defense" period, nearly cost us the war. It will cost us our natural resources if we persist in it.

To analyze the problem of action, the first thing to grasp is that government, no matter how good, can only do certain things. Government can't raise crops, maintain small scattered structures, administer small scattered areas, or bring to bear on small local matters that combination of solicitude, foresight, and skill which we call husbandry. Husbandry watches no clock, knows no season of cessation, and for the most part is paid for in love, not dollars. Husbandry of somebody else's land is a contradiction in terms. Husbandry is the heart of conservation.

The second thing to grasp is that when we lay conservation in the lap of the government, it will always do the things it can, even though they are not the things that most need doing.

The present over-emphasis on game farms, fish hatcheries, nurseries and artificial reforestation, importation of exotic species, predator control, and rodent control is here in point. These are things government can do. Each has an alternative, more or less developed, along naturalistic lines, i.e., management or guidance of natural processes. Research shows these alternatives to be, in general, superior. But they involve husbandry, which government can do only on its own lands. Government lands are a minor fraction of our land area. Therefore government neglects the superior things that need doing, and does the inferior things that it can do. It then imputes to these things an importance and an efficacy they do not merit, thus distorting the growth of public intelligence.

This whole twisted confusion stems from the painless path, from milk-and-water education, from prolonging our reliance on vicarious conservation.

The end result is that ideas once wholly beneficial begin to boomerang on the user, a clear sign, to the ecologist, that some new adjustment is in order. A case in point is the idea of sanctuary. Sanctuaries and refuges have done enormous good; we would have kept few rare species without them;

but for them shootable waterfowl would surely have disappeared. Yet on every hand are signs that we expect too much of them. Most public forests are now shrinking or abolishing their refuges because excess deer and elk, in the absence of natural predators, have become a scourge to the forest and to themselves. Some national parks are being eaten up, and have no recourse except to shoot or to shrink; a pair of sharp horns for any park man to sit on. Many notably successful sanctuaries are now ringed by commercial shooting "clubs"; there is a grave question whether the birds would not be better off without both the sanctuary and the clubs. When administration is in the hands of politicians who care more for votes than for birds, there is not even a question.

Why these kickbacks? The answer, I fear, is that sanctuaries are one of the things government can do, but the growth of private ethics and naturalistic management needed to go with them is beyond the powers of government.

It seems to me that sanctuaries are akin to monasticism in the dark ages. The world was so wicked it was better to have islands of decency than none at all. Hence decent citizens retired to monasteries and convents. Once established, these islands became an alibi for lack of private reform. People said: "We pay the bills for all this virtue. Let goodness stay where it belongs, and not pester practical folks who have to run the world." The present attitude of some duck-hunters offers a close parallel. The more monasteries or sanctuaries, the grimmer the incongruity between inside and outside.

We need more sanctuaries, but some of them will boomerang until they serve a better public. This is particularly true of deer and elk sanctuaries which are too big, duck or goose sanctuaries choked in a noose of limit-shooting clubs, or any sanctuary deprived of its natural predators.

One of the curious evidences that "conservation programs" are losing their grip is that they have seldom resorted to self-government as a cure for land abuse. "We who are about to die," unless democracy can mend its land-use, have not tried democracy as a possible answer to our problem.

I do not here refer to such superficial devices as advisory boards, who offer their wisdom to others, or such predatory devices as pressure groups, who exist to seize what they can. I refer rather to social and economic units who turn the light of self-scrutiny on themselves.

NRA was perhaps a start toward responsible self-scrutiny in industry, but the Supreme Court snapped the rising grouse before he ever got above the alders.

The present Soil Conservation Districts are perhaps a start toward self-scrutiny in farming, but they dare not use their powers for lack of voter-support. These districts are self-governing farm communities which have set themselves up as legal entities. In many states the district is authorized to write land-use regulations with the force of law. So far they dare not. But if farmers once asked: "Why don't we tackle our own erosion-control? I'll pull my cows off the hill if you will," the machinery for action is at hand.

Farmers do not yet ask such rash questions. Why? Probably because they have been led to believe that CCC camps, AAA checks, 4-H clubs, extension, meetings, speeches, and other subsidies and uplifts will do the trick. Those who really know land know this is not true; these milk-and-water measures have indeed retarded the rate of soil-loss, but they have not reversed it. Thus we see that the painless path not only fails to lead us to conservation, but sometimes actually retards the growth of critical intelligence on the whereabouts of alternative routes.

No new device in human affairs is ever an unmixed blessing. The idea here proposed: hitching conservation directly to the producer-consumer relation instead of to the government, entails some serious risks. It would present the professional advertiser with an opportunity for euphemized deception and equivocation vastly larger than cigarettes. The more complex the product or process, the wider the field for the trained hoodwinker.

This brings us to the real and indispensable functions of government in conservation. Government is the tester of fact vs. fiction, the umpire of bogus vs. genuine, the sponsor of research, the guardian of technical standards, and, I hasten to add, the proper custodian of land which, for one reason or another, is not suited to private husbandry. These functions will become real and important as soon as conservation begins to grow from the bottom up, instead of from the top down, as is now the case.

Conservation is a state of health in the land-organism. Health expresses the cooperation of the interdependent parts: soil, water, plants, animals, and people. It implies collective self-renewal and collective self-maintenance.

When any one part lives by depleting another, the state of health is gone. As far as we know, the state of health depends on the retention in each part of the full gamut of species and materials comprising its evolutionary equipment.

Culture is a state of awareness of the land's collective functioning. A culture premised on the destructive dominance of a single species can have but short duration.

The Role of Wildlife in a Liberal Education [1942]

Leopold reflects candidly on the employment prospects of wildlife professionals during and after the war in this presentation to the seventh North American Wildlife Conference, subsequently published in the Transactions. *As in "The State of the Profession," he sees a professional opportunity in all-campus courses, but more important an opportunity, through the window of wildlife, to inculcate an understanding of "land ecology" in the lay student and thus eventually in the community at large. The essay may be read as a statement of the philosophy underlying his own Wildlife Ecology 118 at the University of Wisconsin.*

Most of the wildlife education so far attempted is that designed to teach professionals how to do their job. I here discuss another kind: that aimed to teach citizens the function of wildlife in the land organism.

The two kinds contrast sharply in their war status. Perhaps the output of professionals is now excessive, even if there were no war. On the other hand, wildlife teaching for laymen has the same war status as any other branch of science or of the arts; to suspend teaching it is to suspend culture. Culture is our understanding of the land and its life; wildlife is an essential fraction of both.

The bulk of our funds and brains are invested in professional education. In my opinion it is time to "swap ends" to curtail sharply the output of professionals, and to throw the manpower and dollars thus released into a serious attempt to tell the whole campus, and thus eventually the whole community, what wildlife conservation is all about.

To see our predicament clearly, we must see its history.

When wildlife education started a decade ago, three strong forces impelled us to our present course.

One was the obvious preference for preparing men to earn a salary rather than to live a life.

The second was the depression. The pump-priming policy sucked at the conservation schools like a waterspout. Anyone bearing a sheepskin, wet or dry, could soar into the clouds as a paid expert.

The third was expediency. It is easier to teach wildlife to a professional student in 3 years than to a lay student in a semester or two. Once a professional enrolls he must listen, be the teaching good, bad, or indifferent. On the other hand the lay student elects wildlife courses; if the teaching is not vital, he can elect something else.

To what extent are these three pulls still pulling?

Depression is dead. Expediency is no argument. The question, then, boils down to future jobs. Bureaus are now laying plans for another post-war pump-priming era, but it is a mystery to me where we are to find either the cash or the credit for a repetition of 1933. I do not anticipate a post-war boom in "wildlifers." If I am right, and the market for professionals continues poor, then the deans and the presidents and the donors of wildlife funds will have the option of either shrinking the present schools, or switching their emphasis from professional to liberal teaching.

It is not likely that this switch can be made successfully if postponed until the eleventh hour. The time to start is now.

Fortunately the process of conversion does not call for a complete abandonment of professional output. All-campus teaching cannot be vital without research, and research is not possible without assistants, experimental areas, and definite local projects. This residuum of research can be made to produce a small high-grade annual crop of professionals at the same time that it feeds the all-campus teaching effort with vital local facts and questions.

In my own unit, I began this conversion 3 years ago, when the present overproduction of professionals first became visible. The response from the campus-at-large has been gratifying. I would recommend the change to others, even if there were no war to force the issue.

Liberal education in wildlife is not merely a dilute dosage of technical education. It calls for somewhat different teaching materials and sometimes even different teachers. The objective is to teach the student to see the land, to understand what he sees, and enjoy what he understands. I say land rather than wildlife, because wildlife cannot be understood without understanding the landscape as a whole. Such teaching could well be called land ecology rather than wildlife, and could serve very broad educational purposes.

Perhaps the most important of these purposes is to teach the student how to put the sciences together in order to use them. All the sciences and arts are taught as if they were separate. They are separate only in the classroom. Step out on the campus and they are immediately fused. Land ecology is putting the sciences and arts together for the purpose of understanding our environment.

An illustration of what I mean appears in Figure 1, which traces some of the lines of dependency (or food chains, so called) in an ordinary community. These lines are the arteries of a living thing—the land. In them circulates food drawn from the soil, pumped by a million acts of cooperation and competition among animals and plants. That the land lives is implicit in its survival through eons of time.

Who is the land? We are, but no less the meanest flower that blows. Land ecology discards at the outset the fallacious notion that the wild community is one thing, the human community another.

What are the sciences? Only categories for thinking. Sciences can be taught separately, but they can't be used separately, either for seeing land or doing anything with it. It was a surprise to me to find this was "news" to many well-trained but highly specialized graduate students.

What is art? Only the drama of the land's workings.

With such a synthesis as a starting point, the tenets of conservation formulate themselves almost before the teacher can suggest them. Basic to all conservation is the concept of land-health; the sustained self-renewal of the community. It is at once self-evident from such an over-all view of the community that land-health is more important than surpluses or shortages in any particular land-product. The "famine concept" of conservation is valid mainly for inorganic resources, yet most teachers still apply it to all resources.

There is no need to persuade the student of land ecology that machines to dominate the land are useful only while there is a healthy land to use them on, and that land-health is possibly dependent on land-membership, that is that a flora and fauna too severely simplified or modified may not tick as well as the original. He can see for himself that there is no such thing as good or bad species; a species may get out of hand, but to terminate its membership in the land by human fiat is the last word in anthropomorphic arrogance.

Finally, the student can deduce, if he thinks hard enough, the peculiar nature of human economics. What we call economic laws are merely the impact of our changing wants on the land which supplies them. When that impact becomes destructive of our own tenure in the land, as is so conspicuously the case today, then the thing to examine is the validity of the wants themselves.

I have been sketching the end points, rather than the beginnings, of instruction in land ecology. To reach those end points, the teacher must of course construct a bridge of hard-headed factual materials drawn not only from natural history, but from all land-sciences. This of course raises the

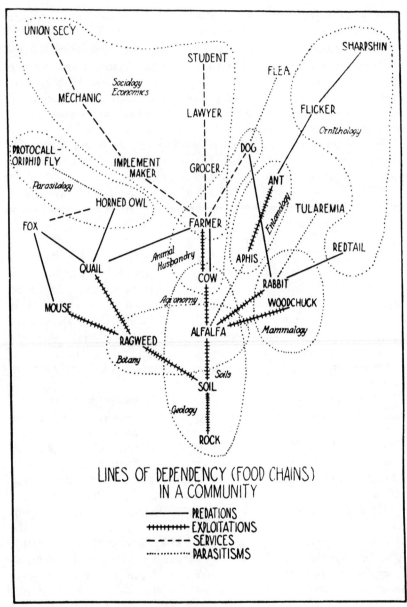

Figure 1 that accompanied Leopold's essay "The Role of Wildlife in a Liberal Education" as it appeared in *Transactions of the Seventh North American Wildlife Conference*, 1942.

question: why is it our job to synthesize and orient; why doesn't agriculture, or geography, or some other bigger and more important discipline do it for us?

My answer is: it is not our job, but it is our opportunity. If this opportunity is real, it is fair to ask: why hasn't it been seen and seized long ago? Why haven't the bigger and more important disciplines synthesized an ecological land concept? I am not sure of the answer, but I think I can see why in zoology and botany. Their pattern of teaching was set by the emergence of the theory of evolution. Some professors are still adding new findings to the evolutionary structure, but in the mind of the average student evolution quits growing, that is dies, when he receives his diploma. There is little opportunity for him to add to his classroom knowledge. Ecology, on the other hand, can lead to lifelong opportunities for study and even experimentation. Therefore, for purposes of a liberal education, ecology is superior to evolution as a window from which to view the world.

What Is a Weed? [1943]

A typescript dated August 2, 1943, this is as fine an example of Leopold's wry, ironic humor in service of a deeply serious concern as anything he wrote. It contains a blank space of several lines that Leopold apparently intended to fill in, but never did. This short, playful essay advances the analogy, implied in others, between wildlife and wildflower conservation. It also underscores the disjunction between industrial and ecological motifs in the philosophy of agriculture.

To live in harmony with plants is, or should be, the ideal of good agriculture. To call every plant a weed which cannot be fed to livestock or people is, I fear, the actual practice of agricultural colleges. I am led to this baleful conclusion by a recent perusal of *The Weed Flora of Iowa*, one of the authoritative works on the identification and control of weed pests.[1]

"Weeds do an enormous damage to the crops of Iowa" is the opening sentence of the book. Granted. "The need of a volume dealing with weeds . . . has long been felt by the public schools." I hope this is true. But among the weeds with which the public schools feel need of dealing are the following:

Black-eyed Susan (*Rudbeckia hirta*) "succumbs readily to cultivation."

A model weed!

Partridge pea (*Cassia Chamaecrista*) "grows on clay banks and sandy fields," where it may be "readily destroyed by cutting."

The inference is that even clay banks must be kept clean of useless blooms. Nothing is said of the outstanding value of this plant as a wildlife food, or of its nitrogen-fixing function.

Flowering spurge (*Euphorbia corollata*) is "common in gravelly soils"

---

1. *The Weed Flora of Iowa*, Bulletin no. 4 (Iowa Geological Survey, 1926), 715 pp.

What Is a Weed? 307

and "difficult to exterminate. To eradicate this plant the ground should be given a shallow plowing and the root-stocks exposed to the sun."

Nothing is said of the wisdom of plowing gravelly soils at all, or of the fact that this spurge belonged to the prairie flora, and is one of the few common relics of Iowa's prairie years. Presumably the public schools are not interested in this.

> Prairie goldenrod (*Solidago regida*), which "though often a very troublesome weed in pastures, is easily killed by cultivation."

The locality troubled by this uncommon and lovely goldenrod is indeed exceptional. The University of Wisconsin Arboretum, in order to provide its botany classes with a few specimens to look at, had to propagate this relic of the prairie flora in a nursery. On my own farm it was extinct, so I hand-planted two specimens, and take pride in the fact that they have reproduced half a dozen new clumps.

> Horsemint (*Monarda mollis*). "This weed is easily exterminated by cultivation," and "should not be allowed to produce seeds."

During an Iowa July, human courage, likewise, might easily be exterminated but for the heartening color-masses and fragrance of this common (and as far as I know) harmless survivor of the prairie days.

> Ironweed (*Veronia Baldwini*) is "frequently a troublesome weed, but it is usually not difficult to exterminate in cultivated fields."

It would be difficult to exterminate from my mind the August landscape in which I took my first hunting trip, trailing after my father. The dried-up cowtracks in the black muck of an Iowa bottomland looked to me like small chasms, and the purple-topped ironweeds like tall trees. Presumably there are still school children who might have the same impressions, despite indoctrination by agricultural authority.

> Peppermint (*Mentha piperita*). "This plant is frequently found along brooks. The effectual means of killing it is to clear the ground of the root-stocks by digging."

One is moved to ask whether, in Iowa, nothing useless has the right to grow along brooks. Indeed why not abolish the brook, which wastes many acres of otherwise useful farmland.

> Water pepper (*Polygonum hydropiper*) "not very troublesome . . . except in low places. Fields that are badly infested should be plowed and drained."

No one can deny that this is a weed, albeit a pretty one. But even after drainage, would not some annual, and perhaps a more troublesome one, follow every plowing? Has Iowa repealed the plant succession? It is also of interest to note that the Iowa wildlife research unit finds *Polygonum hydropiper* to be [Several lines of blank space appear here in the manuscript.]

> Wild rose (*Rosa pratincola*). "This weed often persists," as a relic of the original prairie flora, "in grain fields of northern Iowa. Thorough cultivation for a few seasons will, however, usually destroy the weed."

No comment.

> Blue vervain (*Verbena hastata*) and Hoary vervain (*V. stricta*). The vervains, admittedly weedy, are "easily destroyed by cultivation" and are "frequent in pastures," but nothing is said about why they are frequent.

The obvious reasons are soil depletion and overgrazing. To tell this plain ecological fact to farmers and school children would seem proper in an authoritative volume on weed-control.

> Chicory (*Cichorium Intybus*) "is not often seen in good farming districts except as a wayside weed. Individual plants may be destroyed by close cutting and applying salt to the root in hot dry weather."

School children might also be reminded that during hot dry weather this tough immigrant is the only member of the botanical melting-pot courageous enough to decorate with ethereal blue the worst mistakes of realtors and engineers.

If the spirit and attitude of *The Weed Flora of Iowa* were peculiar to one book or one state, I would hardly feel impelled to challenge it. This publication is, however, only one sample of a powerful propaganda, conducted by many farming states, often with the aid of federal subsidy, and including not only publications but also weed laws and specialized extension workers. That such a propaganda is necessary to protect agriculture is, I think, obvious to all who have ever contended with a serious plant pest. What I challenge is not the propaganda, but the false premises which seem to be common to this and all other efforts to combat plant or animal pests.

The first false premise is that every wild species occasionally harmful to agriculture is, by reason of that fact, to be blacklisted for general persecution. It is ironical that agricultural science is now finding that some of the "worst" weed species perform useful or even indispensable functions. Thus the hated ragweed and the seemingly worthless horseweed are found to prepare the soil, by some still mysterious alchemy, for high-quality high-yield tobacco crops. Preliminary fallowing with these weeds is now recommended to farmers.[2]

The second false premise is the emphasis on weed-control, as against weed-prevention. It is obvious that most weed problems arise from overgrazing, soil exhaustion, and needless disturbance of more advanced successional stages, and that prevention of these misuses is the core of the problem. Yet they are seldom mentioned in weed literature.

These same false premises characterize public predator-control. Because too many cougars or wolves were incompatible with livestock, it was assumed that no wolves or cougars would be ideal for livestock. But the scourge of deer and elk which followed their removal on many ranges has simply transferred the role of pest from carnivore to herbivore. Thus we forget that no species is inherently a pest, and any species may become one.

The same false premises characterize rodent-control. Overgrazing is probably the basic cause of some or most outbreaks of range rodents, the rodents thriving on the weeds which replace the weakened grasses. This relationship is still conjectural, and it is significant that no rodent-control agency has, to my knowledge, started any research to verify or refute it. Still if it is true, we may poison rodents till doomsday without effecting a cure. The only cure is range-restoration.

The same false premises beset the hawk and owl question. Originally rated as all "bad," their early defenders sought to remedy the situation by reclassifying part of them as "good." Hawk-haters, and gunners with a trigger-itch, have had lots of fun throwing this fallacy back in our faces. We should have been better off to assert, in the first place, that good and bad are attributes of numbers, not of species; that hawks and owls are members of the native fauna, and as such are entitled to share the land with us; that no man has the moral right to kill them except when sustaining injury.

It seems to me that both agriculture and conservation are in the process of inner conflict. Each has an ecological school of land-use, and what I may call an "iron heel" school. If it be a fact that the former is the truer, then both have a common problem of constructing an ecological land-practice. Thus, and not otherwise, will one cease to contradict the other. Thus, and not otherwise, will either prosper in the long run.

2. W. M. Lunn, D. E. Brown, J. E. McMurtrey, Jr., and W. W. Garner, "Tobacco Following Bare and Natural Weed Fallow and Pure Stands of Certain Weeds," *Journal of Agricultural Research* 59:11 (1939), 829-846.

Conservation: In Whole or in Part? [1944]

Leopold wrote this report, dated November 1, 1944, for a University of Wisconsin committee on postwar agricultural policies. The typescript, labeled "rough draft," is virtually clean. As its title suggests, the report is a direct and comprehensive statement of a holistic conception of land and a correspondingly integrative approach to conservation. It contains Leopold's most sustained analysis of land health, a concept that figures prominently in "The Land Ethic."

There are two kinds of conservationists, and two systems of thought on the subject.

One kind feels a primary interest in some one aspect of land (such as soil, forestry, game, or fish) with an incidental interest in the land as a whole.

The other feels a primary interest in the land as a whole, with incidental interest in its component resources.

The two approaches lead to quite different conclusions as to what constitutes conservative land-use, and how such use is to be achieved.

The first approach is overwhelmingly prevalent. The second approach has not, to my knowledge, been clearly described. This paper aims to sketch the concept of land-as-a-whole.

Land-Health

Conservation is a state of health in the land.

The land consists of soil, water, plants, and animals, but health is more than a sufficiency of these components. It is a state of vigorous self-renewal in each of them, and in all collectively. Such collective functioning of interdependent parts for the maintenance of the whole is characteristic of an organism. In this sense land is an organism, and conservation deals with its functional integrity, or health.

Conservation: In Whole or in Part? 311

This is almost, but not quite, the same as the familiar "renewable resource" concept. The latter tells us that a particular resource may be healthy or sick, but not that the sickness of one may undermine the health of all.

Conservation is usually thought of as dealing with the *supply* of resources. This "famine concept" is inadequate, for a deficit in the supply in any given resource does not necessarily denote lack of health, while a failure of function always does, no matter how ample the supply. Thus erosion, a malfunction of soil and water, is more serious than "timber famine," because it deteriorates the entire land community permanently, rather than one resource temporarily.

Attitudes

Mass man is unconscious of land-health for three reasons.

First he was, until recently, unable to injure it. He lacked the tools.

Secondly, European civilization developed on a landscape extraordinarily resistant to disorganization, i.e, one which endures very rough usage and severe modification without derangement of function. Thus the oak forests of England became closely grazed sheep downs without losing their soil. The fauna and flora shifted, but did not disintegrate.[1]

Thirdly, science could not, until recently, distinguish fact from fancy in the reaction of land to human use. Thus the Mediterranean countries were permanently deteriorated by overgrazing and erosion before their inhabitants knew what was happening, or why.

As a result of these three historical accidents, the European races acquired machines for dominating land before they had evolved the social inhibitions requisite for their safe use.

In short, the power to injure land-health grew faster than the consciousness that it can be injured.

Land, to the average citizen, is still something to be tamed, rather than something to be understood, loved, and lived with. Resources are still regarded as separate entities, indeed, as commodities, rather than as our cohabitants in the land-community.

Diversity and Stability up to 1840

The Wisconsin land was stable, i.e., it retained its health, for a long period before 1840. The pollens imbedded in peat bogs show that the native plants comprising the prairie, the hardwood forest, and the coniferous forest are

1. E. P. Farrow, *Plant Life on East Anglian Heaths* (Cambridge University Press, 1925).

about the same now as they were at the end of the glacial period, 20,000 years ago. Since that time these major plant communities were pushed alternately northward and southward several times by long climatic cycles, but their membership and organization remained intact. Thus, in one northward push the prairie once reached nearly to Lake Superior; in one southward push the Canadian forest reached to Indiana.

The bones of animals show that the fauna shifted with the flora, but its composition or membership likewise remained intact. The soils not only remained intact, but actually gained in depth and fertility with wind-deposits of loessial soils. With this came a gain in the volume of plant and animal life.

The native Wisconsin community which thus proved its ability to renew itself for 200 centuries was very diverse. It included 350 species of birds, 90 mammals, 174 fishes, 72 amphibians and reptiles, roughly 20,000 insects, about 1500 higher plants, and an unknown but very great number of lower plants and lower animals.

All these creatures were functional members of the land, and their collective activities constituted its inner workings from the glacial epoch to 1840.

These "inner workings" of the community included, as everyone knows, a high proportion of tooth and claw competition, varying in degree from mere jostling to murder. It is hard for the layman, who sees plants and animals in perpetual conflict with each other, to conceive of them as cooperating parts of an organism. Yet the fact remains that throughout geological time up to 1840, the extinction of one species by another occurred more rarely than the creation of new species by evolution, and that occurred very rarely indeed, for we have little evidence of new species appearing during the period of recorded history. The net trend of the original community was thus toward more and more diversity of native forms, and more and more complex relations between them. Stability or health was associated with, and perhaps caused by, this diversity and complexity.

Diversity and Stability since 1840

Since 1840 some members of the native community have been removed. Familiar examples include the buffalo, wild turkey, passenger pigeon, Carolina paroquet, wolverene, marten, and fisher.

Others have been added. These include not only imported birds and mammals like English sparrow, starling, pheasant, Norway rat, and house mouse, but also many wild plants (most weeds are European or Asiatic), many insects good and bad, and many diseases. Domesticated plants, mammals and birds have also been added, and constitute the bulk of the new community. In one measured sample in Columbia County the domestic plus imported wild birds and mammals constitute 99 per cent of the weight of the total present bird and mammal community.[2]

Most of the native species which persist have undergone changes in numerical status or distribution, or both, since 1840. The prairie flora and fauna occupied the best soils, and hence were supplanted early. Later pressures severely curtailed and modified the marsh, bog, forest, and aquatic floras and faunas. Everybody knows of these changes, hence they need not be described.

Losses and Gains

It is necessary to state at this point that this paper is not a nostalgic rehearsal of the glories of primeval Wisconsin. It is an attempt to approach objectively a case of land-illness which nobody understands. The changes we have made in the Wisconsin land are not all inherently or necessarily wasteful. Many of them have enriched and expanded certain elements in the native fauna and flora whilst shrinking others. There is no doubt at all that the introduction of agriculture has increased the numbers, if not the diversity, of many native animals and some native plants. A sketch of these changes has been published.[3]

Symptoms of Illness

Coincident with this period of man-made change in the land community, many symptoms of impaired land-health have become apparent. Most of these are familiar individually, but they are seldom viewed collectively, or as possibly related to each other and to the land as a whole.

Of the various symptoms of illness, soil erosion and abnormal floods are by far the most important. Most critical observers agree that both are getting worse. Much is known of the superficial causes of both, but little of the underlying "physiology" of soil and water.

Less familiar are some of the qualitative deteriorations in land crops. In farm crops, it appears that better varieties and better cultural methods have just about offset the decline in the productivity of the soil. The reason seems

---

2. Aldo Leopold and Paul L. Errington, "Limits of Summer Gain and Winter Loss in Bobwhite Populations at Prairie du Sac, Wisconsin, 1923–1945," unpublished manuscript, 32 pp. [Later emended to "Prairie du Sac Area, Columbia County, unpublished manuscript."]

3. Committee on Wildlife Conservation (Aldo Leopold, Chmn.; L. J. Cole, N. C. Fassett, C. A. Herrick, Chancey Juday, and George Wagner), *The University and Conservation of Wisconsin Wildlife,* Bulletin of the University of Wisconsin Science Inquiry Publication no. 3 (Madison, Feb 1937), 39 pp.

to be plain loss of fertility. It has been discovered recently that decline in soil fertility reduces not only the gross yields of crops, but the nutritional value of the crops, and the welfare of animals which eat them.[4]

The qualitative deterioration of crops applies to trees as well as to agronomic plants. We used to grow 4-log pines; now we do well to grow 2-log pines on the same sites. What, besides fire, has happened to soil? Similar deteriorations have occurred in Europe,[5] and are by no means understood.

All too familiar are those symptoms of land-illness caused by the importation of exotic diseases and pests. There is no mystery about such pains and ailments as the white pine blister rust, chestnut blight, gypsy moth, Dutch elm disease, the corn borer, the Norway rat, the starling, the house mouse, the Canada thistle, and the creeping jenny or German carp. Their ultimate effect on the land, however, presents many unsolved problems, including the damage done by control operations.

Less familiar are the many instances in which native plants and animals, heretofore presumably "well-behaved" citizens of the land community, have assumed all the attributes of pests. The white grub, the cankerworm, the meadowmouse, the fire blight of oaks, and the spruce bud-worm are cases in point.

One of the very recent instances of pest behavior by a heretofore "well-behaved" member of the native community is the irruption of deer in Wisconsin and many other states.[6] While the superficial "causes" of this phenomenon are well known to be a coincidence of lumbering, law enforcement, fire-control, predator-control, and selective harvesting through buck laws, nevertheless it remains a deep mystery why equivalent coincidences never (as far as we know) produced irruptions of hoofed mammals previous to human interference. In all probability some as yet unknown causes lie behind the more superficial ones; possibly fluctuations in the vitamin content of foods.

New plant and animal diseases are now appearing so rapidly that we do not yet know whether they represent some native organism "gone outlaw," or some newly imported pest. Thus the new pine disease, now obliterating plantations of Norway and Jack pine in Oconto and nearby counties, has an unclassified causative agent of unknown origin.

Native members of the community sometimes simply disappear without visible cause, and often despite protective efforts. Prairie chickens, spruce grouse, and certain wildflowers probably belong in this class. Im-

---

4. W. A. Albrecht, "Soil and Livestock," *The Land* 2:4 (1943), 298–305, and other papers by the same author.

5. Aldo Leopold, "Deer and *Dauerwald* in Germany," *Journal of Forestry* 34:4–5 (1936), 366–375, 460–466.

6. Aldo Leopold, "Deer Irruptions," *Wisconsin Conservation Bulletin* 8:8 (1943), 1–11.

ported species may likewise disappear: the Hungarian partridge seems to be on the decline in Wisconsin, after an initial success, without visible cause.

Finally we have unexplained changes in the population behavior of plants and animals; these behaviors are often of considerable economic importance. Thus there is more than a presumption that population cycles have tended to become more violent in all hares and rabbits, in all grouse, and in foxes. Cyclic population behavior has perhaps spread to pheasants and bobwhite quail.

The conservationist who is interested in land as a whole is compelled to view these symptoms collectively, and as probable maladjustments of the land community. Some of them are understood superficially, but hardly any are understood deeply enough to warrant the assertion that they are separate phenomena, unrelated to each other and to the whole. In point of time, nearly all of them are probably new, and fall within the post-1840 period of violent change in the land community. Are they causally related to the period of change, or did the two coincide by accident?

To assert a causal relation would imply that we understand the mechanism. As a matter of fact, the land mechanism is too complex to be understood, and probably always will be. We are forced to make the best guess we can from circumstantial evidence. The circumstantial evidence is that stability and diversity in the native community were associated for 20,000 years, and presumably depended on each other. Both are now partly lost, presumably because the original community has been partly lost and greatly altered. Presumably the greater the losses and alterations, the greater the risk of impairments and disorganizations.

This leads to the "rule of thumb" which is the basic premise of ecological conservation: the land should retain as much of its original membership as is compatible with human land-use. The land must of course be modified, but it should be modified as gently and as little as possible.

This difference between gentle and restrained, as compared with violent and unrestrained, modification of the land is the difference between organic and mustard-plaster therapeutics in the field of land-health.

There are reasons for gentle land-use over and above the presumed risk to the health of the land. Sauer[7] has pointed out that the domesticated plants and animals which we use now are not necessarily those we will need a century hence. To the extent that the native community is extinguished, the genetical source of new domesticated plants and animals is destroyed.

This general concept of land-health as an attribute of the original native community as a whole, and of land-illness as probably related to violent

---

7. Carl O. Sauer, "Theme of Plant and Animal Destruction in Economic History," *Journal of Farm Economics* 20:4 (1938), 765–775.

changes and consequent disorganization, may be called, for short, the "unity concept."

Unity and Land-Use

If the components of land have a collective as well as a separate welfare, then conservation must deal with them collectively as well as separately. Land-use cannot be good if it conserves one component and injures another. Thus a farmer who conserves his soil but drains his marsh, grazes his woodlot, and extinguishes the native fauna and flora is not practicing conservation in the ecological sense. He is merely conserving one component of land at the expense of another.

The conservation department which seeks to build up game birds by extinguishing non-game predators, or to retain excessive deer populations at the expense of the forest, is doing the same thing.

The engineer who constructs dams to conserve water, develop power, or control floods is not practicing conservation if the actual regimen of water which results, either above or below the dam, destroys more values than it creates. I know of no single impoundment of water in which all of the land values affected were weighed in advance. (Unfortunately it must be stated in the same breath, that ecologists competent to weigh all of them do not yet exist.)

Lop-sided conservation is encouraged by the fact that most Bureaus and Departments are charged with the custody of a single resource, rather than with the custody of the land as a whole. Even when their official titles denote a broader mandate, their actual interests and skills are commonly much narrower. The term "land" now brackets a larger span of knowledge than one human mind can compass.

Ironically enough it is the farmer who is, by implication at least, left to unify, as best he can, the conflicts and overlaps of bureaudom. Separatism in bureaus is probably a necessary evil, but this is not the case in agricultural colleges. If the arguments of this paper are valid, the agricultural colleges have a far deeper responsibility for unification of land-use practice than they, or the public, have so far realized.

I will sketch later some of the practical applications of the land-unity concept to land-use and land-users.

Unity and Economics

Some components of the land community are inherently of economic importance (soil, forests, water) while others cannot possibly be, except in a very indirect sense (wildflowers, songbirds, scenery, wilderness areas).

Some components are of economic importance to the community, but of dubious profit to the individual owner (most marshes, most cover on streambanks and steep slopes, most windbreaks).

Some are profitable for the individual to retain if they are still in a productive state, but of dubious profit if they have to be created *de novo*, or if they have to be rebuilt after being damaged (woodlots).

It follows that if conservation on private lands is to be motivated solely by profit, no unified conservation is even remotely possible. Community welfare, a sense of unity in the land, and a sense of personal pride in such unity, must in some degree move the private owner, as well as the public. Conservation cannot possibly "pay" except when the meaning is restricted to components that happen to be profitable. Conservation often pays in the sense that the profitable components can carry the unprofitable ones, just as in any industrial enterprise, a unified purpose involves carrying profitable and unprofitable component enterprises, each necessary to the functioning of the whole.

The fallacious assumption that each separate act of conservation can or must be profitable before its practice can be recommended to farmers is possibly responsible for the meagre fruits of forty years of education, extension, and public demonstration in the conservation field. It is undoubtedly responsible for many dubious claims of profit which are commonly made, or implied, in presenting the subject to the public. It is presumably axiomatic that any "program" saddled with over-claims will backfire in the long run.

Sound conservation propaganda must present land health, as well as land products, as the objective of "good" land-use. It must present good land-use primarily as an obligation to the community. Many constituent parts of it are indeed profitable, and where this is the case, the fact can and should be emphasized. But many constituent parts of it are not, and failure to assert this at once subverts legitimate education to the intellectual level of a cheap "sales" campaign in which only virtues are mentioned.

No one need harbor any illusion that the farmer will immediately undertake the unprofitable components of "good" land-use. But it is probably not illusory to assume that fractional truth is no truth, and that one-resource conservation programs are inherently fractional.

Acts vs. Skills; Law vs. Education

Conservative land-use consists of a system of acts, motivated by a desire, and executed with skill.

Laws and policies must deal almost exclusively with acts, because desires and skills are intangible, and cannot be defined in law, nor created by law. Acts without desire or skill are likely to be futile. Thus, during the CCC epoch many Wisconsin farmers were induced, by subsidy, to perform the acts of soil conservation, but those who lacked desire and skill dropped the acts as soon as the subsidy was withdrawn.

This limitation of conservation law and policy is inherent and unavoidable. It can be offset only by education, which is not precluded from dealing with desires and skills.

Whether education can create these desires and skills is an open question. Certainly it can not do so in time to avoid a much further disorganization of land health than now exists. This paper does not claim to assess the chances for success of the unity concept. It claims only to assess the basic logic of the conservation program.

Farm Practice

Some of the attitudes toward farm land implied in the unity concept have already been set forth in popular form.[8] Summarized in terms of causation, these implications add up rather simply to this: the farmer should know the original as well as the introduced components of his land, and take a pride in retaining at least a sample of all of them. In addition to healthy soil, crops, and livestock, he should know and feel a pride in a healthy sample of marsh, woodlot, pond, stream, bog, or roadside prairie. In addition to being a conscious citizen of his political, social, and economic community, he should be a conscious citizen of his watershed, his migratory bird flyway, his biotic zone. Wild crops as well as tame crops should be a part of his scheme of farm management. He should hate no native animal or plant, but only excess or extinction in any one of them.

Cash outlays for unprofitable components of land are of course not to be expected, but outlays of thought, and to a reasonable extent of spare time, should be given with pride, just as they are now given to equivalent enterprises in human health and civic welfare.

Summary

Conservation means land-health as well as resource-supply. Land-health is the capacity for self-renewal in the soils, waters, plants, and animals that collectively comprise the land.

Stable health was associated geologically with the full native community which existed up to 1840. Impairments are coincident with subsequent

8. Aldo Leopold, *Wildlife Conservation on the Farm* (Racine, Wis.: Wisconsin Agriculturist and Farmer, 1941), 24 pp.

Conservation: In Whole or in Part? 319

changes in membership and distribution. The "inner workings" of land are not understood, but a causal relation between impairments and degree of change is probable. This leads to the rule-of-thumb that changes should be as gentle and as restrained as compatible with human needs.

Land-use is good only when it considers all of the components of land, but its human organization often tends to conserve one at the expense of others.

Some components of land can be conserved profitably, but others not. All are profitable to the community in the long run. Unified conservation must therefore be activated primarily as an obligation to the community, rather than as an opportunity for profit.

Acts of conservation without the requisite desires and skills are futile. To create these desires and skills, and the community motive, is the task of education.

Review of Young and Goldman, *The Wolves of North America* [1945]

Leopold frequently published book reviews—this one in the *Journal of Forestry*—that not only were models of analysis but also reached beyond the work to make some telling point about its implications for society at large. Here in a review of a treatise on wolves, he holds a mirror to contemporary ecological attitudes and even admits, for perhaps the first time in print, that he himself had once thought much differently about the species.

The Wolves of North America. By Stanley P. Young and Edward H. Goldman. Washington, D. C.: American Wildlife Institute, 1944. 660 pp., Illus. $6.

This book is notable, not only as the outstanding contemporary treatise on an outstanding animal, but as a mirror which reflects the thought of our generation on a wide gamut of conservation problems.

The book consists of two parts, treating successively of the ecology of the wolf, and his taxonomy. This review does not purport to cover the taxonomic field, and most lay readers will in any event focus their attention on the wolf's behavior, rather than on his bones.

Viewed as history, the work is a masterly job. It assembles an exhaustive array of interesting quotations on the age-old rivalry between men and wolves as predators on the world's livestock and big-game herds. Some of these historical excerpts go back to the ancients; most of them deal with the American scene. They convey to the reader a vivid picture of wolf troubles and wolf-control stratagems, beginning with the earliest settlers on the Atlantic seaboard, and ending with the motorized cowboy of the modern West.

While populations are especially hard to estimate in so mobile a mammal as the wolf, one gets the impression that wolves were incredibly abundant in the buffalo days, were severely decimated by commercialized poisoning for their fur in the 1870's, regained abundance in the 1880's when cattle and sheep replaced the buffalo as a dependable food supply, held their own for two more decades during a regime of graft-riddled bounty systems, and were finally wiped off the map when the U.S. Biological Survey, in 1914, began its federally supported predator-control campaign, during which bounties were discarded in favor of salaried trappers. The senior author had a large share in organizing this campaign, and has directed it since 1928.

One of the most interesting points in this long and dramatic history is the heavy demand for wolf furs during the commercial poisoning period. It appears that the Russian army at that time used wolfskins for part of its winter uniform, and thus levied tribute on all the world's wolfpacks.

The only fault I can find with Mr. Young's history of the wolf is that his materials are so abundant that he lacks space to evaluate them critically. Some questionable assertions are quoted with the implication that the author accepts them as facts, whereas in some other chapter he implies the contrary. Thus Catesby (p. 175) is allowed to assert, without challenge by the author, that the "Indians . . . had no other dogs" than domesticated wolves prior to the introduction of European dogs. Certainly Lewis and Clark found dogs which were far from wolf-like among all the western Indians.

Viewed as science, *The Wolves of North America* reflects the naturalist of the past, rather than the wildlife ecologist of today. This is demonstrated not in what the authors say, but in what they omit. At no point in the book do they evince any consciousness of the primary ecological enigmas posed by their own work. For example: Why did the heavy wolf population of presettlement days fail to wipe out its own mammalian food supply?

The existence of some compensatory mechanism, whereby the wolf controlled its own numbers, is an almost inevitable deduction from the known facts. The wolf stood at the apex of the animal pyramid; he had no predatory enemies; his efficiency as a killer was dramatically high. What held him down? Diseases or parasites carried from one wolf to another, directly or indirectly? Fighting within his own ranks? Some kind of intraspecific "birth-control"? Or did he indeed wipe out his food supply in long alternating cycles? Such questions are not discussed in this volume. They are extremely pertinent to the modern question: Are we really better off without wolves in the wilder parts of our forests and ranges?

Viewed as literature, this book has much to commend it. Its style is simple, direct, sometimes fluent, never burdened with that curse of modern biology: "scientific" English. The only fault that I can find is that it gives only a few skimpy biographies of the notorious individual wolves of the western ranges, of which we have a dramatic sample in Seton's *Lobo, the King of the Currumpaw*. I suspect that the publisher, rather than the authors, may be responsible for this, for no living man knows more about the famous killers than Stanley P. Young. It is a pity that these biographies should be lost to history. They will be when the present generation of range-raised biologists discards its ropes and saddles.

Viewed as conservation, *The Wolves of North America* is, to me, intensely disappointing. The next to the last sentence in the books asserts: *"There still remain, even in the United States, some areas of considerable size in which we feel that both the red and gray [wolves] may be allowed to continue their existence with little molestation."* Yes, so also thinks every right-minded ecologist, but has the United States Fish and Wildlife Service no responsibility for implementing this thought before it completes its job of extirpation? Where are these areas? Probably every reasonable ecologist will agree that some of them should lie in the larger national parks and wilderness areas; for instance, the Yellowstone and its adjacent national forests. The Yellowstone wolves were extirpated in 1916, and the area has been wolfless ever since. Why, in the necessary process of extirpating wolves from the livestock ranges of Wyoming and Montana, were not some of the uninjured animals used to restock the Yellowstone? How can it be done now, when the only available stocks are the desert wolf of Arizona, and the subarctic form of the Canadian Rockies?

Entirely unmentioned in this book is the modern curse of excess deer and elk, which certainly stems, at least in part, from the excessive decimation of wolves and cougars under the aegis of the present authors and of the Fish and Wildlife Service. None of us foresaw this penalty. I personally believed, at least in 1914 when predator control began, that there could not be too much horned game, and that the extirpation of predators was a reasonable price to pay for better big game hunting. Some of us have learned since the tragic error of such a view, and acknowledged our mistake. One must judge from the present volume that the Fish and Wildlife Service does not see any mistake. Its philosophy seems to be that the rifles can do the necessary trimming of the big-game herds. Yes—so they can—but they seldom do, at least not until the range is ruined and the herds are pauperized.

The publisher of this book is the American Wildlife Institute, a conservation organization which has done invaluable pioneer work in fostering wildlife research. It is disappointing that the Institute should not have encouraged, in this volume, some implementation of the idea that "the wolf may be allowed to continue his existence."

The wolf will certainly disappear from the United States unless the official wildlife agencies exempt certain definite areas of wolf range from official extirpation.

The Outlook for Farm Wildlife [1945]

In lieu of the 1945 meeting of the North American Wildlife Conference, which was cancelled because of World War II, Leopold and other professionals prepared papers for the tenth volume of the *Transactions*. This was Leopold's contribution, offering his prescient vision of the increasing industrialization of agriculture in the years to come and its social as well as ecological consequences. His prediction of an eventual reversal of that trend is hopeful—now as then.

Twenty years have passed since Herbert Stoddard, in Georgia, started the first management of wildlife based on research.

During those two decades management has become a profession with expanding personnel, techniques, research service, and funds. The colored pins of management activity puncture the map of almost every state.

Behind this rosy picture of progress, however, lie three fundamental weaknesses:
1. Wildlife habitat in fertile regions is being destroyed faster than it is being rebuilt.
2. Many imported and also native species exhibit pest behavior. A general disorganization of the wildlife community seems to be taking place.
3. Private initiative in wildlife management has grown very slowly.

In this appraisal of the outlook, I deal principally with the first two items in their bearing on farm wildlife.

Gains and losses in habitat. Wildlife in any settled country is a resultant of gains and losses in habitat. Stability, or equilibrium between gains and losses, is practically nonexistent. The weakness in the present situation may be roughly described as follows: On worn-out soils we are gaining cover but losing food, at least in the qualitative sense. On fertile soils we are losing cover, hence the food which exists is largely unavailable.

Where cover and food still occur together on fertile soils, they often represent negligence or delay, rather than design.

There is a confusing element in the situation, for habitat in the process of going out often yields well.

For example, on the fertile soils of southern Wisconsin, the strongholds of our remaining wildlife are the wood lot, the fencerow, the marsh, the creek, and the cornshock. The wood lot is in process of conversion to pasture; the fencerow is in process of abolition; the remaining marsh is in process of drainage; the creeks are getting so flashy that there is a tendency to channelize them. The cornshock has long been en route to the silo, and the corn borer is speeding up the move.

Using pheasant as an example, such a landscape often yields well while in process of passing out. The marsh, grazed, or drained or both, serves well enough for cover up to a certain point, while the manure spreader substitutes for cornshocks up to a certain point. The rapid shift in the status of plant successions may in itself stimulate productivity.

The situation is complicated further by a "transmigration" of land use. Originally uplands were plowed and lowlands pastured. Now the uplands have eroded so badly that corn yields are unsatisfactory, hence corn must move to the lowlands while pasture must move to the uplands. In order that corn may move to the lowlands, they must be either tiled, drained, or channelized. This, of course, tends to destroy the remaining marsh and natural stream.

The upshot is a good "interim" crop which has a poor future. I don't know how widely a similar situation prevails outside my own state, but I suspect that the basic pattern, with local variations, is widely prevalent.

Runaway populations. Wildlife is never destroyed except as the soil itself is destroyed; it is simply converted from one form to another. You cannot prevent soil from growing plants, nor can you prevent plants from feeding animals. The only question is: What kind of plants? What kind of animals? How many?

Ever since the settlement of the country, there has been a tendency for certain plants and animals to get out-of-hand. These runaway populations include weeds, pests, and disease organisms. Usually these runways have been foreigners (like the carp, Norway rat, Canada thistle, chestnut blight, and white pine blister rust) but native species (like the June beetle and various range rodents) are clearly also capable of pest behavior.

Up to the time of the chestnut blight, these runaways did not threaten wildlife directly on any serious scale, but they now do, and it is now clear that the pest problem is developing several new and dangerous angles:

1. World-wide transport is carrying new "stowaways" to new habitats on an ascending scale. (Example: Anopheles gambiae to Brazil, bubonic plague to western states.)

2. Modern chemistry is developing controls which may be as dangerous as the pests themselves. (Example: DDT.)
3. Additional native species, heretofore law-abiding citizens of the flora and fauna, are exhibiting pest behavior. (Example: excess deer and elk.)

These three new angles must be considered together to appreciate their full import. Mildly dangerous pests like ordinary mosquitoes evoked control measures which severely damaged wildlife; desperately dangerous pests will evoke corresponding control measures, and when these collide with wildlife interests, our squeak of pain will not even be heard.

Moreover, wildlife itself is threatened directly by pests. Sometimes they hit so fast and hard that the funeral is over before the origin of the malady is known. Thus in Wisconsin, we have a new disease known as burn blight, the cause of which is still unknown. It threatens to destroy young Norway pine and jack pine, especially plantations. Oak wilt, the cause of which was only recently discovered, is steadily reducing red and black oaks. Our white pine is already blighted except on artificially-controlled areas. Bud-worm is in the spruce. Hickory can't grow because of a weevil which bites the terminal bud. Deer have wiped out most white cedar and hemlock reproduction. Sawfly has again raided the tamaracks. June beetles began years ago to whittle down the bur and white oaks, and continue to do so. Bag worm is moving up from the south and west and may get our red cedars. Dutch elm disease is headed west from Ohio. What kind of a wood lot or forest fauna can we support if every important tree species has to be sprayed in order to live?

Shrubs are not quite so hard hit, but the shrub flora has its troubles. On the University of Wisconsin Arboretum, an area dedicated to the rebuilding of the original native landscape, the Siberian honeysuckle is calmly usurping the understory of all woods, and threatens to engulf even the marshes.

In Wisconsin wood lots it is becoming very difficult to get oak reproduction even when we fence out the cows. The cottontails won't let a young oak get by. One can't interest the farmer in a wood lot which reproduces only weed trees.

Of the dozen pests mentioned on this page, four are imported, seven are runaway native species, and one is of unknown origin. Of the 12, 6 have become pests in the last few years.

Farm crops and livestock exhibit a parallel list of pests, of which the worst now rampant in my region is the corn borer. The corn borer can be controlled by fall plowing, but what that will do to cornbelt wildlife is something I dislike to think about.

It all makes a pattern. Runaway populations are piling up in numbers and severity. In the effort to rescue one value, we trample another. Wild plants and animals suffer worst because we can't spend much cash on controls or preventives. Everything we lose will be replaced by something else, almost invariably inferior. As Charles Elton has said: "The biological cost of modern transport is high."[1]

In short, we face not only an unfavorable balance between loss and gain in habitat, but an accelerating disorganization of those unknown controls which stabilize the flora and fauna, and which, in conjunction with stable soil and a normal regimen of water, constitute land-health.

The human background. Behind both of these trends in the physical status of the landscape lies an unresolved contest between two opposing philosophies of farm life. I suppose these have to be labeled for handy reference, although I distrust labels:

1. *The farm is a food-factory*, and the criterion of its success is salable products.
2. *The farm is a place to live.* The criterion of success is a harmonious balance between plants, animals, and people; between the domestic and the wild; between utility and beauty.

Wildlife has no place in the food-factory farm, except as the accidental relic of pioneer days. The trend of the landscape is toward a monotype, in which only the least exacting wildlife species can exist.

On the other hand, wildlife is an integral part of the farm-as-a-place-to-live. While it must be subordinated to economic needs, there is a deliberate effort to keep as rich a flora and fauna as possible, because it is "nice to have around."

It was inevitable and no doubt desirable that the tremendous momentum of industrialization should have spread to farm life. It is clear to me, however, that it has overshot the mark, in the sense that it is generating new insecurities, economic and ecological, in place of those it was meant to abolish. In its extreme form, it is humanly desolate and economically unstable. These extremes will some day die of their own too-much, not because they are bad for wildlife, but because they are bad for farmers.

When that day comes, the farmer will be asking us how to enrich the wildlife of his community. Stranger things have happened. Meanwhile we must do the best we can on the ecological leavings.

1. *Journal of Animal Ecology* 13:1 (1944), 87–88.

Review of Farrington, *The Ducks Came Back* [1946]

Ever since his father took him on his first hunting trip in the sloughs along the Mississippi River, Leopold had been painfully aware of the depletion of migratory waterfowl, and he wrote frequently throughout his life about the conundrum. In this review, for the *Journal of Wildlife Management*, of a celebratory book about the Ducks Unlimited organization, Leopold casts a coolly realistic eye on the waterfowl problem, the efforts of Ducks Unlimited to mitigate it, and the peculiar psychology of duck hunters that could lead an author to such unwarranted extravagance. A year later, after four disastrous seasons failed to curb the optimistic propaganda of the Ducks Unlimited public relations machine, Leopold withdrew his membership and his moral support.

The Ducks Came Back: The Story of Ducks Unlimited. By S. Kip Farrington, Jr. New York: Coward-McCann, Inc., 1945. 138 pp.

In this book a persuasive writer (Mr. Farrington) talks to an uncritical audience (the duck hunters) about an important conservation problem (the restoration of waterfowl). His argument is accompanied by soft music, in form of Lynn Bogue Hunt sketches and also excellent photos of the always-photogenic ducks.

The need for a book on the duck crop is great and real. Canada, up to a few years ago, was outdoing her teacher, the U.S., in the art of destroying duck marshes. When the drouths of the 1930's came along, there was a clear choice between restoring some marshes, or saying goodbye to duck shooting. Such work costs money, and obviously the U.S. should share the costs with Canada. Official U.S. money cannot cross the border, but private gifts can. Ducks Unlimited, a non-profit corporation, was set up as the vehicle for such gifts.

It has been remarkably successful, not only in eliciting funds, but in organizing voluntary cooperation from Canadian landowners and officials.

In this book, Mr. Farrington presents to the duck-hunting public the story of the origin, aims, and accomplishments to date of Ducks Unlimited.

I speak from personal knowledge when I say that we duck hunters are curious animals. In our business or professional life we are glad enough to get 6 per cent on our investments, and we feel a lofty disdain for the financial wildcatter. But in the "never-never land" of outdoor sports we insist on bonanzas. We have scant enthusiasm for any wildlife restoration scheme which recognizes difficulties, disappointments and mistakes, or which admits ignorance of wildlife management, or which calls for deferred profits and the curtailment of shooting privileges. We turn a deaf ear to the obvious fact that a century of destruction cannot be undone by wishful check-writing alone. When we went on our economic drunk, we destroyed not only the life of the land, but also something in ourselves. We have yet to learn that this can be restored only by contrition and sacrifice.

I suppose these quirks in the mental makeup of his audience convinced Mr. Farrington that duck-hunters would not invest unless offered a restoration-bonanza. Whatever the reason, this book is the prospectus of a bonanza.

Its major premise is that a large increase in ducks has already been accomplished. This is true if expressed in percentage of the 1930 flights which were near-zero. Begin low enough and almost any change short of extinction is an increase.

Mr. Farrington thinks that the work of Ducks Unlimited accounts for a large part of this increase. It does, but large is a vague word, and Canada is a large place. We should be proud that a start has been made, however small.

Mr. Farrington is sure that predators, drouths, floods, grass-fires, crows, magpies and jackfish kill more ducks than hunters do: he says that "birds shot make a very minor percentage" (of the total loss). This can be demonstrated, with some show of evidence, by assuming, as Mr. Farrington does, that every egg or duckling on the Canadian prairie is a potential greenhead dropping into the blocks at Havana, Illinois. (One could likewise demonstrate, by the same logic, that when a squirrel chews up a pine cone, he destroys a potential forest.) Egg and duckling mortality can and should be reduced, but to assume that it can or should be eliminated is, to put it mildly, a fallacy.

Moreover Mr. Farrington has selected for the attention of his audience the particular causes of mortality which Ducks Unlimited can mitigate with money: drouth losses, predation, and grass fires. He plays down botulism, which still baffles scientific research, and gun powder, the mitigation of which he regards as "straining at a gnat and swallowing a camel."

Having thus proved that further restrictions on gunpowder are quite unnecessary, Mr. Farrington shows what benefits would flow from the relaxation of those that already exist. Legalization of batteries would give useful employment to baymen (page 122), baiting would divert ducks from farm crops (page 123), live decoys would give a little shooting where none is now to be had (page 121), refuges could be converted to much-needed public shooting grounds (page 129). All of these things are true, if there are enough ducks to justify them.

Fortunately for the human race, pressure groups are never quite unanimous. No sooner had Mr. Farrington described Ducks Unlimited as "the conservation miracle of all times," than Mr. A. C. Glassell, President of said miracle, wrote to his state chairmen: "When Ducks Unlimited was organized, we were told—and am sure we believed—that the shortage of ducks was not caused by the gun. Today and in the future the gun is going to be the mighty factor. Our duck crop will soon be exhausted unless we complete our (marsh) building program and can produce 250,000,000 ducks per year. We must do this work in a hurry before they (the ducks) are killed out."

There is no bonanza-philosophy in Mr. Glassell's statement. He sees clearly that with the return of hunters from military service, guns are increasing faster than ducks are, and that the task of building up the annual crop is tough and urgent.

Ducks Unlimited is too important an undertaking to fall victim to extravagant overclaims and outmoded exaggerations. Mr. Farrington's book should be read for what it really is: a charming fantasy. Duck hunters can afford to think that way while facing a fireside bottle, but not while facing the hard realities of 1946.

One thing can always be said to the credit of Ducks Unlimited: while other conservation organizations heaved and pulled at the statute books, it put on its hip boots and began shovelling dirt in the marsh. This is a proud record, and should not be spoiled by claims of non-existent miracles.

Adventures of a Conservation Commissioner [1946]

Among the scores of committees on which Leopold served, the Wisconsin Conservation Commission, the state's policy-making body for forestry, wildlife, and fisheries, provided the greatest challenge. Leopold's tenure lasted from 1943 until his death in 1948. As a commissioner, he was immersed at least monthly, but often weekly or daily, in the politics of conservation. His responsibilities were especially taxing because he had assumed a leadership role during a period of severe overpopulation in the state's deer herd. Addressing fellow professionals at the Midwest Wildlife Conference in Des Moines, Iowa, in December 1946, he gave vent to the frustrations of dealing with an ecological issue in the public arena. The paper is an unpublished typescript with revisions in Leopold's hand.

After helping to set up several Commissions and serving on one, I have come to two conclusions.

The first is that a good Commission can prevent the conservation program from falling below the general level of popular ethics and intelligence.

The second is that no Commission can raise its program much above that level, except in matters to which the public is indifferent. Where the public has feelings, traditions, or prejudices, a Commission must drag its public along behind it like a balky mule, but with this difference: the public, unlike the mule, kicks both fore and aft.

An issue may be so clear in outline, so inevitable in logic, so imperative in need, and so universal in importance as to command immediate support from any reasonable person. Yet that collective person, the public, may take a decade to see the argument, and another to acquiesce in an effective program.

The Migratory Bird Law was such an issue. A handful of national leaders had to drag the mule from 1893, when the first state prohibited spring shooting, until 1916, when the federal law was finally anchored to the Canadian Treaty, a total span of 23 years.

One hears much nowadays about public relations experts who know how to talk gently to the mule; to beguile him into speedier thinking. If there be even one such, I invite him to move to Wisconsin. We will turn our pockets inside out, and give him our shirt besides.

This public we are talking about consists of three groups. Group 1 is the largest; it is indifferent to conservation questions. Group 2 is the smallest; it thinks with its head, but is silent. Group 3 is of intermediate size, and does all its thinking with mouth or pen. Perhaps a Conservation Commissioner would do better not to try to convert Group 3, but to convince Group 2 that there is an issue, and that it should say or do something about it. Perhaps this would shorten the 23 years.

Group 2 always speaks eventually, but the question is how soon, and will there by that time be anything left to speak about?

To illustrate this dilemma, I select a sample issue that is now current in 30 states, and is coming up in others: the issue of excess deer.

The deer issue, despite the fact that it deals with too many rather than too few, is at bottom similar to the migratory bird issue: we are augmenting our present sport by exhausting capital assets needed for the future. In migratory birds the endangered capital was breeding stock; in deer it is food plants. The root of the problem is to convince the public that:

1. Not all "brush" is deer food.
2. Too many deer, by overbrowsing, exterminate the kinds of brush that are food.
3. The worthless kinds move in. There is still lots of brush, but the deer starve.
4. The remedy is to reduce females before starvation occurs.

What I want to discuss is not the biology of deer irruptions, but the psychology of deer hunters when they are confronted by the unpleasant necessity of reducing an overlarge herd. I will use Wisconsin as a case history. Wisconsin is one of 30 states now afflicted with excess deer. A recent survey* of these states indicates that they now contain about a hundred problem areas. Reduction of females has been started on about half of these, but has been carried to the point of relieving overbrowsing on only a tenth. In short, 90 percent of our irrupting herds, of which the Wisconsin herd is one, are still being fed out of capital account. Why? To answer this question I will begin with a thumbnail sketch of what has happened in Wisconsin.

Wisconsin Chronology. In 1940 it began to be clear that the Wisconsin

---

*Aldo Leopold, Lyle K. Sowls, and David L. Spencer, "A Survey of Over-populated Deer Ranges in the United States." Pending publication in *Journal of Wildlife Management*. [Published *JWM* 11:2 (Apr 1947), 162–177.]

deer herd had become too large for its own good. A research project was started.

In 1942 a hard winter brought starvation to many yards. For the most part the deer that died were fawns. The Commission appointed a "Citizen's Deer Committee" to study the problem and recommend action.

In 1943, after an educational campaign, the Commission opened the season on antlerless deer. 62,000 were killed, in addition to 66,000 bucks. There had been little advance opposition, but on the day the "slaughter" started, Group 3 rose in wrath. It hasn't sat down yet.

In 1944 the usual buck law obtained.

In 1945 a special survey of 500 yards showed 54% in poor condition. Nevertheless the usual buck law obtained.

In 1946 a proposed opening on one deer of either sex was voted down by our "Congress" of sportsmen (mostly Group 3, partly Group 2). The Commission acquiesced in this mandate, despite the fact that the research crew reported a further increase in the proportion of overbrowsed yards.

So much for the bare chronology of events. This now brings me to my real subject: the behavior patterns exhibited by Group 3 conservationists since the reduction of 1943, and their effect on Wisconsin policy and administration.

Legislative Reprisals. When Group 3 rose in wrath over the 1943 reduction, it was natural that it should talk loudly to the Legislature. One would suppose that the Legislature would force the Commission's hand on the deer issue and let it go at that. Not so. The Legislature not only enacted mandatory deer feeding and mandatory predator bounties (both reversals of Commission policy), but it quietly killed several important forestry bills recommended by the Commission, and pigeonholed a revision of salary scales that had been previously approved by legislative leaders. Nothing was said publicly to the effect that these actions were reprisals, but they were just that.

These reprisals had an important effect, not only in blockading progress, but in convincing Commissioners that soft-pedalling the deer issue was the price of progress in forestry and other fields of conservation. Hence there has been no follow-up on the 1943 reduction.

This is ironical, for the deer are undoubtedly tearing down the forest faster than any legislature, or any Commission, can rebuild it. In short, Group 3 sportsmen are nullifying a forestry program on which the Commission is spending a million and a half, and on which the U.S. Forest Service and private owners are spending similar sums.

Extra-Legal Reduction. The Commission's educational campaign has inevitably publicized the notion that there are too many deer. The poachers, shiners, and bootleggers of venison have seized upon this as justifying a renewal of illegal deer killing. The meat shortage has added impetus to this trend. The upshot is that the deer herd is being reduced, in some localities, by extra-legal means. It's something like prohibition: the state having failed to provide a legal channel for a reasonable action, the action has cut its own illegal channel, and is going ahead law or no law.

Another way to look at it is that the state government, by failing to face the deer issue squarely, has made itself dependent on the violator for the performance of a necessary biological function. When the state elected to remove the wolf from the forested counties, it automatically inherited the wolf's job of regulating the deer herd. It is now not only shirking that job, but (ironically) it has renewed the bounty on wolves, and as a consequence may soon wipe out the remnant of a few dozen wolves that persists on the Michigan boundary.

Hunter Ethics. Everybody knows that in buck-law states, some hunters shoot everything that might be a buck until they happen to get a buck. The does previously killed are abandoned in the woods. It is axiomatic that such degraded ethics, to the extent that they exist, cancel out the recreational values of deer hunting. It is also axiomatic that states like Wisconsin, in dodging the issue of legalizing excess does, are unwittingly encouraging this degradation. The question is: how many hunters do it?

This question has always been regarded as unanswerable, but I submit that the number can be roughly estimated by comparison with other similar states in which the kill of antlerless deer is legal, and therefore known. Minnesota, for example, has a deer herd of similar size, range area, and population trend. Minnesota's kill since 1942 has averaged 69,000 deer, half of which were bucks. Wisconsin during the same period has averaged 37,000 bucks, close to half of Minnesota's total. It therefore seems highly probable that Wisconsin's illegal antlerless kill is in the five-figure bracket.

This conclusion is sustained by actual post-season counts of carcasses found on sample Wisconsin areas during the closing days of the CCC when large areas were available. I give first areas on which the legal kill, as well as the illegal kill, was known.

Year	Locality	Area Counted	Legal:Illegal Ratio
1942	St. Croix Co.	2060	39:61
1938	Chequamegon NF	?	49:51
1928	T41N R3W	300	27:73

I give next some counts on which only the area is known. These obviously cannot be related to the legal kill, which was unknown.

Year	Locality	Area Counted	Carcasses	Acres per Carcass
1939	Saddle Mound	4045	57	71
?	Bayfield Co.	2110	11	192
1940	Flambeau	4566	11	415
		10721	79	136

The state of hunter ethics, even when both sexes are open, is illustrated by a 1942 count on the Fishlake Forest in Utah where *dressed carcasses* left in the woods added up to 28% of the legal kill. Most of the hunters who abandoned these dressed deer had presumably killed another one bigger, better, or closer to the road.

What I am driving at can be boiled down to three assertions:

1. The average deer hunter loses his scruples in the woods.
2. A buck law, maintained despite a known overpopulation of deer, encourages the killing and abandonment of illegal deer on a far larger scale than has heretofore been publicly admitted.
3. The state, in such a case, is unwittingly contributing to the ethical degradation of its deer hunters.

The Upshot. Here then are three samples of the human behavior patterns with which a Conservation Commission must deal. Collectively they seem to reveal the same natural law of predation as research is discovering in animals: when there is a surplus, something is going to remove it; if one form of predation fails another takes over; if both fail, starvation steps in and finishes the job. This law operates through political as well as biological channels, and it involves both physical and ethical wastage. A Commission can reduce these wastages only as far and as fast as its public learns to face facts.

When the public blocks a sound and necessary reduction in deer, it is all too easy to blame unscrupulous politicians. While our opposition in Wisconsin contains some such, it is clear that the majority consists of solid citizens. How does one explain this?

First, a deer irruption cannot be understood at all except in terms of plant ecology. The public does not know the plants, much less how they react as a community to excess deer or other browsers.

Second, killing antlerless deer after a long period of protection is actually a slaughter, as the opposition says it is. The alternatives of illegal killing, or ultimate starvation, are both invisible, and therefore not objectionable. One can rationalize illegal killing by assuming that only a few people do it, and starvation by blaming the weather.

Thirdly (and this is where the average citizen gets really confused), many localities exhibit no increase in deer; they show only a progressive decline in food plants. The decline in food is invisible, while the failure of deer to increase can be blamed on the slaughter of 1943.

The stage is thus set for a complete circle of self-deceptions. It seems probable that the Wisconsin irruption will run its course before corrective action is taken. We have not learned from other states, nor will other states learn from us. All will end up with impoverished herds, and depleted forests, and (I hope) a fund of painful experience.

Wherefore Wildlife Ecology? [1947]

Midway through his undergraduate wildlife ecology course—perhaps in the spring semester of 1947, the last time he completed the lecture series—Leopold penciled a two-page definition of the course's objectives. This is one of his most succinct and characteristic statements of his role as a teacher.

At the beginning of this course I did not try to define its object, because any attempt at definition would at that time have consisted of meaningless words. I shall now confide in you what the course is driving at.

The object is to teach you how to read land. Land is soil, water, plants and animals. Each of these "organs" of land has meaning as a separate entity, just as fingers, toes, and teeth have. But each has a much larger meaning as the component parts of the organism. No one can understand an animal by learning only its parts, yet when we attempt to say that an animal is "useful," "ugly," or "cruel" we are failing to see it as part of the land. We do not make the error of calling a carburetor "greedy." We see it as part of a functioning motor.

Much can be learned about land with amateur equipment, provided one learns how to think in scientific terms. Hence I am asking you to read the best professional literature, but in the field to use only the eyes, ears, and notebook which everybody carries. The lectures try to connect the two.

What I hope to teach is perhaps ecological research as an outdoor "sport." Yet "sport" is hardly accurate, because in sport one tries to do well what thousands have done better. In ecology one tries to do well what few have ever done at all, at least not in one's home region. The thing I am teaching, then, is amateur exploration, research for fun, in the field of land.

I have an ulterior motive, as everyone has. I am interested in the thing called "conservation." For this I have two reasons: (1) without it, our economy will ultimately fall apart; (2) without it many plants, animals, and places of entrancing interest to me as an explorer will cease to exist. I do not like to think of economic bankruptcy, nor do I see much object in continuing the human enterprise in a habitat stripped of what interests me most.

That there is some basic fallacy in present-day conservation is shown by our response to it. Instead of living it, we turn it over to bureaus. Even the landowner, who has the best opportunity to practice it, thinks of it as something for government to worry about.

I think I know what the fallacy is. It is the assumption, clearly borrowed from modern science, that the human relation to land is only economic. It is, or should be, esthetic as well. In this respect our present culture, and especially our science, is false, ignoble, and self-destructive.

If the individual has a warm personal understanding of land, he will perceive of his own accord that it is something more than a breadbasket. He will see land as a community of which he is only a member, albeit now the dominant one. He will see the beauty, as well as the utility, of the whole, and know the two cannot be separated. We love (and make intelligent use of) what we have learned to understand.

Hence this course. I am trying to teach you that this alphabet of "natural objects" (soils and rivers, birds and beasts) spells out a story, which he who runs may read—if he knows how. Once you learn to read the land, I have no fear of what you will do to it, or with it. And I know many pleasant things it will do to you.

The Ecological Conscience [1947]

Delivered on June 27, 1947, to the Conservation Committee of the Garden Club of America, this address was published later that year in the club's *Bulletin*. Parts of it were revised and incorporated in "The Land Ethic" in *Sand County Almanac*, upon which Leopold was simultaneously working. The body of the essay consists of case studies, one of which summarizes the complaints aired in "Adventures of a Conservation Commissioner," and its "Upshot" incorporates the perspective detailed in "Conservation: In Whole or in Part?"

Everyone ought to be dissatisfied with the slow spread of conservation to the land. Our "progress" still consists largely of letterhead pieties and convention oratory. The only progress that counts is that on the actual landscape of the back forty, and here we are still slipping two steps backward for each forward stride.

The usual answer to this dilemma is "more conservation education." My answer is yes by all means, but are we sure that only the *volume* of educational effort needs stepping up? Is something lacking in its *content* as well? I think there is, and I here attempt to define it.

The basic defect is this: we have not asked the citizen to assume any real responsibility. We have told him that if he will vote right, obey the law, join some organizations, and practice what conservation is profitable on his own land, that everything will be lovely; the government will do the rest.

This formula is too easy to accomplish anything worthwhile. It calls for no effort or sacrifice; no change in our philosophy of values. It entails little that any decent and intelligent person would not have done, of his own accord, under the late but not lamented Babbitian code.

No important change in human conduct is ever accomplished without an internal change in our intellectual emphases, our loyalties, our affections, and our convictions. The proof that conservation has not yet touched these foundations of conduct lies in the fact that philosophy, ethics, and religion have not yet heard of it.

Garden Club
Minneapolis
6-27-47

THE ECOLOGICAL CONSCIENCE

Aldo Leopold

Everyone ought to be dissatisfied with the slow spread of conservation to the land. Our "progress" consists largely of letterhead pieties and convention oratory. The only progress that counts is that on the actual landscape of the back forty, and here we are still slipping two steps backward for each forward stride.

The usual answer to this dilemma is "more conservation education." My answer is yes by all means, but are we sure that only the volume of educational effort needs stepping up? Is something lacking in its content as well? I think there is, and I here attempt to define it.

The basic defect is this: we have not asked the citizen to assume any real responsibility. We have told him that if he will vote right, obey the law, join some organizations, and practice what conservation is profitable on his own land, that everything will be lovely; the government will do the rest.

This formula is too easy to accomplish anything worthwhile. It calls for no effort or sacrifice; no change in our philosophy of values. It entails little that any decent and intelligent person would not have done, of his own accord, under the late but not lamented Babbitian code.

No important change in human conduct is ever accomplished without an internal change in our intellectual emphases, our loyalties, our affections, and our convictions. In our attempt to make conservation easy, we have dodged its spiritual implications. The proof of this error lies in the fact that philosophy, ethics, and religion have not yet heard of it.

[Ecology is the science of communities, and the ecological conscience is therefore the ethic of community life.]

I need a short name for what is lacking in conservation; I will call it the ecological conscience. I will define it further, in terms of some case histories, which I think show the futility of trying to improve the face of the land without improving ourselves. I select three...

Typescript of first page of "The Ecological Conscience," June 27, 1947, heavily edited in Leopold's hand. (Leopold Collection X25 2221, UW Archives)

339

I need a short name for what is lacking; I call it the ecological conscience. Ecology is the science of communities, and the ecological conscience is therefore the ethics of community life. I will define it further in terms of four case histories, which I think show the futility of trying to improve the face of the land without improving ourselves. I select these cases from my own state, because I am there surer of my facts.

Soil Conservation Districts

About 1930 it became clear to all except the ecologically blind that Wisconsin's topsoil was slipping seaward. The farmers were told in 1933 that if they would adopt certain remedial practices for five years, the public would donate CCC labor to install them, plus the necessary machinery and materials. The offer was widely accepted, but the practices were widely forgotten when the five-year contract period was up. The farmers continued only those practices that yielded an immediate and visible economic gain for themselves.

This partial failure of land-use rules written by the government led to the idea that maybe farmers would learn more quickly if they themselves wrote the rules. Hence, in 1937, the Wisconsin Legislature passed the Soil Conservation District Law. This said to the farmers, in effect: "We, the public, will furnish you free technical service and loan you specialized machinery, if you will write your own rules for land-use. Each county may write its own rules, and these will have the force of law." Nearly all the counties promptly organized to accept the proffered help, but after a decade of operation, *no county has yet written a single rule*. There has been visible progress in such practices as strip-cropping, pasture renovation, and soil liming, but none in fencing woodlots or excluding plow and cow from steep slopes. The farmers, in short, selected out those remedial practices which were profitable anyhow, and ignored those which were profitable to the community, but not clearly profitable to themselves. The net result is that the natural acceleration in rate of soil-loss has been somewhat retarded, but we nevertheless have less soil than we had in 1937.

I hasten to add that no one has ever told farmers that in land-use the good of the community may entail obligations over and above those dictated by self-interest. The existence of such obligations is accepted in bettering rural roads, schools, churches, and baseball teams, but not in bettering the behavior of the water that falls on the land, nor in the preserving of the beauty or diversity of the farm landscape. Land-use ethics are still governed wholly by economic self-interest, just as social ethics were a century ago.

To sum up: we have asked the farmer to do what he conveniently could to save his soil, and he has done just that, and only that. The exclusion of cows from woods and steep slopes is not convenient, and is not done. Moreover some things are being done that are at least dubious as conservation practices: for example marshy stream bottoms are being drained to relieve the pressure on worn-out uplands. The upshot is that woods, marshes, and natural streams, together with their respective faunas and floras, are headed toward ultimate elimination from southern Wisconsin.

All in all we have built a beautiful piece of social machinery—the Soil Conservation District—which is coughing along on two cylinders because we have been too timid, and too anxious for quick success, to tell the farmer the true magnitude of his obligations. Obligations have no meaning without conscience, and the problem we face is the extension of the social conscience from people to land.

Paul Bunyan's Deer

The Wisconsin lumberjack came very near accomplishing, in reality, the prodigious feats of woods-destruction attributed to Paul Bunyan. Following Paul's departure for points west, there followed an event little heralded in song and story, but quite as dramatic as the original destruction of the pineries: there sprang up, almost over night, an empire of brushfields.

Paul Bunyan had tired easily of salt pork and corned beef, hence he had taken good care to see that the deer of the original pineries found their way regularly to the stewpot. Moreover there were wolves in Paul's day, and the wolves had performed any necessary pruning of the deer herd which Paul had overlooked. But by the time the brushfields sprang into being, the wolves had been wiped out and the state had passed a buck-law and established refuges. The stage was set for an irruption of deer.

The deer took to the brushfields like yeast tossed into the sourdough pot. By 1940 the woods were foaming with them, so to speak. We Conservation Commissioners took credit for this miracle of creation; actually we did little but officiate at the birth. Anyhow, it was a herd to make one's mouth water. A tourist from Chicago could drive out in the evening and see fifty deer, or even more.

This immense deer herd was eating brush, and eating well. What was this brush? It consisted of temporary short-lived sun-loving trees and bushes which act as a nurse crop for the future forest. The forest comes up under the brush, just as alfalfa or clover come up under oats or rye. In the normal succession, the brush is eventually overtopped by the forest tree seedlings, and we have the start of a new forest.

In anticipation of this well known process, the state, the counties, the U.S. Forest Service, the pulp mills, and even some lumber mills staked out "forests" consisting, for the moment, of brush. Large investments of time, thought, cash, CCC labor, WPA labor, and legislation were made in the expectation that Nature would repeat her normal cycle. The state embarked on a tax subsidy, called the Forest Crop Law, to encourage landowners to hang onto their brushfields until they were replaced by forest.

But we failed to reckon with the deer, and with deer hunters and resort owners. In 1942 we had a hard winter and many deer starved. It then became evident that the original "nurse-trees" had grown out of reach of deer, and that the herd was eating the oncoming forest. The remedy seemed to be to reduce the herd by legalizing killing of does. It was evident that if we didn't reduce the herd, starvation would, and we would eventually lose both the deer and the forest. But for five consecutive years the deer hunters and resort owners, plus the politicians interested in their votes, have defeated all attempts at herd-reduction.

I will not tire you with all the red herrings, subterfuges, evasions, and expedients which these people have used to befog this simple issue. There is even a newspaper dedicated solely to defaming the proponents of herd-reduction. These people call themselves conservationists, and in one sense they are, for in the past we have pinned that label on anyone who loves wildlife, however blindly. These conservationists, for the sake of maintaining an abnormal and unnatural deer herd for a few more years, are willing to sacrifice the future forest, and also the ultimate welfare of the herd itself.

The motives behind this "conservation" are a wish to prolong easy deer hunting, and a wish to show numerous deer to tourists. These perfectly understandable wishes are rationalized by protestations of chivalry to does and fawns. As an unexpected aftermath of this situation, there has been a large increase of illegal killing, and of abandonment of illegal carcasses in the woods. Thus herd-control, of a sort, is taking place outside the law. But the food-producing capacity of the forest has been overstrained for a decade, and the next hard winter will bring catastrophic starvation. After that we shall have very few deer, and these will be runty from malnutrition. Our forest will be a moth-eaten remnant consisting largely of inferior species of trees.

The basic fallacy in this kind of "conservation" is that it seeks to conserve one resource by destroying another. These "conservationists" are unable to see the land as a whole. They are unable to think in terms of community rather than group welfare, and in terms of the long as well as the short view. They are conserving what is important to them in the immediate future, and they are angry when told that this conflicts with what is important to the state as a whole in the long run.

There is an important lesson here: the flat refusal of the average adult to learn anything new, i.e., to study. To understand the deer problem requires some knowledge of what deer eat, of what they do not eat, and of how a forest grows. The average deer hunter is sadly lacking in such knowledge, and when anyone tries to explain the matter, he is branded forthwith as a long-haired theorist. This anger-reaction against new and unpleasant facts is of course a standard psychiatric indicator of the closed mind.

We speak glibly of conservation education, but what do we mean by it? If we mean indoctrination, then let us be reminded that it is just as easy to indoctrinate with fallacies as with facts. If we mean to teach the capacity for independent judgment, then I am appalled by the magnitude of the task. The task is large mainly because of this refusal of adults to learn anything new.

The ecological conscience, then, is an affair of the mind as well as the heart. It implies a capacity to study and learn, as well as to emote about the problems of conservation.

Jefferson Davis' Pines

I have a farm in one of the sand-counties of central Wisconsin. I bought it because I wanted a place to plant pines. One reason for selecting my particular farm was that it adjoined the only remaining stand of mature pines in the County.

This pine grove is an historical landmark. It is the spot (or very near the spot) where, in 1828, a young Lieutenant named Jefferson Davis cut the pine logs to build Fort Winnebago. He floated them down the Wisconsin River to the fort. In the ensuing century a thousand other rafts of pine logs floated past this grove, to build that empire of red barns now called the Middle West.

This grove is also an ecological landmark. It is the nearest spot where a city-worn refugee from the south can hear the wind sing in tall timber. It harbors one of the best remnants of deer, ruffed grouse, and pileated woodpeckers in southern Wisconsin.

My neighbor, who owns the grove, has treated it rather decently through the years. When his son got married, the grove furnished lumber for the new house, and it could spare such light cuttings. But when war prices of lumber soared skyward, the temptation to slash became too strong. Today the grove lies prostrate, and its long logs are feeding a hungry saw.

By all the accepted rules of forestry, my neighbor was justified in slashing the grove. The stand was even-aged; mature, and invaded by heart-rot. Yet any schoolboy would know, in his heart, that there is something wrong about erasing the last remnant of pine timber from a county. When a farmer owns a rarity he should feel some obligation as its custodian, and a community should feel some obligation to help him carry the economic cost of custodianship. Yet our present land-use conscience is silent on such questions.

The Flambeau Raid

The Flambeau was a river so lovely to look upon, and so richly endowed with forests and wildlife, that even the hard-bitten fur traders of the free-booting 1700's enthused about it as the choicest part of the great north woods.

The freebooting 1800's expressed the same admiration, but in somewhat different terms. By 1930 the Flambeau retained only one 50-mile stretch of river not yet harnessed for power, and only a few sections of original timber not yet cut for lumber or pulp.

During the 1930's the Wisconsin Conservation Department started to build a state forest on the Flambeau, using these remnants of wild woods and wild river as starting points. This was to be no ordinary state forest producing only logs and tourist camps; its primary object was to preserve and restore the remnant of canoe-water. Year by year the Commission bought land, removed cottages, fended off unnecessary roads, and in general started the long slow job of re-creating a stretch of wild river for the use and enjoyment of young Wisconsin.

The good soil which enabled the Flambeau to grow the best cork pine for Paul Bunyan likewise enabled Rusk County, during recent decades, to sprout a dairy industry. These dairy farmers wanted cheaper electric power than that offered by local power companies. Hence they organized a cooperative REA and applied for a power dam which, when built, will clip off the lower reaches of canoe-water which the Conservation Commission wanted to keep for recreational use.

There was a bitter political fight, in the course of which the Commission not only withdrew its opposition to the REA dam, but the Legislature, by statute, repealed the authority of the Conservation Commission and made County Commissioners the ultimate arbiters of conflict between power values and recreational values. I think I need not dwell on the irony of this statute. It seals the fate of all wild rivers remaining in the state, including the Flambeau. It says, in effect, that in deciding the use of rivers, the local economic interest shall have blanket priority over state-wide recreational interests, with County Commissioners as the umpire.

The Flambeau case illustrates the dangers that lurk in the semi-honest doctrine that conservation is only good economics. The defenders of the Flambeau tried to prove that the river in its wild state would produce more fish and tourists than the impounded river would produce butterfat, but this is not true. We should have claimed that a little gain in butterfat is less important to the state than a large loss in opportunity for a distinctive form of outdoor recreation.

We lost the Flambeau as a logical consequence of the fallacy that conservation can be achieved easily. It cannot. Parts of every well-rounded conservation program entail sacrifice, usually local, but none-the-less real. The farmers' raid on our last wild river is just like any other raid on any other public wealth; the only defense is a widespread public awareness of the values at stake. There was none.

The Upshot

I have described here a fraction of that huge aggregate of problems and opportunities which we call conservation. This aggregate of case-histories show one common need: an ecological conscience.

The practice of conservation must spring from a conviction of what is ethically and esthetically right, as well as what is economically expedient. A thing is right only when it tends to preserve the integrity, stability, and beauty of the community, and the community includes the soil, waters, fauna, and flora, as well as people.

It cannot be right, in the ecological sense, for a farmer to drain the last marsh, graze the last woods, or slash the last grove in his community, because in doing so he evicts a fauna, a flora, and a landscape whose membership in the community is older than his own, and is equally entitled to respect.

It cannot be right, in the ecological sense, for a farmer to channelize his creek or pasture his steep slopes, because in doing so he passes flood trouble to his neighbors below, just as his neighbors above have passed it to him. In cities we do not get rid of nuisances by throwing them across the fence onto the neighbor's lawn, but in water-management we still do just that.

It cannot be right, in the ecological sense, for the deer hunter to maintain his sport by browsing out the forest, or for the bird-hunter to maintain his by decimating the hawks and owls, or for the fisherman to maintain his by decimating the herons, kingfishers, terns, and otters. Such tactics seek to achieve one kind of conservation by destroying another, and thus they subvert the integrity and stability of the community.

If we grant the premise that an ecological conscience is possible and needed, then its first tenet must be this: economic provocation is no longer a satisfactory excuse for unsocial land-use, (or, to use somewhat stronger words, for ecological atrocities). This, however, is a negative statement. I would rather assert positively that decent land-use should be accorded social rewards proportionate to its social importance.

I have no illusions about the speed or accuracy with which an ecological conscience can become functional. It has required 19 centuries to define decent man-to-man conduct and the process is only half done; it may take as long to evolve a code of decency for man-to-land conduct. In such matters we should not worry too much about anything except the direction in which we travel. The direction is clear, and the first step is to *throw your weight around* on matters of right and wrong in land-use. Cease being intimidated by the argument that a right action is impossible because it does not yield maximum profits, or that a wrong action is to be condoned because it pays. That philosophy is dead in human relations, and its funeral in land-relations is overdue.

Publications of Aldo Leopold
Index

Publications of Aldo Leopold

Aldo Leopold published about five hundred distinct items, including articles, addresses, essays, books, newsletters, reviews, reports, and even a few poems. His publications, including mimeographed serials that he wrote and edited and items that appeared in other mimeographed serials, are listed in chronological order in this bibliography, and from the most to the least precise date of publication. About one hundred items that were reprinted in various books, periodicals, and pamphlets during his lifetime are listed under the citation of original publication.

The basis for this bibliography was a "Preliminary Bibliography of Aldo Leopold," included in the *Wildlife Research News Letter* no. 35 of the University of Wisconsin Department of Wildlife Management, issued May 3, 1948, shortly after Leopold's death. About two hundred items, including original citations, reprints, mimeographed entries, and a few posthumous publications, have been added to the "Preliminary Bibliography." Citations have been verified in the original publication, where possible, and completed or corrected. Unverified citations may not be entirely complete, but there is sufficient evidence, such as a clipping or copy of the item itself, to indicate that it was in fact published. We welcome additions or corrections to the bibliography.

Aldo Leopold is sole author of all cited items unless otherwise noted. Where he is listed as joint author with other individuals, he ordinarily served as principal author, except perhaps for a few committee reports for which another individual is identified as chairman. Titles of articles, essays, poems, etc., are lowercased, while titles of books, periodicals, and pamphlets are capitalized and italicized. Identification of untitled items and other explanatory information is enclosed in brackets.

Abbreviations used in Bibliography

Agr Exp Sta Bull	Annual Report of Agricultural Experiment Station, University of Wisconsin-Madison, Bulletin
AL	Aldo Leopold
Albuq Eve Herald	*The Evening Herald*, Albuquerque
Albuq Morning J	*Albuquerque Morning Journal*
Almanac	*A Sand County Almanac and Sketches Here and There*
Am For & For Life	*American Forests and Forest Life* (formerly *Am Forestry*)
Am Forestry	*American Forestry*

350   Publications of Aldo Leopold

Am Forests	American Forests (formerly Am For & For Life)
Am Game	American Game (formerly Bulletin AGPA)
Am Wildlife	American Wildlife (formerly Am Game)
Audubon Mag	Audubon Magazine (formerly Bird-Lore)
Bulletin AGPA	Bulletin of the American Game Protective Association
Condor	The Condor
IWLA	Izaak Walton League of America
J Forestry	Journal of Forestry
J Mammalogy	Journal of Mammalogy
JWM	The Journal of Wildlife Management
Outdoor Am	Outdoor America (Izaak Walton League of America)
Trans Am Game Conf	Transactions, American Game Conference (1929-1935; formerly National Game Conference—principal papers in Bulletin AGPA)
Trans NAWC	Transactions, North American Wildlife Conference (1936–; formerly Trans Am Game Conf)
Trans Wis Acad	Transactions of the Wisconsin Academy of Sciences, Arts and Letters
WAF	Wisconsin Agriculturist and Farmer (Racine, Wis.)
WAF Booklet	Wildlife Conservation on the Farm (1941)
WCB	Wisconsin Conservation Bulletin
WCD Pub	Wisconsin Conservation Department Publication
Wilson Bull	The Wilson Bulletin
Wis Acad Rev	Wisconsin Academy Review
Wis Sportsman	The Wisconsin Sportsman

1911-1915

Cost of poor mounts to the service, pp. 1-4; [statement concerning qualifications of forest rangers], pp. 105-107; [statement concerning examinations for promotion], pp. 114-116. In *Proceedings of the First Annual Supervisors' Meeting, District 3, Nov 9-14, 1911.* Mimeographed.

The Carson Pine Cone (Carson National Forest, Tres Piedras, N.M.). June 1911-Mar 1914. [A monthly mimeographed bulletin apparently largely written and edited by AL, intended for Forest Service use only and designed to unite and inform members of the Carson National Forest staff. No evidence of issues for May 1912, Apr-June, Aug, Nov 1913, and Feb 1914.] Among items definitely written by AL are the following poems, each reprinted in *The Forest Ranger and Other Verse*, edited by John D. Guthrie, 1919:
    The busy season. July 1911 [unsigned]. Guthrie, p. 116.
    The mystery. Oct 1911 [unsigned]. Guthrie, pp. 45-46.
    Resolutions of a ranger. Jan 1914 [initialed AL, 1/1/14]. Guthrie, pp. 53-54.
    Spare time. Mar 1914 [initialed AL, 12/31/13]. Guthrie, pp. 66-67.
The following open letters were also written by AL:
    Burlington, Iowa, July 15, 1913, To the forest officers of the Carson. July 1913.

Burlington, Iowa, Nov 14, 1913, To the boys on the job. Dec 1913.
Burlington, Iowa, Jan 16, 1914, To the officers of the Carson. Jan 1914.
Santa Fe, N.M., Feb 15, 1914, To the officers of the Carson. Mar 1914.

Game and Fish Handbook. Forest Service, USDA, District 3 (issued by the District Forester, Albuquerque, N.M., Sept 15, 1915, and revised to date). 109pp. Mimeographed; unsigned.

The Pine Cone (Albuquerque, N.M.). 19 issues, Dec 1915–Dec 1920, March 1924, July 1931. [Official bulletin, New Mexico Game Protective Association (except for first issue, published by Albuquerque Game Protective Association); published about three times a year to 1920, Mar 1924, and July 1931; largely written and edited to Mar 1924 by AL, sec. NMGPA; signed contributions cited separately.]

1916

Protectionists take issue with George Willetts. *Albuq Eve Herald* (Mar 1, 1916), 8. [Letter signed by AL et al.]

Forest officers to work in co-operative capacity [letter from R. E. Marsh, R. F. Balthis, and AL]; Resolution: New Mexico GPA proposed predatory animal commission [Miles W. Burford, pres. and AL, sec.]. *The Pine Cone* (New Mexico GPA, Apr 1916), 4.

Game conservation; a warning, also an opportunity. *Arizona* 7:1-2 (1916), 6.

1917

Progressive cattle range management. *The Breeder's Gazette* 71:18 (May 3, 1917), 919.

Forestry and game management. *Yale Forest School News* 5:3 (July 1, 1917), 41.

A new sideline for foresters. *Yale Forest School News* 5:3 (July 1, 1917), 41-42.

Unique punishment for slayers of song birds. *Bulletin AGPA* 6:4 (Oct 1917), 22.

Demise of New Mexican game conserver. *The Sportsmen's Review* 53:12? (c. Dec 1917), 524. [Obituary of Miles W. Burford; unsigned]

1918

The popular wilderness fallacy: An idea that is fast exploding. *Outer's Book—Recreation* 58:1 (Jan 1918), 43-46.

How to build bird houses. *Albuq Eve Herald* (Feb 19, 1918), 5.

How to build bird houses. *Albuq Eve Herald* (Feb 23, 1918), 7.

Do purple martins inhabit bird boxes in the West? *Condor* 20:2 (Mar-Apr 1918), 93.

Birds and Cats. *Albuq Eve Herald* (Apr 6, 1918).

Forestry and game conservation. *J Forestry* 16 (Apr 1918), 404-411.

Make Stinking Lake a game refuge. *Outer's Book—Recreation* 58:4 (Apr 1918), 291. Also in *Bulletin AGPA* 7:1 (Jan 1918), 16.

Pulling together for drainage. *Albuq Eve Herald* (May 13, 1918), sec. 2, p. 3.

Restocking the national forests with elk: Where and how it may be done. *Outer's Book—Recreation* 58:5 (May 1918), 412-415.

Are red-headed woodpeckers moving west? *Condor* 20:3 (May-June 1918), 122.

Transactions of the American Fisheries Society 47:3 (June 1918), 101-102.

Explanation of Rio Grande Valley drainage problems is given by Aldo Leopold. *Albuq Eve Herald* (June 8, 1918), 7.

What about drainage? *Bernalillo County Farm Bureau News* 1:1 (June 1918), 2.
Public recreation—an extravagance or a necessity? *Albuq Morning J* (July 9, 1918), 4.
Putting the "AM" in game warden: The story of how the New Mexico Game Protective Association substituted *Push for Politics* in their state game department. *The Sportsmen's Review* 54:9 (Aug 31, 1918), 173-174.
Forward Albuquerque. Quarterly Bulletin of the Albuquerque Chamber of Commerce no. 1 (Aug 1918), 4pp. [Probably by AL]
America's fire loss exceeds that of any country in the world. *Albuq Morning J* (Nov 1, 1918), 2.
C of C urges early building of ample houses. *Albuq Morning J* (Nov 17, 1918), 4. [Letter to citizens of Albuquerque from secretary, Albuquerque Chamber of Commerce]

1919

Notes on red-headed woodpecker and jack snipe in New Mexico. *Condor* 21:1 (Jan–Feb 1919), 40.
The national forests: The last free hunting grounds of the nation. *J Forestry* 17:2 (Feb 1919), 150-153.
Notes on the behavior of pintail ducks in a hailstorm. *Condor* 21:2 (Mar–Apr 1919), 87.
Forest Service salaries and the future of the national forests. *J Forestry* 17:4 (Apr 1919), 398-401.
Wild lifers vs. game farmers: A plea for democracy in sport. *Bulletin AGPA* 8:2 (Apr 1919), 6-7.
Relative abundance of ducks in the Rio Grande Valley. *Condor* 21:3 (May–June 1919), 122.
Notes on the weights and plumages of ducks in New Mexico. *Condor* 21:3 (May–June 1919), 128-129.
Aldo Leopold says conservation of grass is really first consideration. *Albuq Morning J* (June 28, 1919), 6. [Letter to editor]
Mr. Leopold believes position justifiable. *Albuq Morning J* (July 3, 1919), 6. [Letter to editor]
A breeding record for the red-headed woodpecker in New Mexico. *Condor* 21:4 (July–Aug 1919), 173-174.
City tree planting. *Am Forestry* 25:308 (Aug 1919), 1295. [Unsigned]
Differential sex migration of mallards in New Mexico. *Condor* 21:5 (Sept–Oct 1919), 182-183.
A plea for state-owned ducking grounds. *Wild Life* (Oct 1919), 9.
A turkey hunt in the Datil National Forest. *Wild Life* (Dec 1919), 4-5, 16.
The mystery, pp. 45-46; Resolutions of a ranger, pp. 53-54; Spare time, pp. 66-67; The tourist and the ranger, pp. 80-81; The busy season, p. 116. In *The Forest Ranger and Other Verse*, edited by John D. Guthrie, 1919.

1920

Wanted—national forest game refuges. *Bulletin AGPA* 9:1 (Jan 1920), 8-10, 22.
Determining the kill factor for black-tail deer in the Southwest. *J Forestry* 18:2 (Feb 1920), 131-134.

"Piute forestry" vs. forest fire prevention. *Southwestern Magazine* 2:3 (Mar 1920), 12–13.
The game situation in the Southwest. *Bulletin AGPA* 9:2 (Apr 1920), 3–5.
The forestry of the prophets. *J Forestry* 18:4 (Apr 1920), 412–419.
Range of the magpie in New Mexico. *Condor* 22:3 (May–June 1920), 112.
The "why" & "how" of game refuges. *The Pine Cone* (July 1920), 1. [Drawings with captions, initialed A.L.]
Further notes on differential sex migration. *Condor* 22:4 (July–Aug 1920), 156–157.
Mallard decoys. *Forest and Stream* (Nov 1920), 598–599.
What is a refuge? *All Outdoors* 8:2 (Nov 1920), 46–47 [letter to editor]. Condensed as "The essentials of the game refuge," *The Literary Digest* 68:3 (Jan 15, 1921), 54.
A complaint. *The Game Breeder* (c. Dec 1920), 288–289.

1921

A hunter's notes on doves in the Rio Grande Valley. *Condor* 23:1 (Jan–Feb 1921), 19–21.
A plea for recognition of artificial works in forest erosion control policy. *J Forestry* 19:3 (Mar 1921), 267–273.
Weights and plumage of ducks in the Rio Grande Valley. *Condor* 23:3 (May–June 1921), 85–86.
The wilderness and its place in forest recreational policy. *J Forestry* 19:7 (Nov 1921), 718–721.

1922

The posting problem. *Outdoor Life* 49:3 (Mar 1922), 186–188.
Road-runner caught in the act. *Condor* 24:5 (Sept–Oct 1922), 183.

1923

Wild followers of the forest: The effect of forest fires on game and fish—the relation of forests to game conservation. *Am Forestry* 29:357 (Sept 1923), 515–519, 568.
The 'following' habit in hawks and owls. *Condor* 25:5 (Sept–Oct 1923), 180.
Watershed Handbook. U.S. Forest Service, Southwestern District, Dec 1923. 28pp., illus. Mimeographed. [Unsigned; "first section of the proposed 'Lands Handbook' which will extend and supersede the present 'Uses Handbook.'"] Reissued in substantially revised form, Aug 1, 1933, with a 4p. "Appendix VII: Watershed bibliographies," apparently prepared by AL. Revised version issued Oct 1, 1934.

1924

(With H. B. Jamison and R. Fred Pettit). Report of the quail committee, Albuquerque Game Protective Association. *The Pine Cone*, no. 18 (Mar 1924), 4.
Pioneers and gullies. *Sunset Magazine* 52:5 (May 1924), 15–16, 91–95.
Grass, brush, timber, and fire in southern Arizona. *J Forestry* 22:6 (Oct 1924), 1–10.
Quail production—a remedy for the "song bird list." *Outdoor Am* 3:4 (Nov 1924), 42–43.
Coot caught by turtle. *Condor* 26:6 (Nov–Dec 1924), 226.

1925

The utilization conference. *J Forestry* 23:1 (Jan 1925), 98–100.

A seven-year duck census of the middle Rio Grande Valley. *Condor* 27:1 (Jan–Feb 1925), 8–11.

Recent developments in game management. *USFS Service Bulletin* 9:9 (Mar 2, 1925), 1–2. Mimeographed.

Conserving the covered wagon. *Sunset Magazine* 54:3 (Mar 1925), 21, 56.

Natural reproduction of forests. *Parks and Recreation* 9:2 (Apr 1925), 366–372.

The pig in the parlor. *USFS Service Bulletin* 9:23 (June 8, 1925), 1–2. Mimeographed.

Ten new developments in game management. *Am Game* 14:3 (July 1925), 7–8, 20.

The last stand of the wilderness. *Am For & For Life* 31:382 (Oct 1925), 599–604. Reprinted as a brochure by the American Forestry Association; abstracted as "The vanishing wilderness" in *The Literary Digest* 90:6 (Aug 7, 1926), 54, 56–57; excerpted in *Recreation Resources of Federal Lands*, proceedings, National Conference on Outdoor Recreation (Washington, D.C., 1928), 86–88, 91.

Forestry in Wisconsin. *Report of the Third Annual Convention of the Wisconsin Division of the IWLA* (Oct 14–15, 1925), 82–84.

Wilderness as a form of land use. *The Journal of Land & Public Utility Economics* 1:4 (Oct 1925), 398–404.

A plea for wilderness hunting grounds. *Outdoor Life* 56:5 (Nov 1925), 348–350.

Wastes in forest utilization—what can be done to prevent them. *Empire State Forest Products Association Bulletin* no. 22 (Dec 1925), 6–9. [Proceedings, State-wide Wood Utilization Conference, Syracuse, N.Y., Nov 12, 1925]. Abstracted as "Wastes in utilization" in *Southern Lumberman* 121:1574 (Nov 28, 1925), 39–40.

Forestry and game management. *Colorado Forester* (Ft. Collins: Colorado Agricultural College, 1925), 29–30.

1926

[Untitled address on wilderness conservation]. *Proceedings of the Second National Conference on Outdoor Recreation, January 20–21, 1926* (69th Cong., 1st sess., 1926, S. Doc. 117, pp. 61–65.

Wood preservation and forestry. *Proceedings, 22nd Annual Meeting, American Wood-Preservers' Association* (Jan 26–28, 1926), 30–35. Reprinted in *Railway Engineering and Maintenance* 22:2 (Feb 1926), 60–61; reprinted in *Railway Age* 80:5 (Jan 30, 1926), 346.

On the reputations of forests. *The Forest Worker* 2:3 (May 1926), 17–18. Mimeographed.

The way of the waterfowl: How the Anthony Bill will help ducks and duck hunting; an example of New Mexico's refuge system in actual operation. *Am For & For Life* 32:389 (May 1926), 287–291.

Comment [on Howard R. Flint, "Wasted Wilderness"]. *Am For & For Life* 32:391 (July 1926), 410–411.

Forestry: Its relation to conservation. *Report of the Fourth Annual Convention of the Wisconsin Division of the IWLA* (Sept 9–10, 1926), 129–141. Mimeographed.

Fires and game. *J Forestry* 24:6 (Oct 1926), 726-728.
The next move: A size-up of the migratory bird situation. *Outdoor Life* 58:5 (Nov 1926), 363.
Short lengths for farm buildings. Forest Products Laboratory Report (Nov 8, 1926), 2pp.

1927

Government logging. *USFS Service Bulletin* 11:1 (Jan 3, 1927), 4-5.
The whistling note of the Wilson snipe. *Condor* 29:1 (Jan-Feb 1927), 79-80.
Useless knowledge. *USFS Service Bulletin* 11:6 (Feb 7, 1927), 4-5.
Forest products research and profitable forestry. *J Forestry* 25:5 (May 1927), 542-548.

1928

Pineries and deer on the Gila. *New Mexico Conservationist* 1:3 (Mar 1928), 3.
[Untitled address on game management], pp 138-141, and Report of the Committee on Reforestation [presented by AL, chairman], pp. 153-154. *Official Record, 6th Annual Convention, IWLA* (Apr 18-21, 1928). Mimeographed.
The home builder conserves. *Am For & For Life* 34:413 (May 1928), 276-278, 297.
Mr. Thompson's wilderness. *USFS Service Bulletin* 12:26 (June 25, 1928), 1-2. Mimeographed.
The game survey and its work. *Trans 15th National Game Conf* (Dec 3-4, 1928), 128-132. Reprinted with three additional charts but without the conference discussion in *Am Game* 18 (Apr-May 1929), 45-47.

1929

Fact-finding in wild-life conservation. *Proceedings, 7th National Convention, IWLA* (Apr 18-20, 1929), 262-266.
How the country boy or girl can grow quail. *Wisconsin Arbor and Bird Day Annual* (May 10, 1929), 51-53.
Glues for wood in archery. USFS Forest Products Laboratory Technical Note no. 226 (May 1929), 4pp.
Some thoughts on forest genetics. *J Forestry* 27:6 (Oct 1929), 708-713.
Progress of the game survey. *Trans 16th Am Game Conf* (Dec 2-3, 1929), 64-71.
Report of the Committee on American Wild Life Policy [AL, chairman]. *Trans 16th Am Game Conf* (Dec 2-3, 1929), 196-210. Reprinted as a pamphlet by Michigan Division, IWLA (n.d.), 20pp.
Environmental controls: The forester's contribution to game conservation. *The Ames Forester* 17 (1929), 25-26. Also appeared as "Environmental controls" in *California Fish and Game* [clipping, n.d.], 329-330, in which it is identified as a reprint from *DuPont Promotion News Bulletin* no. 35 (May 24, 1929).
How the country boy or girl can grow quail. *Wisconsin Arbor and Bird Day Annual* (May 10, 1929), 51-53.
Mesa de Los Angeles, p. 18; Ho! Compadres piñoneros!, p. 152. In *Forest Fire and Other Verse*, edited by John D. Guthrie. Portland, Oreg.: Dunham Printing Co., 1929.

1930

Wild game a farm crop. *The Game Breeder* 34:2 (Feb 1930), 39.

Environmental controls for game through modified silviculture. *J Forestry* 28:3 (Mar 1930), 321-326. [Presented at the annual meeting of the SAF, Des Moines, Iowa, Dec 31, 1929]

Game management in the national forests. *Am Forests* 36:7 (July 1930), 412-414.

The decline of the jacksnipe in southern Wisconsin. *Wilson Bulletin* 42:3 (Sept 1930), 183-190.

Game as a side-line for foresters. *Yale Forest School News* 18:4 (Oct 1930), 71.

Game Survey Bulletin (Madison). No. 1, Oct 1930-no. 11, Nov 1931. [A mimeographed serial issued monthly by AL to inform SAAMI members and cooperators of the progress of the game survey]

[Discussion on the American Game Policy]. *Trans 17th Am Game Conf* (Dec 1-2, 1930), 143, 146-147.

The American game policy in a nutshell. *Trans 17th Am Game Conf* (Dec 1-2, 1930), 281-283. Reprinted in pamphlet, *American Game Policy Adopted by the 17th Annual American Game Conference*, by Am Game Assn, pp. 3-4. Reprinted as "Game policy in a nutshell" in *Am Game* 19 (Nov-Dec 1930), 8.

Report to the American Game Conference on an American game policy. *Trans 17th Am Game Conf* (Dec 1-2, 1930), 284-309. [Submitted by Committee on Game Policy, AL, chairman.] Preprinted in pamphlet, *A Proposed American Game Policy to Be Discussed at the Seventeenth Annual American Game Conference*.

Game conditions in the north central states. *Proceedings, 8th Annual Convention, IWLA* (1930), 156-165.

1931

The forester's role in game management. *J Forestry* 29:1 (Jan 1931), 25-31. [Presented at 30th annual meeting of the SAF, Washington, D.C., Dec 29-31, 1930]

Game methods: The American way. *Am Game* 20:2 (Mar-Apr 1931), 20, 29-31.

Game restoration by cooperation on Wisconsin farms. *WAF* 59:16 (Apr 18, 1931), 5, 16.

(With John N. Ball). The quail shortage of 1930. *Outdoor Am* 9:9 (Apr 1931), 14-15, 67.

A history of ideas in game management. *Outdoor Am* 9:11 (June 1931), 22-24, 38-39, 47. [From Leopold's forthcoming book, "Principles of Game Management"]

The role of universities in game conservation. *DuPont Magazine* 25:6 (June 1931), 8-9, 24. Reprinted as "Universities in game conservation," *Outdoor Life* 68:3 (Sept 1931), 33-34; reprinted in *Louisiana Conservation Review* 2:11 (Oct 1932), 15-16.

(With John N. Ball). Grouse in England. *Am Game* 20:4 (July-Aug 1931), 57-58, 63.

Science attacks the game cycle. *Outdoor Am* 10:2 (Sept 1931), 25.

(With John N. Ball). British and American grouse management. *Am Game* 20:5 (Sept-Oct 1931), 70, 78-79.

(With John N. Ball). British and American grouse cycles. *The Canadian Field-Naturalist* 45:7 (Oct 1931), 162-167.
Game range. *J Forestry* 29:6 (Oct 1931), 932-938.
Rebuilding the quail crop. *Outdoor Am* 10:4 (Nov 1931), 38.
Vegetation and birds. *Report of the Iowa State Horticultural Society, 66th Annual Convention, Nov 12-14, 1931*, 66 (1931), 204-206.
Report of the American game policy committee. *Am Game* 20:6 (Nov-Dec 1931), 86. [Unsigned; prior mimeographed version by AL]
The prairie chicken: A lost hope, or an opportunity? *American Field* 116:50 (Dec 12, 1931), 1.
Game food and cover in the cornbelt. *Proceedings, 9th Annual Conference, IWLA* (1931).
Report on a Game Survey of the North Central States. Madison: Sporting Arms and Ammunition Manufacturers' Institute, 1931. 229pp., illus.

1932

Game system deplored as "melting pot." *J Forestry* 30:2 (Feb 1932), 226-227. [Abstract of a statement by AL released by Am Game Assn]
Game and wild life conservation. *Condor* 34:2 (Mar-Apr 1932), 103-106.
Statement of Aldo Leopold [on migratory waterfowl shortage]. U.S. Congress, Senate, Special Committee on Conservation of Wild Life Resources. *Migratory Waterfowl Shortage.* Hearing on the Protection and Preservation of Migratory Waterfowl in the United States (Apr 4-6, 1932), 606-607.
The alder fork: A fishing idyl. *Outdoor Am* 10:10 (May 1932), 11. Reprinted in *Almanac*.
A flight of Franklin's Gulls in northwestern Iowa. *Wilson Bull* 44:2 (June 1932), 116.
Report of the Iowa game survey, chapter one: The fall of the Iowa game range. *Outdoor Am* 11:1 (Aug-Sept 1932), 7-9.
Report of the Iowa game survey, chapter two: Iowa quail. *Outdoor Am* 11:2 (Oct-Nov 1932), 11-13, 30-31.
Results from the American game policy. *Trans 19th Am Game Conf* (Nov 28-30, 1932), 62-66.
[Comment on address by L. W. T. Waller, Jr., The need for educated man power]. *Trans 19th Am Game Conf* (Nov 28-30, 1932), 88-89.
Management of Upland Game Birds in Iowa: A Handbook for Farmers, Sportsmen, Conservationists and Game Wardens. Des Moines: Iowa State Fish and Game Commission, 1932, 35pp., illus. [Author's name absent on title page; indication that "greater part" prepared by AL, with contributions by William Schuenke, pp. 16-17, and Wallace B. Grange, pp. 27-28]

1933

Report of the Iowa game survey, chapter three: Iowa pheasants. *Outdoor Am* 11:3 (Dec-Jan 1933), 10-12, 31.
Weatherproofing conservation. *Am Forests* 39:1 (Jan 1933), 10-11, 48.
How research and game surveys help the sportsman and farmer. [Proceedings] *New*

England Game Conference (Feb 11, 1933). Cambridge: Samuel Marcus Press, for the Massachusetts Fish and Game Association, pp. 51-56.

Report of the Iowa game survey, chapter four: The Hungarian Partridge in Iowa. *Outdoor Am* 11:4 (Feb-Mar 1933), 6-8, 21. [The series was to have run to nine chapters, after which *Outdoor Am* was to have made a bound copy available, but the magazine suspended publication with the Apr-May 1933 issue. Unpublished chapters may be found in the University of Wisconsin Archives.]

Self-spreading game projects in Wisconsin. *Synopsis of 11th Annual Convention, IWLA* (Apr 27-28, 1933), 7-9. Mimeographed.

"Turkish bows" of the New Mexico Indians. *Ye Sylvan Archer* 7:1 (May 1933), 4-6.

The mockingbird in Wisconsin. *Wilson Bull* 45:3 (Sept 1933), 143.

Game as a land crop in the central states. In Central States Forestry Congress, *Proceedings of 4th Annual Conference* (Sept 21-23, 1933), 137-141.

ABC's of winter feeding birds. *Am Game* 22:5 (Sept-Oct 1933), 70, 77-79.

The conservation ethic. *J Forestry* 31:6 (Oct 1933), 634-643. [Fourth Annual John Wesley Powell Lecture, S.W. Div., AAAS, Las Cruces, N.M., May 1, 1933.] Reprinted as "Racial wisdom and conservation," *The Journal of Heredity* 37:9 (Sept 1946), 275-279; reprinted in newsletter of Pan American Section of the International Committee for Bird Preservation, 1000 Fifth Ave., New York City; reprinted in part as "La ética de la conservación," *Boletín del departamento de conservación de Suelos*, Vol. 1 (Ministerio de agricultura y cría, Venezuela, July 1948), 9-10; portions reprinted in *Almanac*.

Game cropping in southern Wisconsin. *Our Native Landscape* (Dec 1933), 2pp. [published by The Friends of Our Native Landscape]

Game Management. New York: Charles Scribner's Sons, 1933. 481pp., illus.

1934

Necessity of game research. *Trans 20th Am Game Conf* (Jan 22-24, 1934), 92-95.

Ecology of jackrabbits. [Review of C. T. Vorhies and W. P. Taylor, *The Life History and Ecology of Jackrabbits in Relation to Grazing in Arizona*. University of Arizona Technical Bulletin no. 49. Tucson, 1933.] *Ecology* 15:1 (Jan 1934), 63-64.

Game Research Newsletter (Chair of Game Management, University of Wisconsin, Madison). No. 1, Mar 10, 1934-No. 17, May 1, 1939. Title changed to *Wildlife Research Newsletter* (Department of Wildlife Management, University of Wisconsin, Madison). No. 18, Oct 25, 1939-No. 34, Nov 13, 1947. [A three- to seven-page mimeographed serial issued several times a year at irregular intervals, apparently written by AL, and intended "for the personal information of the field workers and financial cooperators of the Chair of Game Management, and of faculty members and state and federal officials concerned with wild life research in Wisconsin."]

[Review of N. A. Orde-Powlett. Forestry and sport. *Scottish Forestry Journal* 47, pt. 2, Oct 1933, 93-107.] *J Forestry* 32:4 (Apr 1934), 497.

Conservation economics. *J Forestry* 32:5 (May 1934), 537-544. [Delivered at the Taylor-Hibbard Economics Club, University of Wisconsin, March 1, 1934.] Reprinted in *Am Game* 23:4 (July-Aug 1934), 56, 63; and 23:5 (Sept-Oct 1934), 71, 77-78.

[Review of M. T. Townsend and N. W. Smith. *White-Tailed Deer of the Adirondacks*. Bulletin of the Roosevelt Wild Life Experiment Station 6:2 (Oct 1933).] *J Mammalogy* 15:2 (May 1934), 163-164.

(With Reuben Paulson). Helping ourselves: Being the adventures of a farmer and a sportsman who produced their own shooting ground. *Field and Stream* 39:4 (Aug 1934), 32-33, 56.

The Wisconsin River marshes. *National Waltonian* 2:3 (Sept 1934), 4-5, 11.

The arboretum and the university. *Parks and Recreation* 18:2 (Oct 1934), 59-60. [Address at dedication of University of Wisconsin Arboretum, June 17, 1934]

[Review of Ward Shepard, *Notes on German Game Management, Chiefly in Bavaria and Baden*. Senate Committee on Wild Life Resources, 1934.] *J Forestry* 32:7 (Oct 1934), 774-775.

The game cycle—a challenge to science. *Outdoor Nebraska* 9:4 (Autumn 1934), 4, 14. Also in *Minnesota Conservationist* no. 19 (Dec 1934), 2-3, 14.

An outline plan for game management in Wisconsin, pp. 243-255. In *A Study of Wisconsin: Its Resources, Its Physical, Social and Economic Background*. First Annual Report, Wisconsin Regional Planning Committee, Dec 1934.

Some thoughts on recreational planning. *Parks and Recreation* 18:4 (Dec 1934), 136-137. [From Conference on State and Regional Planning sponsored by Wisconsin Friends of Our Native Landscape]

(With T. H. Beck, chairman, and J. N. Darling). *Report of the President's Committee on Wildlife Restoration*. Washington, D.C., 1934. 27pp.

1935

Whither 1935?—a review of the American Game Policy. *Trans 21st Am Game Conf* (Jan 21-23, 1935), 49-55.

Foreword to "Wildlife cycles in relation to the sun," by Leonard William Wing. *Trans 21st Am Game Conf* (Jan 21-23, 1935), 345.

Wildlife research rapidly growing. *Am Game* 24:1 (Jan-Feb 1935), 5, 13.

"Game conference not accidental," says Aldo Leopold. *Am Game* 24:1 (Jan-Feb 1935), 9. [Quote from AL's acceptance of chairmanship of 22nd Am Game Conf]

Gun and glass hunters. *Am Forests* 41:2 (Feb 1935), 71. [Unsigned editorial]

[Review of Joseph S. Dixon, "A study of the life history and food habits of mule deer in California," *California Fish and Game* 20:3 (July 1934), 181-282, and 20:4 (Oct 1934), 315-354.] *J Mammalogy* 16:1 (Feb 1935), 74-75.

Preliminary report on forestry and game management. *J Forestry* 33:3 (Mar 1935), 273-275. [Prepared by AL but read by W. L. Dutton, USFS; accepted. Proceedings of the 34th Annual Meeting, SAF, Jan 28-30, 1935.]

Coon Valley: An adventure in cooperative conservation. *Am Forests* 41:5 (May 1935), 205-208. Adapted by H. S. Person in *Little Waters: A Study of Headwater Streams & Other Little Waters, Their Use and Relations to the Land* (for Soil Conservation Service, Resettlement Administration, and Rural Electrification Administration, Nov 1935-revised Apr 1936), 67-69.

[Review of S. Charles Kendeigh, "The role of environment in the life of birds," *Ecological Monographs* 4 (July 1934), 299-417.] *Wilson Bulletin* 47:2 (June 1935), 166-167.

Notes on wild life conservation in Germany. *Game Research News Letter* 6 (Sept 16, 1935), 1-3; and 7 (Oct 21, 1935), 1-3.
Why the Wilderness Society? *Living Wilderness* no. 1 (Sept 1935), 6.
[Review of Charles Elton, *Exploring the Animal World*. London, 1933.] *Bird-Lore* 37:5 (Sept-Oct 1935), 364.
Sporting poetry. [Review of Edward S. Parker, *One More Bend*. Dallas, 1935.] *Am Wildlife* 24:6 (Nov-Dec 1935), 90.
[Review of D. Nolte, *Zur Biologie des Rephuhns*. Published under the auspices of the Reichbundes Deutsche Jagerschaft, Berlin, 1934, 105pp.] *Wilson Bulletin* 47:4 (Dec 1935), 300-303.
Forerunners of game management. *The Colorado Forester* (Colorado State College, 1935), 12.
(With W. H. Twenhofel, Noble Clark and G. S. Wehrwein). *The University and the Erosion Problem*. Bulletin of the University of Wisconsin, ser. no. 2097, general ser. no. 1881, Science Inquiry, n.d. 46pp. [According to Noble Clark, the bulletin was largely written by AL, except for the first section on "Nature and extent of erosion" by W. H. Twenhofel. It was published in 1935, the first in the Science Inquiry series, according to the foreword in publication no. 2.]
Wild life research in Wisconsin. *Trans Wis Acad* 29 (1935), 203-208.

1936

Remarks [on wildlife management by private agencies.] *Proc NAWC*, February 3-7, 1936 (Senate Committee Print, 74th Cong., 2d sess., 1936), 156-158. Reprinted as "Wildlife management on private and state lands" in *Am Forests* 42:3 (Mar 1936), 120-121.
Farmer-sportsman set-ups in the north central region. *Proc NAWC*, February 3-7, 1936 (Senate Committee Print, 74th Cong., 2d sess., 1936), 279-285.
Study influence of the sun on wildlife cycles. Game Management section in *Today's Science for Tomorrow's Farming*, Agr Exp Sta Bull no. 435 (March 1936), 29-30.
Threatened species: A proposal to the Wildlife Conference for an inventory of the needs of near-extinct birds and animals. *Am Forests* 42:3 (Mar 1936), 116-119.
Naturschutz in Germany. *Bird-Lore* 38:2 (Mar-Apr 1936), 102-111.
Wildlife conference. *J Forestry* 34:4 (Apr 1936), 430-431.
Deer and *Dauerwald* in Germany: I. History. *J Forestry* 34:4 (Apr 1936), 366-375.
Deer and *Dauerwald* in Germany: II. Ecology and policy. *J Forestry* 34:5 (May 1936), 460-466.
[Review of Horace Mitchell, *Raising Game Birds*. Philadelphia, 1936.] *Am Wildlife* 25:3 (May-June 1936), 40.
[Review of Walter P. Taylor, *Ecology and Life History of the Porcupine (Erethizon epixanthum) as Related to the Forests of Arizona and the Southwestern United States*. University of Arizona Bulletin VI-5, 1935.] *J Forestry* 34:6 (June 1936), 632-633.
[Review of Robert H. Connery, *Governmental Problems in Wild Life Conservation*. Studies in History, Economics, and Public Law no. 411. New York: Columbia University Press, 1935.] *J Forestry* 34:6 (June 1936), 635-636.
Franklin J. W. Schmidt. *Wilson Bull* 48:3 (Sept 1936), 181-186.
Farm game management in Silesia. *Am Wildlife* 25:5 (Sept-Oct 1936), 67-68, 74-76.

The conservation quandary [review of Thomas Barbour and Margaret Dewar Porter, *Notes on South African Wild Life Conservation Parks and Reserves: A Report Prepared for the American Committee for International Wild Life Protection*. Special Publ. Amer. Committee for International Wild Life Protection no. 7, 1935. 34pp.]. *Geographical Review* 26:4 (Oct 1936), 694-695.

Quail population studies in Iowa and Wisconsin [review of Paul L. Errington and F. N. Hamerstrom, Jr., *The Northern Bob-White's Winter Territory*. Iowa State College of Agriculture, Research Bull. 201, 1936]. *Ecology* 17:4 (Oct 1936), 680-681.

Notes on game administration in Germany. *Am Wildlife* 25:6 (Nov-Dec 1936), 85, 92-93.

[Review of *Upland Game Restoration*. Western Cartridge Co. and Winchester Repeating Arms Co., 1936.] *Outdoor Am* 2:2 (Dec 1936), 11.

(With Gardiner Bump, George C. Embody, Carl L. Hubbs, and Herbert L. Stoddard). *Wildlife Crops: Finding Out How to Grow Them*. Washington, D.C.: American Wildlife Institute, 1936, 23pp. [A research and demonstration program prepared by the Technical Committee of the American Wildlife Institute, AL, chairman]

1937

How to build a game crop? The university sets out to find the answer. *Wis Sportsman* 1:5 (Dec-Jan 1937), 2-3.

[Review of A. Freiherr von Vietinghoff-Reisch, *Naturschutz—eine national-politische Kulturaufgabe*, 1936.] *J Forestry* 35:1, (Jan 1937), 87-88.

The thick-billed parrot in Chihuahua. *Condor* 39:1 (Jan-Feb 1937), 9-10. Reprinted in *Almanac* as "Guacamaja."

[Review of A. E. Parkins and J. R. Whitaker, *Our Natural Resources and Their Conservation*. New York, 1936.] *Bird-Lore* 39:1 (Jan-Feb 1937), 74-75.

Killing technique of the weasel. *J Mammalogy* 18:1 (Feb 1937), 98-99.

Second report of Game Policy Committee. *J Forestry* 35:2 (Feb 1937), 228-232. [Unsigned—AL, chairman; Proceedings of the 36th Annual Meeting, SAF, Dec 14-16, 1936.] Abstracted by H. H. Chapman as "Approval of report of committee on forests and wildlife and of the principles stated therein." *S.A.F. Affairs* 3:4 (Apr 1937), 68-69.

(With L. J. Cole, N. C. Fassett, C. A. Herrick, Chancey Juday, and George Wagner). *The University and Conservation of Wisconsin Wildlife*. Bulletin of the University of Wisconsin, ser. no. 2211, general ser. no. 1995, Science Inquiry Publication no. 3, Feb 1937. 39 pp. [Report of the Committee on Wildlife Conservation, AL, chairman]

The research program. *Trans NAWC* (Mar 1-4, 1937), 104-107. Reprinted in *Am Wildlife* 26:2 (Mar-Apr 1937), 22, 28. Partially quoted in "Complains of Neglect of Wildlife Research," *St. Louis Post Dispatch* (Tuesday, Mar 2, 1937), 9A.

Conservationist in Mexico. *Am Forests* 43:3 (Mar 1937), 118-120, 146.

Farm game population increased in trials at Riley; Study response of prairie chickens and sharptail grouse to fall and winter feeding. In Poultry and Game Birds section of *Findings in Farm Science*, Agr Exp Sta Bull no. 438 (Mar 1937), 61-64.

White-winged scoter in Missouri. *Wilson Bull* 49:1 (Mar 1937), 49-50.

Conservation in the world of tomorrow. *The Milwaukee Journal* (Sunday, Apr 4, 1937), sec. 3, p. 5. [Guest columnist for Gordon MacQuarrie, "Right off the reel"]

Conservation of wildlife, pp. 52-54; Bibliography, pp. 56-57; Appendix A: Biography of a covey, pp. 58-61; Appendix E: Chronology of Wisconsin wildlife conservation, pp. 69-71. In Wisconsin Department of Public Instruction, *Teaching of Conservation in Wisconsin Schools*, Curriculum Bulletin 1:1 (May 1937). [Unsigned]

The effect of the winter of 1935-36 on Wisconsin quail. *American Midland Naturalist* 18:3 (May 1937), 408-416.

1936 pheasant nesting study. *Wilson Bull* 49:2 (June 1937), 91-95.

The wildlife program of the university. *Wis Sportsman* 1:10 (June 1937), 8.

[Review of A. Vietinghoff-Reisch, "Forstlicher Naturschutz und Naturschutz im nationalen Lebensraume Deutschlands," *Zeitschr. f. Weltforstwirtschaft* 3 (1936), 868-885.] *J Forestry* 35:8 (Aug 1937), 794-795.

Marked birds. *Wis Sportsman* 2:2 (Sept 1937), 4.

Right off the reel [guest column re sportsman conduct]. *The Milwaukee Journal* (c. Sept 1937).

Marshland elegy. *Am Forests* 43:10 (Oct 1937), 472-474. Reprinted in *Almanac*.

[Review of Margaret Morse Nice, *Studies in the Life History of the Song Sparrow*. Vol. 1, *A Population Study of the Song Sparrow*. Transactions of the Linnaean Society of New York no. 4 (Apr 1937).] *The Canadian Field-Naturalist* 51:8 (Nov 1937), 126.

[Review of Rudolf Bennitt and Werner O. Nagel, *A Survey of the Resident Game and Furbearers of Missouri*. The University of Missouri Studies 12:2 (Apr 1, 1937), 215pp.] *J Mammalogy*, 18:4 (Nov 1937), 520-521.

Conservation blueprints. *Am Forests* 43:12 (Dec 1937), 596, 608.

Compare value of grains for winter game feeding; Game cover may be provided cheaply; Marked birds tell the story of their movements. Game Management section of *What's New in Farm Science*, pt. 1, Agr Exp Sta Bull no. 439 (Dec 1937), 49-52. [Unsigned]

The Chase Journal: An early record of Wisconsin wildlife. *Trans Wis Acad* 30 (1937), 69-76.

Teaching wildlife conservation in public schools. *Trans Wis Acad* 30 (1937), 77-86.

1938

Chukaremia. *Outdoor Am* 3:3 (Jan 1938), 3. [Editorial]

(With Orville S. Lee and Harry G. Anderson). Wisconsin pheasant movement study, 1936-37. *JWM* 2:1 (Jan 1938), 3-12.

Wildlife research—is it a practical and necessary basis for management? *Trans 3rd NAWC* (Feb 14-17, 1938), 42-45, 55. [Part of a forum discussion with W. P. Taylor, Rudolf Bennitt, and H. H. Chapman]

(With Harry G. Anderson). The 1936 cotton-tail scarcity in Wisconsin. *J Mammalogy* 19:1 (Feb 1938), 110-111.

Haymowers, fires and WPA men called more perilous than crows. *The Wisconsin State Journal* (Madison, Mar 10, 1938) [letter to the editor]. Excerpted in *Wis Sportsman* (Mar 1938), 4.

Conservation esthetic. *Bird-Lore* 40:2 (Mar–Apr 1938), 101–109. Condensed in *Conservation* 4:3 (May–June 1938), 18–21. Reprinted with minor revision in *Almanac*.

Letter to a wildflower digger. *The Wisconsin State Journal* (June 7, 1938).

(With F. B. Trenk, S. A. Wilde, A. J. Riker, Noble Clark and G. S. Wehrwein). *The University and Wisconsin Forestry*. Bulletin of the University of Wisconsin, ser. no. 2334, general ser. no. 2118, Science Inquiry Publication no. 7, June 1938. 56pp. [Report of the Committee on Forestry. The manuscript was prepared in final form by Noble Clark and Niemen Hoveland.]

Whither Missouri? *The Missouri Conservationist* 1:1 (July 1938), 6. [Address at dedication of the Ashland Wildlife Area, Boone County, Mo., Apr 26, 1938.]

Wildlife conservation on the farm. *WAF* 65:23 (Nov 5, 1938), 5. Reprinted as "Wildlife conservation" in *Wis Sportsman* 3:6 (Feb 1939), 2; reprinted as "Winter cover" in WAF Booklet.

Wildlife conservation on the farm. *WAF* 65:24 (Nov 19, 1938), 18 [unsigned]. Reprinted as "Winter food" in WAF Booklet.

Ups and downs of quail furnish clues to best management; Why do game birds nest in hayfields? Mature, well-developed pheasants survive best. In Game Management section of *What's New in Farm Science*, pt. 1, Agr Exp Sta Bull no. 442 (Nov 1938), 48–51. [Unsigned]

Feed the song birds. *WAF* 65:25 (Dec 3, 1938), 5. Reprinted in WAF Booklet.

Woodlot wildlife aids. *WAF* 65:27 (Dec 31, 1938), 4. Reprinted abridged as "Woodlot wildlife" in WAF Booklet.

Report on Huron Mountain Club. Printed by Huron Mountain Club, Michigan, 1938. 18pp. Reprinted in *Report of Huron Mountain Wildlife Foundation 1955–1966*, Jan 1967, pp. 40–57.

1939

(With Ellwood B. Moore and Lyle K. Sowls). Wildlife food patches in southern Wisconsin. *JWM* 3:1 (Jan 1939), 60–69.

The farm pond attracts game. *WAF* 66:3 (Feb 11, 1939), 7. Reprinted as "The farm pond" in WAF Booklet.

Farmer-sportsman, a partnership for wildlife restoration. *Trans 4th NAWC* (Feb 13–15, 1939), 145–149, 167–168 [part of forum discussion]. Reprinted, slightly cut, as "Game policy—model 1930," *Bird-Lore* 41:2 (Mar–Apr 1939), 94–97.

The Farmer as a Conservationist. Stencil Circular 210, Extension Service, College of Agriculture, University of Wisconsin, Madison (Feb 1939), 1–8. [Presented at Wisconsin Farm and Home Week, University of Wisconsin, Feb 1939.]

Burned marsh means a loss. *WAF* 66:8 (Apr 22, 1939), 22. Abridged as "The marsh" in WAF Booklet.

(With Rudolf Bennitt, chairman, et al.) Report of the Committee on Professional Standards [of the Wildlife Society]. *JWM* 3:2 (Apr 1939), 153–155.

Academic and professional training in wildlife work. *JWM* 3:2 (Apr 1939), 156–161. [Prepared by AL with the assistance of and endorsed by the other members of the Committee on Professional Standards of the Wildlife Society; adopted by the society, February 13, 1939.]

Dane County management areas. *The Passenger Pigeon* 1:4 (Apr 1939), 49–50. Reprinted in [Wisconsin] *Wildlife* 1:6 (Memorial issue, c. May 1939), 7.
Game areas of Wisconsin [map]. *The Passenger Pigeon* 1:4 (Apr 1939), 55. Reprinted in *The Passenger Pigeon* 2:9 (Sept 1940), 105.
Wildlife conservation. *Wis Sportsman* 3:8 (Apr 1939), 2. [Re crows]
Plant evergreens for bird shelter. *WAF* 66:9 (May 6, 1939), 5. Reprinted as "Wildlife conservation" in *Wis Sportsman* 3:11 (July 1939), 2; and as "Evergreens for cover" in WAF Booklet.
(With John T. Curtis). Wild flower corners. *WAF* 66:12 (June 17, 1939), 16. Abridged in WAF Booklet.
The farmer as a conservationist. *Am Forests* 45:6 (June 1939), 294–299, 316, 323. [Revised version of Stencil Circular 210, Feb 1939.]
A biotic view of land. *J Forestry* 37:9 (Sept 1939), 727–730 [part of a "Symposium on Land Use," at joint meeting of the Society of American Foresters and the Ecological Society of America, Milwaukee, June 21, 1939]. Reprinted condensed in *The Council Ring* (National Park Service monthly mimeographed publication), 1:12 (Nov 1939), 4pp; reprinted as a pamphlet by Forest and Bird Protection Society of New Zealand [1940], 12pp.
Wild feeds on farms. *WAF* 66:23 (Nov 18, 1939), 19. Abridged as "Wild foods" in WAF Booklet.
The Hungarian partridge pioneers; Can prairie chickens winter on buds? What is the yield of wild food crops? Rabbits range at least a mile. In Wildlife Management section of *What's New in Farm Science*, pt. 1, Agr Exp Sta Bull no. 446 (Nov 1939), 21–23. [Unsigned]
Look for bird bands. *WAF* 66:24 (Dec 2, 1939), 19. Abridged in WAF Booklet.

1940

New Year's inventory checks missing game. *WAF* 67:3 (Jan 27, 1940), 10. Abridged as "Stories in the snow" in WAF Booklet.
[Obituary of Royal N. Chapman]. *JWM* 4:1 (Jan 1940), 104.
[Review of A. F. Gustafson, H. Ries, C. H. Guise, and W. J. Hamilton, Jr. *Conservation in the United States* (Ithaca, 1939). 445pp.] *Ecology* 21:1 (Jan 1940), 92–93.
Windbreaks aid wildlife. *WAF* 67:5 (Mar 9, 1940), 15. Reprinted as "Windbreaks" in WAF Booklet.
When geese return spring is here. *WAF* 67:7 (Apr 6, 1940), 18 [unsigned]. Reprinted as "When the geese return" in WAF Booklet.
(With Victor H. Cahalane, chairman, William L. Finley, and Clarence Cottam). Report of the [American Ornithologists' Union] Committee on Bird Protection, 1939. *The Auk* 57:2 (Apr 1940), 279–291.
[Letter to Dr. Schmitz regarding "fox squirrel dens"]. *J Forestry* 38:4 (Apr 1940), 375.
Farm arboretum adds to home beauty. *WAF* 67:10 (May 18, 1940), 4.
History of the Riley Game Cooperative, 1931–1939. *JWM* 4:3 (July 1940), 291–302. Excerpted by Russ Pyre in "Hook, line and sinker," *The Wisconsin State Journal* (Sunday, Nov 3, 1940).
Origin and ideals of wilderness areas. *The Living Wilderness* 5 (July 1940), 7.

Quail population shrinks during winter. *WCB* 5:7 (July 1940), 39-40. [Attributed to W. E. Scott, but written by AL]
(With Rudolf Bennitt, chairman, et al.) Report of the Committee on Professional Standards [of the Wildlife Society]. *JWM* 4:3 (July 1940), 338-341.
Song of the Gavilan. *JWM* 4:3 (July 1940), 329-332. Reprinted in *Almanac*.
The state of the profession. *JWM* 4:3 (July 1940), 343-346. [Address of the president of the Wildlife Society, Washington, D.C., March 18, 1940.]
Pheasant damage checked. *WAF* 67:17 (Aug 24, 1940), 14.
Exit orchis: A little action now would save our fast disappearing wildlife. [Wisconsin] *Wildlife* 2:2 (Aug 1940), 17. Reprinted in *Am Wildlife* 29:5 (Sept-Oct 1940), 207.
[Review of Milton B. Trautman, *The Birds of Buckeye Lake, Ohio*. University of Michigan Museum of Zoology Miscellaneous Publications no. 44 (May 7, 1940), 466pp.] *Wilson Bull* 52:3 (Sept 1940), 217-218.
[Review of *The Status of Wildlife in the United States*. Report of the Special (Senate) Committee on the Conservation of Wildlife Resources. S. Rept. 1203, 76th Cong., 3rd sess., 1940. 457pp.] *J Forestry* 38:10 (Oct 1940), 823.
A house divided. *Wis Sportsman* (Oct 1940).
Birds earn their keep on Wisconsin farms. *WAF* 67:24 (Nov 30, 1940), 18.
Half a duck apiece. *Am Forests* 46:11 (Nov 1940), 509. [Editorial]
Wisconsin wildlife chronology. *WCB* 5:11 (Nov 1940), 8-20. Reprinted as WCD Pub no. 301, 15pp. Summarized in *The Milwaukee Journal* (Sunday, Jan 19, 1941), 6.
Cover plantings need winter protection. *WAF* 67:26 (Dec 28, 1940), 11.
Escudilla. *Am Forests* 46:12 (Dec 1940), 539-540. Reprinted in *Maryland Conservationist* 18:1 (Winter 1941), 20-21; reprinted in *Almanac*.
(With Robert A. McCabe). Snow-killed bobwhite covey. *Wilson Bull* 52:4 (Dec 1940), 280.
Spread of the Hungarian partridge in Wisconsin. *Trans Wis Acad* 32 (1940), 5-28.

1941

Pheasant planting requires skill. *WAF* 68:2 (Jan 25, 1941), 19. Reprinted in WAF Booklet.
[Review of E. G. Cheyney and T. Shantz-Hansen, *This is Our Land: The Story of Conservation in the United States*. St. Paul, 1940. 337pp.] *J Forestry* 19:1 (Jan 1941), 72.
Houses for birds make friends. *WAF* 68:5 (Mar 8, 1941), 28.
(With F. N. Hamerstrom, Jr.). John S. Main. *Wilson Bull* 53:1 (Mar 1941), 31-32.
(With Robert McCabe). Other records of snow-killed bob-white coveys. *Wilson Bull* 53:1 (Mar 1941), 44.
Bur oak is badge of Wisconsin. *WAF* 68:7 (Apr 5, 1941), 10. Abridged in WAF Booklet; reprinted revised in *Almanac* as "Bur oak."
Bluebirds welcome. *WAF* 68:8 (Apr 19, 1941), 16. Reprinted in WAF Booklet.
(With Victor H. Cahalane, chairman, et al.) Report of the [American Ornithologists' Union] Committee on Bird Protection, 1940. *The Auk* 58:2 (Apr 1941), 292-298.
Pest-hunts. *The Passenger Pigeon* 3:5 (May 1941), 42-43.
Wild life likes water. *WAF* 68:13 (June 28, 1941), 10. Reprinted in WAF Booklet.

Bob white members can be increased. *WAF* 68:15 (July 26, 1941), 19.
Wilderness as a land laboratory. *The Living Wilderness* 6 (July 1941), 3. Reprinted in *Outdoor Am* 7:2 (Dec 1941), 7; condensed in *Forest and Bird* [New Zealand] 65 (Aug 1942), 2.
Fifth column of the fence row. *WAF* 68:17 (Aug 23, 1941), 11.
Wildlife Conservation on the Farm. Wisconsin Agriculturist and Farmer, Racine, Wis. [c. Sept 1941]. 24pp. [Booklet of articles reprinted from *WAF* series, several shortened arbitrarily to fit space requirements; see individual entries.]
Sky dance of spring. *Wildlife Conservation on the Farm* [c. Sept 1941; apparently not published in *WAF*]. Rewritten as "Sky dance" for *Almanac*.
Faville prairie preserve. *Wild Flower* 18:4 (Oct 1941), 67-68. [Revision of "Exit orchis," *Wildlife* 2:2 (Aug 1940), 17.]
Feed the birds early. *WAF* 68:24 (Nov 29, 1941), 10. [Unsigned]
Fur crop in danger. *WAF* 68:25 (Dec 13, 1941), 19.
Farmers and rabbits. *WAF* 68:26 (Dec 27, 1941), 19.
Riley game cooperative proves a success; History of Faville Grove shows wildlife changes. In Game Management section of *What's New in Farm Science*, pt. 1, Agr Exp Sta Bull no. 453 (Dec 1941), 58-60. [Unsigned]
Cheat takes over. *The Land* 1:4 (Autumn 1941), 310-313. Reprinted condensed in *Conservation* 8:3 (May-June 1942), 27-30; reprinted revised in *Almanac*.
Lakes in relation to terrestrial life patterns, pp. 17-22. In *A Symposium on Hydrobiology*, by James G. Needham et al. Madison: University of Wisconsin Press, 1941. [Presented at institute at the University of Wisconsin, Sept 4-6, 1940.]
Wilderness values. *1941 Yearbook, Park and Recreation Progress* (National Park Service, 1941), 27-29. Reprinted in *The Living Wilderness* 7 (Mar 1942), 24-25.

1942

Farming in color. *WAF* 69:2 (Jan 24, 1942), 4.
A raptor tally in the Northwest. *Condor* 44:1 (Jan-Feb 1942), 37-38.
Wild ducks need more "pond-room." *WAF* 69:7 (Apr 4, 1942), 25.
The role of wildlife in a liberal education. *Trans 7th NAWC* (Apr 8-10, 1942), 485-489. Reprinted in *Michigan Conservation* 12:1 (Jan 1943), 8.
The grizzly—a problem in land planning. *Outdoor Am* 7:6 (Apr 1942), 11-12. [Presented at 20th Annual IWLA convention, Chicago, Mar 27, 1942.]
(With Victor H. Cahalane, chairman, et al.) Report of the [American Ornithologists' Union] Committee on Bird Protection, 1941. *The Auk* 59:2 (Apr 1942), 286-299.
(With Wm. L. Finley). Substitute statement on pole-trapping of raptors. *The Auk* 59:2 (Apr 1942), 300. [Minority opinion appended to "Report of the Committee on Bird Protection, 1941."]
The plover is back from Argentine. *WAF* 69:10 (May 16, 1942), 10. Reprinted as "Back from the Argentine" in *Almanac*.
The last stand. *Outdoor Am* 7:7 (May-June 1942), 8-9. Reprinted in *The Living Wilderness* 8 (Dec 1942), 25-26; reprinted in *WCB* 9:2 (Feb 1944), 3-5.
Odyssey. *Audubon Mag* 44:3 (May-June 1942), 133-135. Reprinted in *Almanac*.
"Packratting." *Wildlife News* 2:1 (June 20, 1942), 11. [Letter to editor]
Half-excellent [review of George T. Renner, *Conservation of Natural Resources*. New York: John Wiley & Sons, 1942]. *The Land* 2:2 (July 1942), 111-112.

Wildlife conservation. *The Wisconsin State Journal* (Madison, Sunday, Aug 23, 1942), sec. 1, p. 6. [Guest editorial]
"Control" of the golden eagle in Texas. *Wilson Bull* 54:3 (Sept 1942), 218.
Land-use and democracy. *Audubon Mag* 44:5 (Sept–Oct 1942), 259–265.
Introduction to *The Ducks, Geese and Swans of North America: A Vade Mecum for the Naturalist and the Sportsman*, by Francis H. Kortright. Washington, D.C.: American Wildlife Institute, 1942, v.

1943

Flambeau: The story of a wild river. *Am Forests* 49:1 (Jan 1943), 12–14, 47. Reprinted as "The Flambeau" in *WCB* 8:3 (Mar 1943), 13–17; reprinted revised as "Flambeau" in *Almanac*.
Wildlife in American culture. *JWM* 7:1 (Jan 1943), 1–6. Previously mimeographed in *Proceedings, 7th Annual Midwest Wildlife Conference, Des Moines* (Dec 4–6, 1941), 19–25; reprinted in *Pacific Discovery* 2 (July–Aug 1949), 12–15; reprinted in *Almanac*.
Obituary: P.S. Lovejoy. *JWM* 7:1 (Jan 1943), 125–128. Reprinted as "Lovejoyiana" in *Michigan Conservation* 12:4 (May 1943), 10–11.
A lesson from the woodlands. *WCB* 8:2 (Feb 1943), 25–27. Reprinted as "A mighty fortress" in *Almanac*.
Pines above the snow. *WCB* 8:3 (Mar 1943), 27–29. Reprinted in *Almanac*.
[Review of Sherman Strong Hayden, *The International Protection of Wild Life*. Columbia University Studies in History, Economics, and Public Law no. 491. New York, 1942.] *The Geographical Review* 33:2 (Apr 1943), 340–341.
[Review of Norman F. Smith, A study of the spread of forest cover into wild-land openings. *Michigan Academy of Sciences, Arts and Letters, Papers* no. 28 (1942), 269–277]. *J Forestry* 41:5 (May 1943), 381–382.
The excess deer problem. *Audubon Mag* 45:3 (May–June 1943), 156–157.
[Review of Ira N. Gabrielson, *Wildlife Refuges*. New York, 1943.] *J Forestry* 41:7 (July 1943), 529–531.
Deer irruptions. *WCB* 8:8 (Aug 1943), 1–11. [Published with permission of the Wisconsin Academy of Sciences, Arts and Letters]
(With J. R. Jacobson, Henry C. Kuehn, Miss Joyce Larkin, John O. Morland, Dr. E. G. Ovitz, Howard Quirt, and Mrs. Harry E. Thomas). Majority report of the Citizens' Deer Committee to Wisconsin Conservation Commission. *WCB* 8:8 (Aug 1943), 19–22 [AL, chairman]. Reprinted in *Wisconsin's Deer Problem*, WCD Pub no. 321, 20–23.
Home range. *WCB* 8:9 (Sept 1943), 23–24. Reprinted in *Almanac*.
(With Theodore M. Sperry, William S. Feeney, and John A. Catenhusen). Population turnover on a Wisconsin pheasant refuge. *JWM* 7:4 (Oct 1943), 383–394. Reprint issued with Arboretum cover as Journal Paper no. 4 of the University of Wisconsin Arboretum.
[Review of Helen M. Martin, ed., *They Need Not Vanish*. Michigan Department of Conservation, Lansing, 1942.] *J Forestry* 41:12 (Dec 1943), 924.
[Review of William R. Van Dersal, *The American Land*. New York, 1943.] *J Forestry* 41:12 (Dec 1943), 928.
Facts on pheasants appear from Arboretum study. In Poultry and Game section of

What's New in Farm Science, pt. 1, Agr Exp Sta Bull no. 461 (Dec 1943), 13–15 [unsigned]. Reprinted in *WCB* 9:5 (May 1944), 11–13.

Deer irruptions. *Trans Wis Acad* 35 (1943), 351–366. [Compiled by AL for the Natural Resources Committee, Wisconsin Academy of Sciences, Arts and Letters: AL, Ernest F. Bean, Norman C. Fassett.] Previously published with permission in *WCB* 8:8 (Aug 1943), 1–11; reprinted in *Wisconsin's Deer Problem*, WCD Pub no. 321, 3–11.

1944

[Review of H. M. Bell and E. J. Dyksterhuis, Fighting the mesquite and cedar invasion on Texas ranges. *Soil Conservation* 9:5 (Nov 1943), 111–114.] *J Forestry* 42:1 (Jan 1944), 63.

Post-war prospects. *Audubon Mag* 46:1 (Jan–Feb 1944), 27–29.

The present winter and our native game birds. *WCB* 9:2 (Feb 1944), 25–26.

[Contribution to a symposium prepared by W. E. Scott, "Does inbreeding cause the cycle on game animals?"] *WCB* 9:2 (Feb 1944), 9.

[Review of Montague Stevens, *Meet Mr. Grizzly*. Albuquerque, 1943.] *J Forestry* 42:3 (Mar 1944), 222.

(With Irven Buss). Cliff swallows to order. *WCB* 9:4 (Apr 1944), 21–22.

(With W. F. Grimmer). The crow. *WCB* 9:5 (May 1944), 10. [Excerpt from a report on crow damage]

What next in deer policy? *WCB* 9:6 (June 1944), 3–4, 18–19.

[Review of Edward H. Graham, *Natural Principles of Land Use*. New York, 1944.] *Soil Conservation* 10:2 (Aug 1944), 38–39.

(With sixteen others). Six points of deer policy. *WCB* 9:11 (Nov 1944), 10. [Reprinted from mimeographed statement prepared by Walter P. Taylor, Aldo Leopold, and Thomas A. Schroeder, "Experience with deer irruptions," May 1, 1944, submitted following 9th North American Wildlife Conference for additional signatures]

(With Miles D. Pirnie and William Rowan). Introduction to *The Canvasback on a Prairie Marsh*, by H. Albert Hochbaum. Washington, D.C.: American Wildlife Institute, 1944, xi–xii.

1945

[Review of Stanley P. Young and Edward H. Goldman, *The Wolves of North America*. Washington, D.C.: American Wildlife Institute, 1944.] *J Forestry* 43:1 (Jan 1945), 928–929.

The outlook for farm wildlife. *Trans 10th NAWC* (Feb 26–28, 1945), 165–168. [Conference canceled due to government ban on all conventions]

(With Hans Peter Thomsen). War status of predators in Norway. *J Mammalogy* 26:1 (Feb 1945), 88–89. ["Translated from Ukens nytt fra Norge, Vol. 1, no. 298, October 21, 1943, p. 8"]

Deer, wolves, foxes and pheasants. *WCB* 10:4 (Apr 1945), 3–5.

[Review of Sally Carrighar, *One Day on Beetle Rock*. New York, 1944.] *J Forestry* 43:4 (Apr 1945), 301–302.

[Review of Durward L. Allen, *Michigan Fox Squirrel Management*. Michigan

Department of Conservation, Game Div., Pub. no. 100, 1943.] *J Forestry* 43:6 (June 1945), 462.
Wildlife explorations at Prairie du Sac. *WCB* 10:7-8 (July-Aug 1945), 3-5.
The green lagoons. *Am Forests* 51:8 (Aug 1945), 376-377, 414. Reprinted in *Almanac*.

1946

[Review of A. G. Tansley, *Our Heritage of Wild Nature*. Cambridge University Press, 1945.] *J Forestry* 44:3 (Mar 1946), 215-216.
(With Robert A. McCabe). [Review of W. L. McAtee, ed., *The Ring-Necked Pheasant and Its Management in North America*. American Wildlife Institute, 1945.] *Wilson Bull* 58:2 (June 1946), 126-127.
[Review of S. Kip Farrington, Jr., *The Ducks Came Back: The Story of Ducks Unlimited*. New York: Coward-McCann, Inc., 1945.] *JWM* 10:3 (July 1946), 281-283.
The deer dilemma. *WCB* 11:8-9 (Aug-Sept 1946), 3-5.
Leopold explains opposition to deer hunting restrictions. *The Milwaukee Journal* (Sunday, Sept 1, 1946), sec. 3, p. 6.
Erosion as a menace to the social and economic future of the Southwest. *J Forestry* 44:9 (Sept 1946), 627-633. [Introduction by H. H. Chapman, who submitted this paper originally read by AL at a meeting of the New Mexico Association for Science in 1922.]
(With W.F. Grimmer). Introduction: The history and future of the pheasant in Wisconsin, pp. 15-25. In Irven O. Buss, *Wisconsin Pheasant Populations*. WCD Pub no. 326, A-46 (1946).

1947

(With Sara Elizabeth Jones). A phenological record for Sauk and Dane Counties, Wisconsin, 1935-1945. *Ecological Monographs* 17:1 (Jan 1947), 81-122. Reprint issued with Arboretum cover as Journal Paper no. 8 of the University of Wisconsin Arboretum.
Summarization of the Twelfth North American Wildlife Conference. *Trans 12th NAWC* (San Antonio, Feb 3-5, 1947), 529-536. Reprinted in *The Pennsylvania Game News* 17:12 (Mar 1947), 14-15, 30-31; condensed in *National Parks Magazine* 21:89 (Apr-June 1947), 26-28.
The distribution of Wisconsin hares. *Trans Wis Acad* 37 (1945; issued April 10, 1947), 1-14.
(With Lyle K. Sowls and David L. Spencer). A survey of over-populated deer ranges in the United States. *JWM* 11:2 (Apr 1947), 162-177. Summary printed as "National deer survey" in *WCB* 12:5 (May 1947), 27-30.
[Review of E. M. Queeny, *Prairie Wings: Pen and Camera Flight Studies*. New York: Ducks Unlimited, Inc., 1946]. *JWM*, 11:2 (Apr 1947), 190-191.
On a monument to the passenger pigeon. In *Silent Wings*. Madison: Wisconsin Society for Ornithology, May 11, 1947, pp. 3-5 [revised from talk delivered at annual meeting of the Wisconsin Society for Ornithology, Appleton, April 6, 1946]. Excerpted in *National Humane Review* (Aug 1948); reprinted revised as "On a monument to the pigeon" in *Almanac*.

The ecological conscience. *The Bulletin of the Garden Club of America* (Sept 1947), 45-53 [address at Conservation Meeting, Minneapolis, June 1947]. Reprinted in *WCB* 12:12 (Dec 1947), 4-7, and in WCD Pub no. 343 (n.d.), 4pp.; condensed in *Plants and Gardens* 3:4 (Winter 1947), 210-211; reprinted in *The Journal of Soil and Water Conservation* 3:3 (July 1948), 109-112; excerpted in *The Missouri Conservationist* 9:6 (June 6, 1948), 2.

Mortgaging the future deer herd. *WCB* 12:9 (Sept 1947), 3.

(With Irven O. Buss). [Review of Frank C. Edminister, *The Ruffed Grouse: Its Life Story, Ecology and Management*. New York: Macmillan Company, 1947.] *Condor* 49:6 (Nov-Dec 1947), 246-247.

(With Olaus J. Murie, chairman, et al.) Wilderness and aircraft. *The Living Wilderness* 12:22 (Autumn 1947), 1-6. [Slightly abridged from multigraphed report of the NAS-NRC-DBA Committee on Aircraft vs. Wilderness.]

Game management. *Encyclopaedia Britannica* (1947), 3pp.

1948

Bone test tells age of cottontails; Check pheasant hens' egg records, chicks' ages; Pheasants winter in same spot each year; Prairie du Sac quail numbers drop; Young muskrats identified by pelts. Wildlife Management and Fur Farming section of *What's New in Farm Science*, pt. 1, Agr Exp Sta Bull no. 474 (Jan 1948), 46-49. [Unsigned]

Why and how research? *Trans 13th NAWC* (Mar 8-10, 1948), 44-48. [Read by Robert A. McCabe in the absence of Professor Leopold because of illness.]

Charles Knesal Cooperrider, 1889-1944. *JWM* 12:3 (July 1948), 337-339.

1949-

A Sand County Almanac and Sketches Here and There. New York: Oxford University Press, 1949. xiii + 226pp., illus. Reissued in an enlarged edition, *A Sand County Almanac with Other Essays on Conservation from Round River*. New York, Oxford University Press, 1966. xv + 269pp., illus. Original edition issued in paperback by Oxford University Press, 1968. Enlarged edition issued in paperback by Sierra Club/Ballentine Books, 1970.

Luna B. Leopold, ed. *Round River: From the Journals of Aldo Leopold*. New York: Oxford University Press, 1953. xiii + 173pp., illus. Reissued in paperback, 1972.

(With Alfred E. Eynon). Avian daybreak and evening song in relation to time and light intensity. *Condor* 63:4 (July-Aug 1961), 269-293. [Written by Eynon, based on field data and unpublished manuscripts of Aldo Leopold.]

Dear Herbert. In *Memoirs of a Naturalist*, by Herbert L. Stoddard, Sr. Norman: University of Oklahoma Press, 1969, pp. 226-229. [Letter from Leopold to Stoddard dated March 26, 1934.]

Some fundamentals of conservation in the Southwest. *Environmental Ethics* 1 (Summer 1979), 131-141. [Edited from typescript, c. Mar 1923.]

Foreword. In *Companion to a Sand County Almanac: Interpretive & Critical Essays*, edited by J. Baird Callicott. Madison: University of Wisconsin Press, 1987, pp. 281-288. [Edited from typescript dated July 31, 1947, of foreword to "Great Possessions," an earlier version of *Sand County Almanac*, which Leopold submitted to Alfred A. Knopf on September 5, 1947.]

Index

Abert, J. W., 175, 184
Abraham, 184
Actinosis, 51
"Adventures of a Conservation Commissioner," 16, 330-35, 338
Agricultural Adjustment Administration (AAA), 198, 300
Agricultural college, 113, 166, 261, 264, 306, 316
Agriculture: AL's views on, 22-24, 103, 106, 128, 167, 236, 299-300, 306-9, 310, 326; biotic, 24, 30n26, 191, 237, 263-65, 272-73, 316-19; and energy, 269-70; and esthetics, 191, 213-16 passim, 258-65; industrialization of, 24, 297, 326; and wildlife, 50, 185, 274-75, 323-26; and wildlife management, 150-55 passim, 159, 166, 169-72, 203-8, 236, 279, 323-26
—regions: Mexico, 239-40; Southwest, 87, 100, 106-8, 128; Wisconsin, 198, 203, 218-23, 255-65, 274-75, 312-13, 324, 340-41
Alaska, 52, 130, 233
Albrecht, W. A., 314n4
Albuquerque Chamber of Commerce: AL as secretary of, xiii, 20, 22, 49; relation to boosterism, 98-105 passim
Albuquerque Game Protective Association, 47-48
Alhambra, 185
Alpine flora, 233
Amazon, 10, 123-24, 289
American Game Conference, 150
American Game Policy, xiv, 19, 150-55, 156-63 passim. See also Wildlife management; Wildlife policy
"American Game Policy, Report to the American Game Conference on an," 19, 150-55, 203
American Game Protective Association, 62, 131
American Ornithologists' Union (A.O.U.), 168
American Wildlife Institute, 322
Ammunition industry, 63, 166
Anopheles gambiae, 324
Antelope, 47, 51, 52, 59, 233, 285
Apache National Forest, xiii, 40, 90. See also Blue River
Appalachians, 139, 241
"Arboretum and the University, The," 6, 209-11. See also Wisconsin, University of: Arboretum
Argali, 59
Aristotle, 9, 12,
Arizona, 87-89 passim, 114-22, 128. See also Southwest; *individual forests, rivers, other place names*
Artificial propagation: of fish, 257; overemphasis on, 18, 298; of pheasants, 204, 207; and violence, 271. See also Game farming
Atlantic tidal marsh, 251
Audubon, John James, 236
Audubon Society, 166, 296

Babbitt (Sinclair Lewis's character), 167, 168, 188, 189, 209, 338
Bache, Richard, 245

371

Bag worm, 325
Balance of nature, 7, 39, 91, 267
Barnes, Will C., 240
Basswood, 274
Bates, Carlos G., 118
Bear, 242; black, 121. *See also* Grizzly
Beaver, 174, 177, 278
Beech, 271
Beetle, lady, 266
Beloit College, 235
Bennett, H. H., 218
Benton, Thomas Hart, 263
Berry, Wendell, 24
Bible, 6, 13, 169, 182, 209; evidence of landscape in, 71-77, 109, 184
Big Bend, international park at, 243
Billboards, 102, 182
Biological Survey, U.S., 48, 232, 321
Biotic community, 6-8, 12, 237, 267-68. *See also* Community; Food chains; Pyramid
"Biotic View of Land, A," 7, 15, 24, 266-73
Birch: bog, 261-62; white, 222; yellow, 290
Black-eyed Susan, 306
Bluegill, 259
Bluegrass, 183
Blue River (Arizona), 92-93, 106, 107-8, 110
Bobcat, 47
Bog flora, 210, 233, 261-62, 272, 313, 318
Boone, Daniel, 174, 183, 278
Boosterism, AL on, 98-105, 126-27, 131-32, 189, 194
Botulism, 328
Boulder Dam, 257
Boxwood: in Bible, 75-77 *passim*
Brazil, 324
Bryant, William Cullen, 6, 96
Bubonic plague, 324
Buffalo, 50-51, 257, 312, 320-21
Bunyan, Paul, 341, 344
Burlington, Iowa, xiii, xv
Burn blight, 325
Burroughs, John, 6, 97, 137
Bursum Bill, 102
"Busy Season, The," 17, 40
Butler, Samuel, 73-74

California, 68-70 *passim*, 121, 138, 241
Canada, 109, 130, 260, 277, 327, 328
Cane-lands, 183
Cankerworm, 314
Canute, King, 185
Capitalism, 188
Carhart, Arthur, xiv, 78
Caribou, 232
Carlisle, Thomas, 104
Carp, 257, 324; German, 314
Carrying capacity, 282-86 *passim*
Carson, Rachel, ix
Carson National Forest, 5, 40, 41-46, 82
Carson Pine Cone, x, xiii, 40, 41-42
Cat, as predator, 207
Catesby, Mark, 321
Cave Creek (Arizona), 93
Ceanothus, 83, 115
Cedar: in Bible, 75-76, 77; red, 35, 325; in Southwestern ruins, 89; white, 257, 271, 325
Channelization of streams, 223, 227-28, 250, 252, 324, 345
Chemical controls, 325
Chequamegon National Forest, 333
Chestnut, 59; blight, 59, 314, 324
Chickadee, 9, 34
Chicory, 308
Chihuahua Sierra (Mexico), 239-44, 289
China, 91, 182, 213
Christianity, 6, 9, 182. *See also* Bible; *individual prophets*
"Chukaremia," 18, 245-46
Citizen's Deer Committee (Wisconsin), 332
Citizenship, AL's views on, 105, 147, 295-300 *passim*, 318, 338
Civic values, in Albuquerque, 98-105
Civilian Conservation Corps (CCC): AL and, xiv; AL's critique of, 197, 198, 200; at Coon Valley, 220, 221; mentioned, 230, 231, 251, 261, 264, 271, 300, 318, 333-34, 340, 342
Civilization, AL's views on, 49-52, 94-97, 98, 110, 126, 127, 129, 134, 142, 173, 179-80, 182, 183-85, 192, 209-10, 213-14, 244, 253-54, 281-86, 295, 300, 311, 326, 345-46
Civil Works Administration (CWA), 196, 197
Clements, Frederick E., 4, 5, 6, 14, 27

Climate: and cycles, 238; and watershed stability, 89-91, 213
Clover, 265
Colonia Pacheco, Chihuahua, 240
Colorado, 233
Colorado River, 174, 177-78
Columbia River, 250, 252
Commoner, Barry, ix
Communism, 188
Community: concept, 6-8, 209-10, 237, 293, 303, 310-19 *passim*, 337; consciousness, and ethics: 14-16, 191-92, 340. *See also* Biotic community
Condor, 232, 271, 272
Condor, The, 18, 19, 60, 164, 165
Connecticut, 33
Conservation: AL's attitude on protection vs. use, 164-68; definition of, 187-88, 202, 212, 255, 257, 263, 298, 300, 310-11; evolution of ideas in, 189-92, 215-16, 230-32, 236-37, 253, 261, 266-67, 308-9, 311, 320-22; history of, 12, 17, 29*n*18, 55, 186-87, 193, 230-31, 244, 330; holistic approach to, 213, 310-19; motives for, 257-58, 317-19; in Old Testament, 74; standards of, 82-85, 231-32
—movement: factions in, 164-68 *passim*, 214, 218, 234; organizations, 231-34 *passim*, 327-29
—related concerns: economics, 193-202, 297, 316-17; engineering, 249-54; esthetics, 216, 231, 236; ethics, 181-92, 309, 330, 338-46; farming, 215-16, 218-23, 255-65, 272-73, 298-99, 308-9, 318; intelligent consumption, 143-47, 165, 191-92, 295-97, 300; politics, 281-86, 296, 298-300, 330-35; public attitudes, 330-35
"Conservation: In Whole or in Part," 8, 310-19, 338
"Conservation Economics," 22, 193-202, 215
"Conservation Ethic, The," 13-14, 22, 181-92, 193, 215, 235, 266
"Conservationist in Mexico," 12, 239-44
"Conserving the Covered Wagon," 25-26, 128-32
Consumption. *See* Conservation, related concerns, intelligent consumption
Contingent possession, 202

Controls, 237, 325
Coolidge, Calvin, 132
"Coon Valley: An Adventure in Cooperative Conservation," 16, 218-23
Cooperative Wildlife Research Unit Program, 279
Cooper, James Fenimore, 227
Cooper Ornithological Society, 60
Copeland Report, 194
Corn borer, 314, 325
Coronado, Francisco, 103, 176, 184, 241, 242
Coronado National Forest, 115
Cottonwood, 175, 240
Cougar, 257, 289, 322
Couzens, S. W., 175, 184
Cowles, Henry C., 4, 6, 27
Coyote, 47, 51, 241, 289
Craftsmanship, 94, 110
Crane, 271; sandhill, 233
Creeping jenny, 314
"Criticism of the Booster Spirit, A," 16, 22, 98-105
Crook National Forest, 115
Crow, 328
Curlew, 233
Curry, John Steuart, 263
Cycles: climatic, 90-91, 312; erosion, 119; of seasons, 274; in wildlife populations, 256-57, 261-62, 315

Darling, F. Fraser, 276, 277
Darling, Jay N., 234
Darwin, Charles, 14, 282
Dauerwald, 271
David (prophet), 72-77 *passim*, 276
Davis, Jefferson, 343
DDT, 325
DeBeaux, Oscar, 215*n*1
Decatur, Stephen, 98
Deer: overpopulation of, xv, 241-42, 299, 316, 322, 325; population mechanisms in, 285
—regions: Chihuahua Sierra, 241-42, 244, 257; Germany, 11, 226, 228, 271, 278; Southwest, xiv, 18, 51, 59, 81; Wisconsin, xv, 197, 257, 261-62, 325, 330-35, 341-43
—species: mule, 59; Sonora, 59, 232; whitetail, 59

Index

Deforestation: in ancient world, 91, 109; in Spain, 39
Delaware Valley, 35
Democracy, AL's concept of, 62-67 *passim*, 105, 182, 259, 285, 295-300 *passim*
Depression (1930s), xiv, 20, 193, 240, 302
Disease: biotic system and, 269; in forest soils, 227, 228, 293; in plants and animals, 51, 238, 314, 324-25
Diversity, and ecosystem stability, 7-8, 15, 29n10, 311-15
Doniphan, Alexander, 176
Douglas, Andrew E., 89, 91
Ducks, 51, 61, 232, 264, 278, 296, 327-29; pintail, 60-61; wood, 50. *See also* Waterfowl
Ducks Came Back, The (review), 327-29
Ducks Unlimited, 327-29
Dutch elm disease, 314, 325

Eagle, 201, 231; in Bible, 77
"Ecological Conscience, The," 15-16, 338-46
Ecological Society of America, 7, 266, 272
Ecology: in AL's writing, 3-8; biotic (ecosystem) concept of, 6-7, 266-73, 302-5; presentation to laymen, 224-25, 277, 301-5, 306-9; role of, in history, 183-85; as a sport, 278, 336-37; as a worldview, 3-8, 13, 16, 28, 209-11, 212, 217, 253. *See also* Fire; Land health; Landscape history; Organicism; Predators; Soil erosion; Wildlife management
—regions: Germany, 227-28; Iowa, 306-9; Sierra Madre, 239-44, 289; Southwest, 89-93, 114-122, 174-79, *passim*, 239-44 *passim*; Wisconsin, 185, 209-11, 247-48, 252, 256, 261-62, 274-75, 311-15, 324, 325, 334-35, 341-43
—related concerns: economics, 167, 185-88; engineering, 6-7, 252-54; esthetics, 8-12, 96, 191, 215-16, 227-29 *passim*, 236, 239, 277, 303, 326, 337; ethics, 12-16, 182, 192, 338-46; politics, 281-86, 330-35
"Ecology and Politics," 20, 281-86
Economics: and determinism, 94, 222; and ecology, 13, 185-88, 266-67; and ethics, 12, 13-14, 16, 345; of land use and conservation, 20, 22, 188-89, 191-92, 193-202, 215-16, 255-56, 258, 295-300 *passim*, 303, 316-17, 325; of resources in Southwest, 86-94, 98-105, 106-10, 112; and self interest, 16, 22, 186, 215, 272-73, 317, 340; unearned increment and, 86, 100; wilderness and, 125-27, 131-32; of wildlife management, 150-55, 165-68; of wood utilization, 143-47
Ecosystem: AL's biotic view of, 7, 237, 266-73, 303-4; origin of term, 7, 29n9
Education, AL's views on, 19-20, 188, 214, 234, 247, 258, 264, 279-80, 286, 295-96, 301-5, 306-9, 318, 319, 336-37, 338, 343
Egret, 192
Elephant Butte Reservoir, 93
Elk, 59, 81, 289, 299, 322, 325
Elm, Chinese, 265
Elton, Charles, 6, 14, 224-25, 326
Emergency Conservation Committee, 168
Energy, in ecosystems, 7, 268-72
Engineering: in Germany, 227-28; as world view, 6-7, 184, 209, 249-54, 257
"Engineering and Conservation," 6-7, 249-54
Eric the Red, 124
Espejo, Antonio, 184
Esthetics: AL's concept of, 8-12; and biota as a whole, 213, 327; and ecology, 10, 276-77; and economics, 213, 337; and ethics, 14, 15, 345; and evolution, 10, 229; of German landscape, 226-29; of landscape, 35-36, 123-24, 128, 148-49, 173-80 *passim*, 188, 191, 197, 212, 216, 239-40, 248, 259-60, 263-65, 294, 303, 337; land (term), 10, 214, 215; and professions, 43, 252-53; of structures, 104-5; in Western history, 9, 29n12-15, 214; of wildflowers, 247-48, 306-8; and wildlife management, 158-59, 163, 169-72, 236, 276-77
Ethics: AL's concept of, 8, 12-16; and conservation, 94-97, 181-92, 309, 338-46; definition of, 181-82, 345; evolution of, 14, 181-82, 215, 345-46; of sportsmen, 64, 161, 243-44, 333-34; as suspension of laws of predation, 283-84. *See also* Land ethic

Europe: esthetics in, 9, 29n15, 36; forestry in, 38, 119, 186; human population in, 284; resistance to abuse, 109, 202, 213, 170, 288, 311; wildlife management in, 141, 156-63, 231, 275. See also Germany; Scotland

Evolution: and ecology, 305; and elaboration of biota, 7-8, 11, 268-70, 312; and esthetics, 229, 277; and ethics, 8, 14-15, 182, 215

Exotic species, 18, 167-68, 222, 245-46, 271, 298, 312-14 passim, 323-26. See also individual species

Exploring the Animal World (review), 6, 224-25

Ezekiel, 72-76 passim, 94, 182

Famine concept, 303, 311; and timber, 24, 37, 70, 258

"Farmer as a Conservationist, The," 12, 23-24, 255-65

Farming. See Agriculture

Farrington, S. Kip, Jr., 327-29

Fascism, 188

Feeding station, 206

Fir: in Bible, 75-77 passim; white, 44, 144-45

Fire: control of, xiii, 18, 40, 198; destruction by, 37-38, 69; ecology of, 4-5, 69-70, 115-21, 175, 179, 185; and game, 51, 70, 197; "light-burning," 68-70; in Old Testament, 72-74, 109; and range management, 83, 85, 91, 92; in Sierra Madre, 240; use of, to maintain prairie, 265; in Wisconsin, 209, 210, 221, 256, 260, 264

Fish and Wildlife Service, U.S., 322. See also Biological Survey, U.S.

Fisher, 232, 312

Fishlake National Forest (Utah), 334

Flambeau River (Wisconsin), 344-45

Floods, 93, 107, 313. See also Soil erosion

Food chains, 6, 268-69, 303-4

Food patch, 206, 222

Forbes, S. A., 6

Forest Crop Law (Wisconsin), 198, 342

Forest policy: and administrative standards, 41-46, 82-85; and erosion, 118-19, 122; and fire, 51, 68-70; and private land use, 12, 38-39, 198-202, 216, 256-61 passim, 294; and wilderness recreation, 24-25, 78-81, 130-32, 133, 138-40. See also Forestry and associated references

Forest Products Laboratory, xiv, 20, 22, 143, 148

Forest ranger, 40, 46

Forestry: biotic approach to, 258-60, 271-72, 287-89, 292-93, 311, 314-16; and economics, 69, 143-47, 186-87, 192, 195, 197, 293-94, 297; and game management, 53-59, 185; in Germany, 226-29 passim, 271-72, 292-93; in Lake States, 209-11, 221, 247-48, 263, 290-94, 325, 332, 341-45; in Old Testament, 71-77. See also Conservation; Ecology; Fire; Forest policy; Forest Service; Grazing; Landscape history; Land-use policy; National forests; Range management; Recreation; Soil erosion; Watershed management; Wilderness; individual national forests and tree species

"Forestry and Game Conservation," 17, 53-59

"Forestry of the Prophets, The," 13, 71-77

Forest Service, U.S.: AL in, ix, xiii-xiv, 17, 24-25, 40, 41-46, 68, 78, 82, 106, 143, 148, 173, 223. See also Forestry and associated references

"Foreword": to Almanac, 3; to "Great Possessions," 27

Fox, 47, 207, 208, 315

France, population in World War II, 284-85

Franciscan, The (Albuquerque), 104-5

Galisteo River (New Mexico), 110, 175, 176

Gallinule, 264

Game. See Wildlife; individual species

"Game and Wildlife Conservation," 19, 164-68

Game Breeder, 156

Game conservation, history of, 17, 55. See also Wildlife management

Game farming, 18, 56, 63-67, 170, 195. See also Artificial propagation

Game Management, x, xiv, 203, 235

Game management. See Wildlife management

"Game Methods: The American Way," 19, 156-63

Game protective associations, xiii, 4, 47–48
Game Survey of the North Central States, Report on a, x, 164–65
Garden Club of America, 338
Gavilan, Rio, xiv
Genetical deterioration of human species, 282
Georgia, 222, 241, 323
Germany: AL's 1935 trip to, xiv, 11–12, 26–27, 226–29, 239; forestry in, 271–72, 292–93, 314; human population in, 259, 284; near-pogram in, 182; roads in, 133
Gila National Forest: AL and, xiv, 148; wilderness in, xiv, 25, 26, 81; watersheds of, 92, 118
Gila River, 107, 174, 177, 179, 251
Glaciation, 35, 259, 265, 312; hypothesis about pheasants, 167
Glassell, A. C., 329
Goethe, J. W. von, 226
Goldenrod, prairie, 307
Goldman, Edward H., 320–22
Goose, 236, 299; blue, 165
Gopher, gray, 266
Grama, side-oats, 240
Grand Canyon, 174, 177
"Grand-Opera Game," 10, 12, 18, 169–72
Grapes, wild, 264
"Grass, Brush, Timber, and Fire in Southern Arizona," 4–5, 68, 114–22
Graves, Henry, 68
Grazing: and erosion, 5, 92, 108–10, 114–18, 174, 176–78, 183–84, 220, 240; in the Holy Land, 77, 109; and pests, 308, 309; policy, xiii, 40, 44, 45, 83, 118–19, 121, 199, 243; and wilderness, 81; and wildlife, 50. *See also* Range management; Soil erosion; Watershed management
Great Plains, dust storms in, 22
"Great Possessions": book manuscript, xv; essay, 8
Greece, 181, 182
Greeley, William B., 68, 131
Grizzly, 4, 228, 230–33 *passim,* 236, 271, 272, 289
Grouse: cycles in, 256–57, 261–62, 315; in Gila, 81; management of, 157–59, 238; ruffed, 233, 274, 343; sharptail, 261; spruce, 314
Grub, white, 314

Gulf of California, 178
Guthrie, John D., 40, 148
Gypsy moth, 314

Hadley, Arthur T., 96
Hair-trigger equilibrium, 112, 202, 213
Hamilton, Alexander, 249
Hanno (of Carthage), 124, 125
Hare, 238, 315
Harmony of nature, 11
Hawk, 225, 228, 237, 345; goshawk, 271; red-tailed, 208, 266
Hazelnut, 264
Hazlitt, William, 77
Health. *See* Land health
Heath hen, 64
"Helping Ourselves," 16, 203–8
Hemlock, 325
Heron, 201, 228, 345
Hickory, 263, 264; weevil, 325
Highest use, doctrine of, 17, 25, 53, 78–81 *passim,* 133, 136
Hitler, 284
"Ho! Compadres Piñoneros!," 10, 12, 148–49
"Home Builder Conserves, The," 22, 143–47
Honeysuckle, Siberian, 325
Hoover, Herbert, 141, 196, 249
Horsemint, 307
Horseweed, 308–9
Hudson, W. T., 161, 162
Hudson's Bay Company, 277
Hunt, Lynn Bogue, 327
Hunter psychology: deer, 331–35, 342–43; duck, 327–29
Hunting: AL's attitude toward, 60–61, 137, 161–63, 164–68, 190, 245–46, 278, 307; as management tool, 322; and predators, 50; production of game for, 150–55; regulation of, 56–58, 65–67, 190; in Sierra Madre, 243; and wilderness, 52, 124–25, 137–38, 141
Huntington, Ellsworth, 89
Husbandry, 236, 298–99
Hutchinson, G. E., 7

Idaho, 139
Indians: and dogs, 321; and fire, 51, 68–70, 115; Incas, 109; of Mississippi Valley,

183; mounds of, 201; Piutes, 68-70, 71; prehistoric, of Southwest, 89, 96, 121, 242-43; and prehistoric equilibrium, 267, 282, 289; of Sierra Madre, 239-43 passim; of Southwest, 71, 102, 176, 179
Individualism, 285-86
Industrialist, 249
Iowa: AL in, xiii, xv, 33; weed flora of, 306-8; wildlife in, 168
Ironweed, 307
Irruptions: of deer, 241-42, 314, 331-35, 341; as symptom of land sickness, 287-88
Isaiah, 71-77 passim, 182, 276
Izaak Walton League, 166

Jackfish, 328
Jackson, Wes, 24
"January Thaw," 20, 274
Javelina, 59
Jefferson, Thomas, 22, 249
Joel (prophet), 71-74 passim
John, St. (of the Apocalypse), 281
June beetle, 324, 325
Juniper, 83, 90, 115-120 passim, 176-77, 240; alligator, 115, 120

Kaibab National Forest, and deer, 241, 271
Kenton, Simon, 174, 183
Kentucky, 183
Keokuk Dam, 50
Kingfisher, 35, 345
Kitchener, Horatio H., 249

Ladyslipper, 201, 258, 279; white, 261-62; yellow, 247-48
Lake States: northern hardwood forest, 290-94; and wilderness, 135, 138-40 passim, 290
Land, AL's concept of, 7, 212, 266-73 passim, 280, 310-19 passim, 336-37
Land, holistic approach to, 310-19, 341-43
Land ethic (term), 10, 212, 215. See also Ethics
"Land Ethic, The," 7, 12, 14-16, 27, 30n24, 181, 266, 310, 338
Land health: AL's concept of, 7, 8, 10, 15, 264, 282, 287, 303, 310-19 passim, 326; and energy circuits, 268-73; pathology as reverse of, 217, 270; in Sierra Madre, 239; wilderness and, 27, 244, 288-89
"Land Pathology," 10-11, 14, 22-23, 212-17
Landscape history: AL as teacher of how to read, 336-37; in Germany, 226-29; human society and, 213-15, 272-73; in Michigan, 290-94; in Mississippi Valley, 183; in Sierra Madre, 239-44; in Southwest, 70, 89-93, 107-9, 114-22, 173-80, 183-85, 239-44 passim; in Wisconsin, 185, 209-11, 220, 247, 251, 252, 255-65 passim, 274-75, 311-15, 324, 341-44
"Land Use and Democracy," 22, 295-300
Land-use policy: AL's biotic approach to, 250, 258, 299-300, 308-9, 316-18, 340-41, 344-45; and integration of uses, 197-202, 213, 218-23, 260; and private lands, 22, 110-13 passim, 187, 215-17; and public ownership, 111, 113, 188, 215, 259-60; and public vs. private ownership, 22, 194-97. See also Land
Las Cruces, New Mexico, 181
"Last Stand, The," 12, 290-94
Lawrenceville School (New Jersey): AL at, xiii, 8, 12, 37; tramp in vicinity of, 33-36
Leopold, Aldo: biography, ix-x, xiii-xv, 3-28 passim, 30n19; characterized as prophet, 30n20; writings, ix, x-xi, 3, 28, 349. See also individual essays and headnotes
Leopold, Estella Bergere (wife), xiii, 10, 15
Leopold Desk Company, 12
"Letter to a Wildflower Digger," 16, 247-48
Lewis and Clark, 125, 321
Ligon, J. Stokely, 48
Lindeman, Raymond L., 7, 29n9
Lord, Russell, 281
Los Lunas, New Mexico, 60-61
Lowdermilk, Walter C., 251
Lumber Code, 195, 216
Lumber industry, 37-39, 143-47 passim, 185-87 passim, 192, 256, 290-94 passim, 297, 341-42

McCabe, T. T., 164-68 passim
Machiavelli, 101
MacMillan, Donald B., 124
MacMillan, Lake (New Mexico), 93
Madre de Dios, Rio, 123-24

Magpie, 328
"Maintenance of Forests, The," 12–13, 17, 37–39
Manzanita, 115, 116
Maple: Norway, 292; sugar, 290, 292
Marks, Jeanette, 95
Marsh flora, 210, 313, 317
"Marshland Elegy," 11
Marten, 312
Massachusetts, 168
Mather Field, 85
Meadowlark, 266
"Means and Ends in Wildlife Management," 12, 21, 235–38
Mechanism, concept of, 6, 13, 14, 24, 26, 94–95, 180, 181–92 *passim*, 209, 213–14, 253, 257–58; as iron-heel attitude, 6, 210, 211, 309
Mediterranean lands, 91, 109, 124, 213, 311; in Bible, 71–77, 109
Merganser, 228
Mexico, AL's trips to, xiv, 11, 27, 239–44, 257
Michigan, 51, 189, 290–94
Middle West (U.S.), 109
Migratory Bird Law, Federal, 62, 330
Mill, John Stuart, 24
Milwaukee, Wisconsin, 266
Mimbres Valley (New Mexico), 110
Mining, 146; in Southwest, 44, 68, 69, 86, 87
Mink, 207
Minnesota, 333
Mississippi River: and erosion, 195, 219–20, 250, 252; and waterfowl, 50
Missouri, 233, 246, 279
Montana, 139, 322
More Game Birds in America Foundation, 156, 164–66 *passim*
Mormon colonies, 240
Mosquito, control of, 251–52, 325
Mountain lion, 47, 241, 242, 271
Mountain mahogany, 115, 240
Mt. Taylor, 176
Mouse: 288; house, 312, 314; jumping, 275; meadow, 275, 314
Muskrat, 264
Muir, John, 6, 72, 95–96, 247, 276
Munns, E. N., 91
Murie, Adolph, 289

Music, as metaphor for ecological system, 11–12, 169, 191, 229
Mussolini, 284
Myrtle, in Bible, 76–77

National Academy of Science, 276
National Conference on Outdoor Recreation (1924), 25, 132
National forests: 101, 239, 243, 294; AL's philosophy of administration, 41–46, 82–85; and erosion, 88, 92–93, 178, 196; and fire, 51, 68–70; and range management, 50, 53, 111; and timber, 87–88; and wilderness, 78–81, 125, 130–32, 133, 138–39; and wildlife, 50–52, 53–59, 67, 232. *See also* Forest Service; *individual forests*
National parks, 10, 101, 230, 243, 299; and threatened species, 232, 322; and wilderness, 80, 125, 131, 239. *See also individual parks*
National Recovery Administration (NRA), 299
National Research Council, 233
National Wildlife Federation, 150
Naturschutz, 227, 272
Nebraska, 233
New Deal: AL and, xiv, 20; critique of conservation programs of, 22, 193–94, 197–202, 218–23, 251
New England, 51, 81, 109, 185
New Jersey, AL in, xiii, 8, 33–36
New Mexico, 41, 47–48, 50, 51, 54, 59, 60–61, 98–105, 181, 233. *See also* Southwest; *individual cities, forests, rivers*
New Mexico Wool Growers Association, 48
"New Year's Inventory Checks Missing Game," 20, 274–75
North American Wildlife Conference, 20, 233, 301
Northwest (U.S.), 109, 138
"Notes on the Behavior of Pintail Ducks in a Hailstorm," 18, 60–61
Noumenon (pl. noumena), concept of, 10, 11, 29n16

Oak, 83, 115, 197, 201, 222, 251, 263, 264, 292–93; in Bible, 75, 77
—diseases: fire blight, 314; wilt, 325

—species: bur, 325; live, 240, 241; white, 274, 325
Obligation, to community, 110, 317, 340-41, 343
Odysseus, 181, 182
Olmsted, Frank H., 107, 251
Organicism, concept of, 6-8, 13, 14, 23-24, 26, 27, 94-95, 209, 212, 217, 240, 250, 301, 336; and health, 287-88, 310, 312; and principle of wholeness, 259
Ortega y Gasset, José, 173, 192
Osage orange, 261
Otero, E. M., 47, 48
Otter, 232, 271, 275, 345
Ouspensky, P. D., 6, 10, 94-95, 137
"Outlook for Farm Wildlife, The," 18, 24, 323-26
Ovis mexicanus, 59
Owl: 263, 345; barred, 274; in Germany, 11, 228, 271; great-horned, 9, 35, 208; long-eared, 275; screech, 274
Ozarks, and wilderness, 139

Palouse, 297
Paradigms: ecological, 4-7, 13-14; management, 53; nature, 94-96
Parakeet, Carolina, 236, 312
Parkman, Francis, 227
Parks, AL's attitude toward, 136-37, 216. *See also* National parks
Partridge: Hungarian, 167, 197, 236, 245, 315; spruce, 232
Partridge pea, 306
Passenger pigeon, 312
Pathology, symptoms of, in land, 213-14, 217, 313-15
Pattie, James Ohio, 173-79, 184
Paul, St. (apostle), 73
Paulson, Reuben, 16, 203-8 *passim*
Peattie, Donald C., 277
Peccary, 232
Pennsylvania, 241
Peppermint, 307
Peru, 277
Pest: behavior, 323-26; concept of, 306-9, 314. *See also* Weed
Pheasant, 10, 18, 64, 167, 169, 236, 238, 245, 271, 274, 312, 315, 324; management of, 204-8, 222, 278. *See also* Exotic species

Philippines, 241
Philosophic worldviews, 6-7, 16, 28, 94-97, 209-11, 212, 237, 253, 266-67
Physiology: conception of, 6, 27; of landscape, 244, 288, 289, 313
"Pig in the Parlor, The," 25, 133
Pinchot, Gifford, 17, 24-25, 78
Pine: 40, 83, 144, 197, 221, 240, 241, 256, 314, 343, 344; disease, 314; jack, 251, 314, 325; piñon, 83, 116, 148; red (Norway), 24, 258, 261, 314, 325; white, 24, 59, 144, 258; yellow, 89, 90, 115, 116, 120, 226
Pine Cone, The, xiii, 47
Piñon jay (piñonero), 10, 148-49
"Pioneeers and Gullies," 13, 22, 106-13, 128
Pittman-Robertson program, 278
" 'Piute Forestry' vs. Forest Fire Prevention," 4, 5, 17, 68-70
Plum, wild, 264
Poplar, yellow, 59
"Popular Wilderness Fallacy, The," 17-18, 49-52, 53
Population, animal: mechanisms for limitation of, 282-86, 321
Population, human: relation to technological culture, 165, 182, 282-86; and violence of land use, 270
Porcupine Mountains (Michigan), 290-94
Prairie chicken, 50, 168, 169, 205, 206, 274-75, 314
Prairie dog, 288
Prairie flora, 201, 233, 237, 257, 271, 272, 318; in Iowa, 306-9; in Wisconsin, 220, 258, 264-65, 311-13 *passim*
Predators: AL's attitude toward, 4, 11, 18, 28n4, 47-48, 64, 154, 161, 207-8, 230, 241-42, 257, 309, 320-22; attitudes of public toward, 161-62, 225, 237, 263, 267, 275, 309, 328; and biotic community, 225, 268, 271-72, 345; control of, 18, 47-48, 50, 67, 207-8, 225, 238, 298, 309, 316, 333; and esthetics, 161; in Europe, 161-62, 228; human, and ethics, 283-84, 333-34; preservation of, 230-34, 241-42, 322; in Sierra Madre, 241-42. *See also individual species*
Prescott National Forest, 82, 90, 115, 116
Progressive era, 12, 16, 29n18

Professionalism: AL and, 16-17, 150, 197-98, 201, 235, 246, 252; in engineering, 249-54 *passim;* in forestry, 17, 18, 37-39, 40, 41-46, 82-85; in wildlife management, 17-21, 53-59, 224, 235-38, 246, 276-80, 301-2, 323

Public interest: as criterion for conservation policy, 201-2, 219, 340-41; mechanisms for protecting, on private lands, 14, 110-11, 113, 215-17, 260-61, 343

Puerco, Rio, 110, 174, 175

Pyramid, biotic: as image of land, 7, 267-70; implications for land use, 270-73

Quack grass, 265

Quail: bobwhite, 10, 169, 196, 203-6 *passim,* 238, 245, 263, 264, 268, 313; in Georgia, 222, 278; management of, 51, 170-72, 191, 222, 236-37; masked bobwhite, 232; population mechanisms in bobwhite, 282-83, 315

Rabbits, 203, 206, 208, 325; cycles in, 256-57, 261-62, 315

Rabies, 51

Raccoon, 258, 263, 274

Ragweed, 288, 308-9

Rail, 264

Range management: on national forests, 53, 83, 84, 88, 91-93; on private lands, 110-12. *See also* Grazing; Soil erosion

Rat, Norway, 312, 314, 324

Rattlesnake, 95-96

Reclamation Service, U.S., 107, 113

Recreation: AL and, xiii, 24; economics of, 67, 196, 199, 243, 344-45; non-lethal, 278; and science, 289; and wilderness, 78-81, 132, 133, 141

Red-winged blackbird, 259, 264

Refuges, 54, 57, 167, 172, 197, 206, 207, 232, 233, 278, 288, 298-99

Religion, 95-96, 182. *See also* Bible; Ethics

Reservoirs: economics of, 93, 107, 108, 119, 316; siltage of, 92-93, 122, 179, 196, 250, 252, 270; and wildlife, 18, 50

Resource, concept of, 311

Restoration, of environments, 209-11, 220-23, 247-48, 278, 325

Revolt of the Masses, The (Ortega), 173, 192

Riley Game Cooperative, 203-8

Rio Grande River, 60-61, 91, 102, 174, 175, 177

"River of the Mother of God, The," 10, 12, 25-26, 123-27

Rodents: control of, 237, 279, 298; irruptions of, 238, 288; as symptom of abused soil, 272, 324

"Role of Wildlife in a Liberal Education, The," 20, 301-5

Roosevelt, Franklin, xiv

Roosevelt, Theodore, 131

Roosevelt Reservoir, 93, 119

Rose, wild, 308

Round River, x

Rural Electrification Administration (REA), 344

Russia, 321

Sagehen, 232

Salmon, 250

Sand County Almanac, A, ix-xi *passim,* xv, 3, 8, 12, 14, 24, 27, 28, 181, 212, 266, 274, 287, 338

San Felipe, New Mexico, 174

San Jose River (New Mexico), 176

San Simon Valley (Arizona), 240

Santa Fe, New Mexico, xiii, 10, 101, 173, 174

Santa Fe Trail, 174, 176

Santa Rita del Cobre, New Mexico, 177

Sapello Valley (New Mexico), 92, 110

Sauer, Carl O., 315

Sawfly, 325

Scabies, 51

Schorger, A. W., 279

Scotland, grouse management in, 157-59, 169

Sears, Paul, 296

Selective logging, 293, 294

Sequoia, 89

Seton, Ernest Thompson, 321

"Shack, the" (AL's farm), xiv, 15, 343

Shakespeare, 196, 229

Sheep: bighorn, 59, 232; desert bighorn, 232, 272; Mexican mountain, 59; mountain, 51, 236, 289

Shooting preserve, 204-8 *passim,* 299

Siberia, wilderness in, 130, 135

Sierra Madre. *See* Chihuahua Sierra

Silphium, 265
Sitgreaves, Lorenzo, 184
Sitgreaves National Forest, 90
Skill, AL's concept of, 83, 86, 89, 110, 257-58, 298, 317-18
Skunk, 47, 208
Snakeweed, 240
Socialism, 188
Society of American Foresters, 7, 266
Socorro, New Mexico, 174, 175
Soil Conservation Districts, 299-300, 340-41; Wisconsin law, 340
Soil Conservation Service, 218, 228, 252
Soil Erosion Service, 218-23 *passim*
Soil erosion: control techniques, 111-13, 198, 219, 221-22, 300; and economics, 38, 188-89, 195-96, 199, 200; and land illness, 214, 269, 287, 311, 313; research on, 251; in Sierra Madre (absence of), 239-44 *passim;* in Southwest, xiv, 83, 87-93, 106-113, 115, 117-18, 122, 173-80, 183-86, 239; in Wisconsin, 209-10, 218-23, 252, 256, 324, 340-41. *See also* Ecology; Grazing; Landscape history; Range management; Soil fertility; Watershed management
Soil fertility: and energy flows, 269-72, 323-24; and land-illness, 287, 313-14; restoration of, 237, 288, 308-9
Solomon (prophet), 71-76 *passim*
"Some Fundamentals of Conservation in the Southwest," 5-6, 13, 14, 26, 86-97, 114
Songbirds, 8, 16, 51, 96, 162, 191, 195, 197, 264, 316
"Song of the Gavilan," 11
South Africa, 244, 285
South America, 123-24
Southeast (U.S.), 109, 278
Southern Pacific Railroad, 68, 106
Southwest: AL in, xiii-xiv, 10, 13, 17, 18, 22, 128, 148; boosterism, 98-105; environmental change, 114-22, 173-80; erosion, 106-113, 183-85; fire, 68-70; forestry, 40, 41-46, 82-85; resources and conservation, 86-93, 96; and Sierra Madre, compared, 239-44 *passim;* wilderness, 79, 80-81, 130, 138; wildlife, 47-48, 59, 60-61, 148-49, 233, 241
Spain, 39
Sparrow, English, 312

Spessart Forest, 292-93
Sporting Arms and Ammunition Manufacturers' Institute, xiv
Sportsmanship: distortions of, 245-46; ethics of, 13, 137, 161, 162
Sportsman, The, 169, 171
Spruce, 226, 227; bud-worm, 314, 325
Spurge, flowering, 306-7
Squirrel: 264, 328; flying, 258, 274; ground, 288
Stability: and beauty, 239; and diversity, 8, 29n10, 311-15; and society, 286. *See also* Land health
"Standards of Conservation," 17, 82-85, 106
Starling, 312, 314
"State of the Profession, The," 12, 19-20, 276-80, 301
Stevenson, Robert Louis, 77
Stoddard, Herbert L., 169, 323
Stony Mountain (New Jersey), 8, 33-36 *passim*
Succession, in *Bible,* 77
Sumerian tribes, 124
Sunflower, 265
Sustained yield, 57, 186
Swallow, bank, 265
Swamp flora, 233
Swan, trumpeter, 232-33
Sweet William, 263
Sycamore, 240

Tamarack, 210, 271, 275, 325
Tansley, Arthur G., 7
Taos Canyon (New Mexico), 93
Technocracy, 188
Tern, 345
Texas, 130, 279
Thistle, 265; Canada, 314, 324
"Threatened Species," 4, 18, 230-34
Threshold, concept of, 284
Thrush, hermit, 290
Tibet, 59
Titmouse, 271
Tobacco, 73, 288, 309
Tonto National Forest, 115, 118-19
Tools, AL's concept of, 12, 43, 166, 179, 182, 185, 190. 217, 236, 254, 269, 282, 286, 311; relation to predation and ethics, 282-83

"To the Forest Officers of the Carson," 17, 41–46
Tourism, 103, 189, 239, 242, 243, 252
"Tramp in November, A," 8–9, 33–36
Tree-ring analysis, 89–91, 120
Tres Piedras, New Mexico, 5
Tristram (E.A. Robinson poem), 185
Trout, 81, 92, 177, 191, 197, 220, 240, 271; brook, 257; brown, 257
Turkey, wild, 81, 121, 233, 241, 242, 244, 312
Turner, Frederick Jackson, 26, 134
Tusayan National Forest, 90

Udall, Stewart, ix
Ukraine, 260
Ulysses, 124
Uniformity, as operational policy, 45–46
Utah, 173–74, 334
Utility, 13, 24–26 *passim*, 64, 259, 266–67, 307; and beauty, 10, 212–13, 271, 337

Van Hise, Charles R., 256
Varmints. *See* Predators
"Varmint Question, The," 4, 17, 18, 47–48
Vassar, 272
Vermin. *See* Predators; Pest
Vervain, blue, 308
Villa, Pancho, 240
Violence, and land use, 270–73, 292–93, 315–16
"Virgin Southwest, The," 5, 12, 173–80
Vitamin content of foods, 314

Walnut, 263; black, 59
Warbler, Brewster's, 233
War Mothers of America memorial, 100–101
Waterfowl, migratory, 50, 187, 192, 195, 251, 264, 327–29, 330, 331; management of, 18, 62, 153, 199, 231, 299. *See also individual species*
Water lily, 259, 264
Water pepper, 307–8
Watershed Handbook, xiv
Watershed management: AL and, xiv, 4–6, 28n5; in Southwest, 83, 87–93, 106–13, 118–19; substitutes for, 250–52. *See also* Grazing; Soil erosion
Weaver, J. E., 251, 271
Weed, 38, 306–9, 312, 324. *See also* Pest

Weed Flora of Iowa, The (review), 306–9
Wells, H. G., 125
"What is a Weed?," 12, 306–9
"Wherefore Wildlife Ecology?," 20, 336–37
White Mountains (Arizona), 240
White pine blister rust, 59, 314, 324, 325
White, Stewart Edward, 228
Whitman, Walt, 191, 196
"Why the Wilderness Society?," 26
Wilderness: AL's advocacy of idea, 24–27, 31n27; definition, 24, 79, 135–36; and economic growth, 125–27, 131–32, 136–37; and roads, 26, 126–27, 128–30, 133, 139–40; system, ix, 25, 27, 79, 125, 138–40, 289
—regions: Amazonia, 123–24; Appalachia, 139; Europe, 141; Germany, 26–27, 226–29; Lake States, 138–40 *passim*, 290, 294; Mexico, 27, 239–44; Northwest, 138; Ozarks, 139; Siberia, 130, 135; Southwest, 25–26, 130, 138
—values: culture, 26, 128–30, 134, 137–38, 141–42; land use, 134–42, 316–17; psychological needs, 26, 124–25; recreation, 24, 78–81, 132, 133, 141; science, 27, 287–89; standard of normality, 272, 287–89; and wildlife, 17–18, 49–52, 153, 231, 232, 236, 322
"Wilderness" (German essay), 11–12, 26–27, 226–29
"Wilderness" (in *Almanac*), 26, 27, 287
"Wilderness and Its Place in Forest Recreational Policy," 24–25, 78–81
"Wilderness as a Form of Land Use," 26, 134–42
"Wilderness as a Land Laboratory," 27, 287–89
Wilderness Society, xiv, 26
Wildflowers, 8, 16, 96, 126, 247–48, 191, 195, 198, 231, 233, 255–65 *passim*, 264, 296, 306–8, 314, 316; management of, 272, 279
Wildlife: behavior, 60–61, 148–49; and civilization, 49–52; and education, 301–5, 336–37; and game (terms), 232; history, 279; ownership, 19, 54; and soil, 323–24; values, 236. *See also* Wildlife management *and associated references*
Wildlife Ecology 118, xv, 20, 281–86, 301–5, 336–37

Index   383

Wildlife management: AL and profession of, ix, xiii, 4, 12, 17-21, 150-55, 164-68, 276-80; American vs. European system of, 18-19, 27, 66, 152, 156-63, 169; approaches of wildlifers and game farmers, 62-67; biotic approach to, 263-65, 266-73, 287-89, 313-19 *passim*, 323-26; history of ideas in, 190-91, 236-37, 280, 309; objectives of, 58-59, 231, 235-38, 276-77; practice of, 169-72, 203-8, 222; and public attitudes, 327-29, 330-35; research on, 166, 203, 236-38, 272, 278-79, 302. *See also* Artificial propagation; Ecology; Game farming; Wildlife; Wildlife policy
—related concerns: agriculture, 50, 153-55, 159-61, 166, 187, 199, 236, 255-65, 272-73, 274-75, 279, 309, 313, 318, 323-26; economics, 49-50, 59, 150-55, 165-68, 187, 266-67; education, 279-80, 301-2; esthetics, 158-59, 163, 169, 172, 195, 236, 276-77; ethics, 18, 59, 64, 161-62, 309, 330, 333-34, 345; forestry, 50-52, 53-59, 153, 271-72, 340-42; geography, 194-95. *See also* Agriculture; Forestry
—sub-fields: exotic species, 245-46; fur species, 277-78; in Germany, 228; non-game species, 18, 65, 155; pest species, 47-48, 309, 323-36; threatened species, 58-59, 190, 230-34, 279, 322. *See also* Predators; Songbirds; Waterfowl; *individual species*
Wildlife Management, Journal of, 277
Wildlife Management Institute, 150, 233
Wildlife policy: and private land use, 201, 278; and private vs. public ownership, 19, 66-67, 152; and public attitudes, 245-46, 330-35; and research, 278-79; for threatened species, 58-59, 201, 230-34, 271-72, 279, 322. *See also* American Game Policy; Wildlife management
Wildlife Review, 277
"Wild Lifers vs. Game Farmers: A Plea for Democracy in Sport," 18, 62-67
Wildlife Society, xiv, 12, 19, 276, 279
Wild rivers, 344-45
Willow, 83, 111-12, 221, 222, 241

Wilson, James, 131
Wilson, Woodrow, 189
Windbreaks, 260-61, 317
Wisconsin: AL in, ix, xiv-xv, 17, 23, 193, 212, 218, 235, 247, 249, 266, 274, 301, 310, 330, 336; conservation on farmland in, 198, 203-8, 218-23, 255-65; deer problems in, 241, 331-35, 341-43; drainage in, 251, 259; erosion in, 252; grouse management in, 157-59; land history, 311-15; landscape restoration, 209-11, 247-48; land-use case histories, 340-45; wildlife in, 168, 194, 231, 274-75, 324-25
Wisconsin Agriculturist and Farmer, xv, 20, 274
Wisconsin Conservation Commission: AL and, xv, 16, 330-35, 341; programs of, 204, 207
Wisconsin Industrial Commission Act, 201
Wisconsin River, 343
Wisconsin Shooting Preserve Law, 204-8 *passim*
Wisconsin, University of: AL at, ix, xiv-xv, 17, 20, 193, 203, 209, 218, 249, 281, 301, 310; Arboretum, 209-11, 247-48, 261-62, 325; Science Inquiry, 217; wildflower management at, 272. *See also* Wildlife Ecology 118
Wolf, 4, 47, 232, 241-42, 257, 271, 289, 320-22, 333, 341; desert, 322; sub-arctic, 322. *See also* Predators
Wolverine, 232, 312
Wolves of North America, The (review), 4, 320-22
Wood, Grant, 263
Woodpecker, 271; ivory-billed, 232, 272; pileated, 343
Wood utilization, 143-47
Woolsey, T. S., 91
World War I, xiii, 20, 62, 100-101
World War II: AL and, xv, 22, 24, 323; and conservation, 295-300 *passim;* and education, 301-2; and ethical implications of ecology, 281-86; and logging in Lake States, 290-94 *passim,* 343; and postwar policies, 310, 329
Wright, Sewall, 285-86
Wyoming, 322

384 Index

Yale University, xiii, 4
Yellow-legs, 264
Yellowstone National Park, 196, 289, 322

Young, Stanley P., 320–22

Zuni Reservoir, 93